Rapid Bioassessment Protocols
For Use in Streams and Wadeable Rivers:

Periphyton, Benthic Macroinvertebrates, and Fish
Second Edition

Washington, DC 20460

NOTICE

This document has been reviewed and approved in accordance with U.S. Environmental Protection Agency policy. Mention of trade names or commercial products does not constitute endorsement or recommendation for use.

This entire document, including data forms and other appendices, can be downloaded from the website of the USEPA Office of Wetlands, Oceans, and Watersheds:

FOREWORD

In December 1986, U.S. EPA's Assistant Administrator for Water initiated a major study of the Agency's surface water monitoring activities. The resulting report, entitled "Surface Water Monitoring: A Framework for Change" (U.S. EPA 1987), emphasizes the restructuring of existing monitoring programs to better address the Agency's current priorities, e.g., toxics, nonpoint source impacts, and documentation of "environmental results." The study also provides specific recommendations on effecting the necessary changes. Principal among these are:

1.   To issue guidance on cost-effective approaches to problem identification and trend assessment.

2.   To accelerate the development and application of promising biological monitoring techniques.

In response to these recommendations, the Assessment and Watershed Protection Division developed the rapid bioassessment protocols (RBPs) designed to provide basic aquatic life data for water quality management purposes such as problem screening, site ranking, and trend monitoring, and produced a document in 1989 (Plafkin et al. 1989). Although none of the protocols were meant to provide the rigor of fully comprehensive studies, each was designed to supply pertinent, cost-effective information when applied in the appropriate context.

As the technical guidance for biocriteria has been developed by EPA, states have found these protocols useful as a framework for their monitoring programs. This document was meant to have a self-corrective process as the science advances; the implementation by state water resource agencies has contributed to refinement of the original RBPs for regional specificity. This revision reflects the advancement in bioassessment methods since 1989 and provides an updated compilation of the most cost-effective and scientifically valid approaches.

DEDICATION

All of us who have dealt with the evaluation and diagnosis of perturbation to our aquatic resources owe an immeasurable debt of gratitude to *Dr. James L. Plafkin*. In addition to developing the precursor to this document in 1989, Jim was a driving force within EPA to increase the use of biology in the water pollution control program until his untimely death on February 6, 1990. Throughout his decade-long career with EPA, his expertise in ecological assessment, his dedication, and his vision were instrumental in changing commonly held views of what constitutes pollution and the basis for pollution control programs. Jim will be remembered for his love of life, his enthusiasm, and his wit. As a small token of our esteem, we dedicate this revised edition of the RBPs to his memory.

ACKNOWLEDGMENTS

Dr. James L. Plafkin of the Assessment and Watershed Protection Division (AWPD) in USEPA's Office of Water, served as principal editor and coauthor of the original Rapid Bioassessment Protocols document in 1989. Other coauthors of the original RBPs were consultants to the AWPD, Michael T. Barbour, Kimberly D. Porter, Sharon Gross and Robert M. Hughes. Principal authors of this revision are Michael T. Barbour, James (Sam) Stribling, Jeroen Gerritsen, and Blaine D. Snyder. Many others also contributed to the development of the original RBP document. Special thanks goes to the original Rapid Bioassessment Workgroup. The Workgroup, composed of both State and USEPA Regional biologists (listed in Chapter 1), was instrumental in providing a framework for the basic approach and served as primary reviewers of various drafts. Dr. Kenneth Cummins and Dr. William Hilsenhoff provided invaluable advice on formulating certain assessment metrics in the original RBP approach. Dr. Vincent Resh also provided a critical review that helped strengthen the RBP approach. While not directly involved with the development of the RBPs, Dr. James Karr provided the framework (Index of Biotic Integrity) and theoretical underpinnings for "re-inventing" bioassessment for water resource investigations. Since 1989, extensive use and application of the IBI and RBP concept has helped to refine specific elements and strengthen the overall approach. The insights and consultation provided by these numerous biologists have provided the basis for the improvements presented in this current document.

This revision of the RBPs could not have been accomplished without the support and oversight of Chris Faulkner of the USEPA Office of Water. Special thanks go to Ellen McCarron and Russell Frydenborg of Florida DEP, Kurt King of Wyoming DEQ, John Maxted of Delaware DNREC, Dr. Robert Haynes of Massachusetts DEP, and Elaine Major of University of Alaska, who provided the opportunity to test and evaluate various technical issues and regional specificity of the protocols in unique stream systems throughout the United States. Editorial and production support, report design, and HTML formatting were provided by a team of Tetra Tech staff — Brenda Fowler, Michael Bowman, Erik Leppo, James Kwon, Amanda Richardson, Christiana Daley, and Abby Markowitz. Technical assistance and critical review was provided by Dr. Jerry Diamond of Tetra Tech.

A Technical Experts Panel was convened by the USEPA to provide an in-depth review and recommendations for revisions to this document. This group of esteemed scientists provided not only useful comments, but assisted in revising sections of the document. In particular, Drs. Jan Stevenson and Loren Bahls revised the periphyton chapter; and Dr. Phil Kaufmann provided assistance on the habitat chapter. The Technical Experts Panel included:

> Dr. Reese Voshell, Virginia Tech University (Chair)
>
> Dr. Loren Bahls, University of Montana
>
> Dr. David Halliwell, Aquatic Resources Conservation Systems
>
> Dr. James Karr, University of Washington
>
> Dr. Phil Kaufmann, Oregon State University
>
> Dr. Billie Kerans, Montana State University
>
> Dr. Jan Stevenson, University of Louisville

Dr. Charles Hawkins (Utah State University) and Dr. Vincent Resh (University of California, Berkeley) served as outside readers.

Much appreciation is due to the biologists in the field (well over a hundred) who contributed their valuable time to review both the original and current documents and provide constructive input. Their help in this endeavor is sincerely appreciated.

TABLE OF CONTENTS

Rapid Bioassessment Protocols for Use in Streams and Wadeable Rivers: Periphyton, Benthic Macroinvertebrates, and Fish, Second Edition

v

Rapid Bioassessment Protocols for Use in Streams and Wadeable Rivers: Periphyton, Benthic Macroinvertebrates, and Fish, Second Edition

vii

LIST OF FIGURES AND TABLES

FIGURES

Rapid Bioassessment Protocols for Use in Streams and Wadeable Rivers: Periphyton, Benthic Macroinvertebrates, and Fish, Second Edition

ix

TABLES

LIST OF ACRONYMS

Acronym	Full Name (acronym stands for)
AFDM	Ash Free Dry Mass
ANOVA	Analysis of Variance
APHA	American Public Health Association
ASTM	American Society of Testing and Materials
AUSRIVAS	Australian River Assessment System
AWPD	Assessment and Watershed Protection Division
BEAST	Benthic Assessment of Sediment
BMP	Best Management Practices
CBWD	Chesapeake Bay and Watershed Programs
CWA	Clean Water Act
DEC	Department of Environmental Conservation
DEM	Department of Environmental Management
DEM	Division of Environmental Management
DEP	Department of Environmental Protection
DEQ	Department of Environmental Quality
DHEC	Department of Health and Environmental Control
DNR	Department of Natural Resources
DNREC	Department of Natural Resources and Environmental Control
DQO	Data Quality Objectives
EDAS	Ecological Data Application System
EMAP	Environmental Monitoring and Assessment Program
EPA	Environmental Protection Agency
EPT	Ephemeroptera, Plecoptera, Trichoptera
GIS	Geographic Information System
GPS	Global Positioning System
HBI	Hilsenhoff Biotic Index
IBI	Index of Biotic Integrity
ICI	Invertebrate Community Index
ITFM	Intergovernmental Task Force on Monitoring
ITIS	Integrated Taxonomic Information Service

Acronym	Full Name (acronym stands for)
IWB	Index of Well Being
MACS	Mid-Atlantic Coastal Systems
MBSS	Maryland Biological Stream Survey
MIWB	Modified Index of Well Being
NAWQA	National Water Quality Assessment Program
NPDES	National Pollutant Discharge Elimination System
NPS	nonpoint source pollution
PASS	Preliminary Assessment Scoresheet
PCE	Power Cost Efficiency
POTWS	Publicly Owned Treatment Works
PTI	Pollution Tolerance Index
QA	Quality Assurance
QC	Quality Control
QHEI	Qualitative Habitat Evaluation Index
RBP	Rapid Bioassessment Protocols
RDMS	Relational Database Management System
RM	River Mile
RPS	Rapid Periphyton Survey
SAB	Science Advisory Board
SCI	Stream Quality Index
SOP	Standard Operating Procedures
STORET	Data Storage and Retrieval System
SWCB	State Water Control Board
TCR	Taxonomic Certainty Rating
TMDL	Total Maximum Daily Load
TSN	Taxonomic Serial Number
USDA	United States Department of Agriculture
USEPA	United States Environmental Protection Agency
USGS	United States Geological Survey
WPA	Watershed Protection Approach
WQD	Water Quality Division

Rapid Bioassessment Protocols for Use in Streams and Wadeable Rivers: Periphyton, Benthic Macroinvertebrates, and Fish, Second Edition

xiii

This Page Left Intentionally Blank

1 THE CONCEPT OF RAPID BIOASSESSMENT

1.1 PURPOSE OF THE DOCUMENT

The primary purpose of this document is to describe a practical technical reference for conducting cost-effective biological assessments of lotic systems. The protocols presented are not necessarily intended to replace those already in use for bioassessment nor is it intended to be used as a rigid protocol without regional modifications. Instead, they provide options for agencies or groups that wish to implement

> *Biological assessment* is an evaluation of the condition of a waterbody using biological surveys and other direct measurements of the resident biota in surface waters.

rapid biological assessment and monitoring techniques. This guidance, therefore, is intended to provide basic, cost-effective biological methods for states, tribes, and local agencies that (1) have no established bioassessment procedures, (2) are looking for alternative methodologies, or (3) may need to supplement their existing programs (not supersede other bioassessment approaches that have already been successfully implemented).

The Rapid Bioassessment Protocols (RBPs) are essentially a synthesis of existing methods that have been employed by various State Water Resource Agencies (e.g., Ohio Environmental Protection Agency [EPA], Florida Department of Environmental Protection [DEP], Delaware Department of Natural Resources and Environmental Control [DNREC], Massachusetts DEP, Kentucky DEP, and Montana Department of Environmental Quality [DEQ]). Protocols for 3 aquatic assemblages (i.e., periphyton, benthic macroinvertebrates, fish) and habitat assessment are presented. All of these protocols have been tested in streams in various parts of the country. The choice of a particular protocol should depend on the purpose of the bioassessment, the need to document conclusions with confirmational data, and available resources. The original Rapid Bioassessment Protocols were designed as inexpensive screening tools for determining if a stream is supporting or not supporting a designated aquatic life use. The basic information generated from these methods would enhance the coverage of broad geographical assessments, such as State and National 305(b) Water Quality Inventories. However, members of a 1986 benthic Rapid Bioassessment Workgroup and reviewers of this document indicated that the Rapid Bioassessment Protocols can also be applied to other program areas, for example:

! Characterizing the existence and severity of impairment to the water resource

! Helping to identify sources and causes of impairment

! Evaluating the effectiveness of control actions and restoration activities

! Supporting use attainability studies and cumulative impact assessments

! Characterizing regional biotic attributes of reference conditions

Therefore, the scope of this guidance is considered applicable to a wider range of planning and management purposes than originally envisioned, i.e., they may be appropriate for priority setting, point and nonpoint-source evaluations, use attainability analyses, and trend monitoring, as well as initial screening.

1.2 HISTORY OF THE RAPID BIOASSESSMENT PROTOCOLS

In the mid-1980's, the need for cost-effective biological survey techniques was realized because of rapidly dwindling resources for monitoring and assessment and the extensive miles of un-assessed stream miles in the United States. It was also recognized that the biological data needed to make informed decisions relevant to the Nation's waters were greatly lacking across the country. It was further recognized that it was crucial to collect, compile, analyze, and interpret environmental data rapidly to facilitate management decisions and resultant actions for control and/or mitigation of impairment. Therefore, the principal conceptual underpinnings of the RBPs were:

> ! Cost-effective, yet scientifically valid, procedures for biological surveys
>
> ! Provisions for multiple site investigations in a field season
>
> ! Quick turn-around of results for management decisions
>
> ! Scientific reports easily translated to management and the public
>
> ! Environmentally-benign procedures.

The original RBPs were developed in two phases. The first phase centered on the development and refinement of the benthic macroinvertebrate protocols. The second phase involved the addition of analogous protocols pertinent to the assessment of fish assemblages.

The benthic macroinvertebrate protocols were originally developed by consolidating procedures in use by various State water quality agencies. In 1985, a survey was conducted to identify States that routinely perform screening-level bioassessments and believed that such efforts were important to their monitoring programs. Guidance documents and field methods in common use were evaluated in an effort to identify successful bioassessment methods that used different levels of effort. Original survey materials and information obtained from direct personal contacts were used to develop the draft protocols.

Missouri Department of Natural Resources (DNR) and Michigan Department of Natural Resources both used an approach upon which the screening protocol (RBP I) in the original document was based. The second (RBP II) was more time and labor intensive, incorporating field sampling and family-level taxonomy, and was a less intense version of RBP III. The concept of family-level taxonomy was based on the approach used by the Virginia State Water Control Board (SWCB) in the late 1980s. The third protocol (RBP III) incorporated certain aspects of the methods used by the North Carolina Division of Environmental Management (DEM) and the New York Department of Environmental Conservation (DEC) and was the most rigorous of the 3 approaches.

In response to a number of comments received from State and USEPA personnel on an earlier version of the RBPs, a set of fish protocols was also included. Fish protocol V was based on Karr's work (1981) with the Index of Biological Integrity (IBI), Gammon's Index of Well Being (1980), and standard fish population assessment models, coupled with certain modifications for implementation in different geographical regions. During the same time period as the development of the RBPs, Ohio EPA developed precedent-setting biological criteria using the IBI and Index of Well Being (IWB), as well as a benthic macroinvertebrate index, called the Invertebrate Community Index (ICI), and published methods and supporting documentation (Ohio EPA 1987). A substantial database on their use for site-specific fish and benthic macroinvertebrate assessments exists, and has been published (DeShon 1995, Yoder 1995, Yoder and Rankin 1995a,b). In the intervening years since 1989, several other states have followed suit with similar methods (Davis et al. 1996).

A workgroup of State and USEPA Regional biologists (listed below) was formed in the late 1980's to review and refine the original draft protocols. The Rapid Bioassessment Workgroup was convened from 1987 through 1989 and included biologists using the State methods described above and biologists from other regions where pollution sources and aquatic systems differed from those areas for which the draft protocols were initially developed.

USEPA
James Plafkin[1], Assessment and Watershed Protection Division (AWPD), USEPA
Michael Bilger[2], USEPA Region I
Michael Bastian[2], USEPA Region VI
William Wuerthele, USEPA Region VIII
Evan Hornig[2], USEPA Region X

STATES
Brenda Sayles, Michigan DNR
John Howland[2], Missouri DNR
Robert Bode, New York DEC
David Lenat, North Carolina DEM
Michael Shelor[2], Virginia SWCB
Joseph Ball, Wisconsin DNR

The original RBPs (Plafkin et al. 1989) have been widely distributed and extensively tested across the United States. Under the direction of Chris Faulkner, Monitoring Branch of AWPD the AWPD of USEPA, a series of workshops has been conducted across the Nation since 1989 that have been directed to training and discussions on the concept and approach to rapid bioassessment. As a result of these discussions and the opportunity of applying the techniques in various stream systems, the procedures have been improved and refined, while maintaining the basic concept of the RBPs. This document reflects those improvements and serves as an update to USEPA's Rapid Bioassessment Protocols.

1.3 ELEMENTS OF THIS REVISION

Refinements to the original RBPs have occurred from regional testing and adaptation by state agency biologists and basic researchers. The original concept of large, composited samples, and multimetric analyses has remained intact for the aquatic assemblages, and habitat assessment has remained integral to the assessment. However, the specific methods for benthic macroinvertebrates have been refined, and protocols for periphyton surveys have been added. A section on conducting performance-based evaluations, i.e., determining the precision and sensitivity of methods, to enable sharing of comparable data despite certain methodological differences has been added. Various technical issues, e.g., the

---

[1]   deceased
[2]   no longer with state agency or USEPA department relevant to water resource assessments of ecosystem health.

---

testing of subsampling, selection of index period, selection and calibration of biological metrics for regional application have been refined since 1989. Many of these technical issues, e.g., development of reference condition, selection of index period and selection/calibration of metrics, have been discussed in other documents and sources (Barbour et al. 1995, Gibson et al. 1996, Barbour et al. 1996a). This revision draws upon the original RBPs (Plafkin et al. 1989) as well as numerous other sources that detail relevant modifications. This document is a compilation of the basic approaches to conducting rapid bioassessment in streams and wadeable rivers and focuses on the periphyton, benthic macroinvertebrates, and fish assemblages and assessing the quality of the physical habitat structure.

2 APPLICATION OF RAPID BIOASSESSMENT PROTOCOLS (RBPS)

2.1 A FRAMEWORK FOR IMPLEMENTING THE RAPID BIOASSESSMENT PROTOCOLS

The Rapid Bioassessment Protocols advocate an integrated assessment, comparing habitat (e.g., physical structure, flow regime), water quality and biological measures with empirically defined reference conditions (via actual reference sites, historical data, and/or modeling or extrapolation). Reference conditions are best established through systematic monitoring of actual sites that represent the natural range of variation in "minimally" disturbed water chemistry, habitat, and biological conditions (Gibson et al. 1996). Of these 3 components of ecological integrity, ambient water chemistry may be the most difficult to characterize because of the complex array of possible constituents (natural and otherwise) that affect it. The implementation framework is enhanced by the development of an empirical relationship between habitat quality and biological condition that is refined for a given region. As additional information is obtained from systematic monitoring of potentially impacted and site-specific control sites, the predictive power of the empirical relationship is enhanced. Once the relationship between habitat and biological potential is understood, water quality impacts can be objectively discriminated from habitat effects, and control and rehabilitation efforts can be focused on the most important source of impairment.

2.2 CHRONOLOGY OF TECHNICAL GUIDANCE

A substantial scientific foundation was required before the USEPA could endorse a bioassessment approach that was applicable on a national basis and that served the purpose of addressing impacts to surface waters from multiple stressors (see Stribling et al. 1996a). Dr. James Karr is credited for his innovative thinking and research in the mid-1970's and early 1980's that provided the formula for developing bioassessment strategies to address issues mandated by the Clean Water Act. The USEPA convened a few key workshops and conferences during a period from the mid-1970's to mid-1980's to provide an initial forum to discuss aspects of the role of biological indicators and assessment to the integrity of surface water. These workshops and conferences were attended by National scientific authorities who contributed immensely to the current bioassessment approaches advocated by the USEPA. The early RBPs benefitted from these activities, which fostered attention to biological assessment approaches. The RBPs embraced the multimetric approach described in the IBI (see Karr 1981, Karr et al. 1986) and facilitated the implementation of bioassessment into monitoring programs across the country.

Since the publication of the original RBPs in 1989, U.S. Environmental Protection Agency (USEPA) has produced substantial guidance and documentation on both bioassessment strategies and implementation policy on biological surveys and criteria for water resource programs. Much of this effort was facilitated by key scientific researchers who argued that bioassessment was crucial to the underpinnings of the Clean Water Act. The work of these researchers that led to these USEPA documents resulted in the national trend of adapting biological assessment and monitoring approaches for detecting problems, evaluating Best Management Practices (BMPs) for mitigation of nonpoint source impacts, and monitoring ecological health over time. The chronology of the crucial USEPA guidance, since the mid-1980's, relevant to bioassessment in streams and rivers is presented in Table 2-1. (See Chapter 11 [Literature Cited] for EPA document numbers.)

Table 2-1. Chronology of USEPA bioassessment guidance (relevant to streams and rivers).

Year	Document Title	Relationship to Bioassessment	Citation
1987	Surface Water Monitoring: A Framework for Change	USEPA calls for efficacious methods to assess and determine the ecological health of the nation's surface waters.	USEPA 1987
1988	Proceedings of the First National Workshop on Biological Criteria (Lincolnwood, Illinois)	USEPA brings together agency biologists and "basic" researchers to establish a framework for the initial development of biological criteria and associated biosurvey methods.	USEPA 1988
1989	Rapid Bioassessment Protocols for Use in Streams and Rivers: Benthic Macroinvertebrates and Fish	The initial development of cost-effective methods in response to the mandate by USEPA (1987), which are to provide biological data on a national scale to address the goals of the Clean Water Act.	Plafkin et al. 1989
1989	Regionalization as a Tool for Managing Environmental Resources	USEPA develops the concept of ecoregions and partitions the contiguous U.S. into homogeneous regions of ecological similarity, providing a basis for establishment of regional reference conditions.	Gallant et al. 1989
1990	Second National Symposium on Water Quality Assessment: Meeting Summary	USEPA holds a series of National Water Quality Symposia. In this second symposium, biological monitoring is introduced as an effective means to evaluating the quality of water resources.	USEPA 1990a
1990	Biological Criteria: National Program Guidance for Surface Waters	The concept of biological criteria is described for implementation into state water quality programs. The use of biocriteria for evaluating attainment of "aquatic life use" is discussed.	USEPA 1990b
1990	Macroinvertebrate Field and Laboratory Methods for Evaluating the Biological Integrity of Surface Waters	This USEPA document is a compilation of the current "state-of-the-art" field and laboratory methods used for surveying benthic macroinvertebrates in all surface waters (i.e., streams, rivers, lakes, and estuaries).	Klemm et al. 1990
1991	Biological Criteria: State Development and Implementation Efforts	The status of biocriteria and bioassessment programs as of 1990 is summarized here.	USEPA 1991a
1991	Biological Criteria Guide to Technical Literature	A limited literature survey of relevant research papers and studies is compiled for use by state water resource agencies.	USEPA 1991b
1991	Technical Support Document for Water Quality–Based Toxics Control	USEPA describes the approach for implementing water quality-based toxics control of the nation's surface waters, and discusses the value of integrating three monitoring tools, i.e., chemical analyses, toxicity testing, and biological surveys.	USEPA 1991c
1991	Biological Criteria: Research and Regulation, Proceedings of the Symposium	This national symposium focuses on the efficacy of implementing biocriteria in all surface waters, and the proceedings documents the varied applicable approaches to bioassessments.	USEPA 1991d

Chapter 2: Application of Rapid Bioassessment Protocols (RBPs)

Table 2-1. Chronology of USEPA bioassessment guidance (relevant to streams and rivers) (Continued).

Year	Document Title	Relationship to Bioassessment	Citation
1991	Report of the Ecoregions Subcommittee of the Ecological Processes and Effects Committee	The SAB (Science Advisory Board) reports favorably that the use of ecoregions is a useful framework for assessing regional fauna and flora. Ecoregions become more widely viewed as a basis for establishing regional reference conditions.	USEPA 1991e
1991	Guidance for the Implementation of Water Quality–Based Decisions: The TMDL Process	The establishment of the TMDL (total maximum daily loads) process for cumulative impacts (nonpoint and point sources) supports the need for more effective monitoring tools, including biological and habitat assessments.	USEPA 1991f
1991	Design Report for EMAP, the Environmental Monitoring and Assessment Program	USEPA's Environmental Monitoring and Assessment Program (EMAP) is designed as a rigorous national program for assessing the ecological status of the nation's surface waters.	Overton et al. 1991
1992	Procedures for Initiating Narrative Biological Criteria	A discussion of the concept and rationale for establishing narrative expressions of biocriteria is presented in this USEPA document.	Gibson 1992
1992	Ambient Water-Quality Monitoring in the U.S. First Year Review, Evaluation, and Recommendations	Provide first-year summary of task force efforts to develop and recommend framework and approach for improving water resource quality monitoring.	ITFM 1992
1993	Fish Field and Laboratory Methods for Evaluating the Biological Integrity of Surface Waters	A compilation of the current "state-of-the-art" field and laboratory methods used for surveying the fish assemblage and assessing fish health is presented in this document.	Klemm et al. 1993
1994	Surface Waters and Region 3 Regional Environmental Monitoring and Assessment Program: 1994 Pilot Field Operations and Methods Manual for Streams	USEPA focuses its EMAP program on streams and wadeable rivers and initiates an approach in a pilot study in the Mid-Atlantic Appalachian mountains.	Klemm and Lazorchak 1994
1994	Watershed Protection: TMDL Note #2, Bioassessment and TMDLs	USEPA describes the value and application of bioassessment to the TMDL process.	USEPA 1994a
1994	Report of the Interagency Biological Methods Workshop	Summary and results of workshop designed to coordinate monitoring methods among multiple objectives and states. [Sponsored by the USGS]	Gurtz and Muir 1994
1995	Generic Quality Assurance Project Plan Guidance for Programs Using Community Level Biological Assessment in Wadeable Streams and Rivers	USEPA develops guidance for quality assurance and quality control for biological survey programs.	USEPA 1995a
1995	The Strategy for Improving Water Quality Monitoring in the United States: Final Report of the Intergovernmental Task Force on Monitoring Water Quality	An Intergovernmental Task Force (ITFM) comprised of several federal and state agencies draft a monitoring strategy intended to provide a cohesive approach for data gathering, integration, and interpretation.	ITFM 1995a
1995	The Strategy for Improving Water Quality Monitoring in the United States: Final Report of the Intergovernmental Task Force on Monitoring Water Quality, Technical Appendices	Various issue papers are compiled in these technical appendices associated with ITFM's final report.	ITFM 1995b

Table 2-1. Chronology of USEPA bioassessment guidance (relevant to streams and rivers) (Continued).

Year	Document Title	Relationship to Bioassessment	Citation
1995	Environmental Monitoring and Assessment Program Surface Waters: Field Operations and Methods for Measuring the Ecological Condition of Wadeable Streams	A revision and update of the 1994 Methods Manual for EMAP.	Klemm and Lazorchak 1995
1996	Biological Assessment Methods, Biocriteria, and Biological Indicators: Bibliography of Selected Technical, Policy, and Regulatory Literature	USEPA compiles a comprehensive literature survey of pertinent research papers and studies for biological assessment methods. This document is expanded and updated from USEPA 1991b.	Stribling et al. 1996a
1996	Summary of State Biological Assessment Programs for Wadeable Streams and Rivers	The status of bioassessment and biocriteria programs in state water resource programs is summarized in this document, providing an update of USEPA 1991a.	Davis et al. 1996
1996	Biological Criteria: Technical Guidance for Streams and Small Rivers	Technical guidance for development of biocriteria for streams and wadeable rivers is provided as a follow-up to the Program Guidance (USEPA 1990b). This technical guidance serves as a framework for developing guidance for other surface water types.	Gibson et al. 1996
1996	The Volunteer Monitor's Guide to Quality Assurance Project Plans	USEPA develops guidance for quality assurance for citizen monitoring programs.	USEPA 1996a
1996	Nonpoint Source Monitoring and Evaluation Guide	USEPA describes how biological survey methods are used in nonpoint-source investigations, and explains the value of biological and habitat assessment to evaluating BMP implementation and identifying impairment.	USEPA 1996b
1996	Biological Criteria: Technical Guidance for Survey Design and Statistical Evaluation of Biosurvey Data	USEPA describes and define different statistical approaches for biological data analysis and development of biocriteria.	Reckhow and Warren-Hicks 1996
1997	Estuarine/Near Coastal Marine Waters Bioassessment and Biocriteria Technical Guidance	USEPA provides technical guidance on biological assessment methods and biocriteria development for estuarine and near coastal waters.	USEPA 1997a
1997	Volunteer Stream Monitoring: A Methods Manual	USEPA provides guidance for citizen monitoring groups to use biological and habitat assessment methods for monitoring streams. These methods are based in part on the RBPs.	USEPA 1997b
1997	Guidelines for Preparation of Comprehensive State Water Quality Assessments (305[b] reports)	USEPA provides guidelines for states for preparing 305(b) reports to Congress.	USEPA 1997c
1997	Biological Monitoring and Assessment: Using Multimetric Indexes Effectively	An explanation of the value, use, and scientific principles associated with using a multimetric approach to bioassessment is provided by Drs. Karr and Chu.	Karr and Chu 1999
1998	Lake and Reservoir Bioassessment and Biocriteria Technical Guidance Document	USEPA provides technical guidance on biological assessment methods and biocriteria development for lakes and reservoirs.	USEPA 1998

Chapter 2: Application of Rapid Bioassessment Protocols (RBPs)

Year	Document Title	Relationship to Bioassessment	Citation
1998	Environmental Monitoring and Assessment Program Surface Waters: Field Operations and Methods for Measuring the Ecological Condition of Wadeable Streams	A revision and update of the 1995 Methods Manual for EMAP.	Lazorchak et al. 1998

2.3 PROGRAMMATIC APPLICATIONS OF BIOLOGICAL DATA

States (and tribes to a certain extent) are responsible for identifying water quality problems, especially those waters needing Total Maximum Daily Loads (TMDLs), and evaluating the effectiveness of point and nonpoint source water quality controls. The biological monitoring protocols presented in this guidance document will strengthen a state's monitoring program if other bioassessment and monitoring techniques are not already in place. An effective and thorough biological monitoring program can help to improve reporting (e.g., 305(b) reporting), increase the effectiveness of pollution prevention efforts, and document the progress of mitigation efforts. This section provides suggestions for the application of biological monitoring to wadeable streams and rivers through existing state programs.

2.3.1 CWA Section 305(b)—Water Quality Assessment

Section 305(b) establishes a process for reporting information about the quality of the Nation's water resources (USEPA 1997c, USEPA 1994b). States, the District of Columbia, territories, some tribes, and certain River Basin Commissions have developed programs to monitor surface and ground waters and to report the current status of water quality biennially to USEPA. This information is compiled into a biennial *National Water Quality Inventory* report to Congress.

Use of biological assessment in section 305(b) reports helps to define an understandable endpoint of relevance to society—the biological integrity of waterbodies. Many of the better-known and widely reported pollution cleanup success stories have involved the recovery or reappearance of valued sport fish and other pollution-intolerant species to systems from which they had disappeared (USEPA 1980). Improved coverage of biological integrity issues, based on monitoring protocols with clear bioassessment endpoints, will make the section 305(b) reports more accessible and meaningful to many segments of the public.

Biological monitoring provides data that augment several of the section 305(b) reporting requirements. In particular, the following assessment activities and reporting requirements are enhanced through the use of biological monitoring information:

> ! Determine the status of the water resource (Are the designated/beneficial and aquatic life uses being met?).

> ! Evaluate the causes of degraded water resources and the relative contributions of pollution sources.

> ! Report on the activities underway to assess and restore water resource integrity.

> ! Determine the effectiveness of control and mitigation programs.

Rapid Bioassessment Protocols for Use in Streams and Wadeable Rivers: Periphyton, Benthic Macroinvertebrates, and Fish, Second Edition

2-5

!   Measure the success of watershed management plans.

2.3.2 CWA Section 319—Nonpoint Source Assessment

The 1987 Water Quality Act Amendments to the Clean Water Act (CWA) added section 319, which established a national program to assess and control nonpoint source (NPS) pollution. Under this program, states are asked to assess their NPS pollution problems and submit these assessments to USEPA. The assessments include a list of "navigable waters within the state which, without additional action to control nonpoint source of pollution, cannot reasonably be expected to attain or maintain applicable water quality standards or the goals and requirements of this Act." Other activities under the section 319 process require the identification of categories and subcategories of NPS pollution that contribute to the impairment of waters, descriptions of the procedures for identifying and implementing BMPs, control measures for reducing NPS pollution, and descriptions of state and local programs used to abate NPS pollution. Based on the assessments, states have prepared nonpoint source management programs.

Assessment of biological condition is the most effective means of evaluating cumulative impacts from nonpoint sources, which may involve habitat degradation, chemical contamination, or water withdrawal (Karr 1991). Biological assessment techniques can improve evaluations of nonpoint source pollution controls (or the combined effectiveness of current point and nonpoint source controls) by comparing biological indicators before and after implementation of controls. Likewise, biological attributes can be used to measure site-specific ecosystem response to remediation or mitigation activities aimed at reducing nonpoint source pollution impacts or response to pollution prevention activities.

2.3.3 Watershed Protection Approach

Since 1991, USEPA has been promoting the Watershed Protection Approach (WPA) as a framework for meeting the Nation's remaining water resource challenges (USEPA 1994c). USEPA's Office of Water has taken steps to reorient and coordinate point source, nonpoint source, surface waters, wetlands, coastal, ground water, and drinking water programs in support of the watershed approach. USEPA has also promoted multi-organizational, multi-objective watershed management projects across the Nation.

The watershed approach is an integrated, inclusive strategy for more effectively protecting and managing surface water and ground water resources and achieving broader environmental protection objectives using the naturally defined hydrologic unit (the watershed) as the integrating management unit. Thus, for a given watershed, the approach encompasses not only the water resource, such as a stream, river, lake, estuary, or aquifer, but all the land from which water drains to the resource. The watershed approach places emphasis on all aspects of water resource quality—physical (e.g., temperature, flow, mixing, habitat); chemical (e.g., conventional and toxic pollutants such as nutrients and pesticides); and biological (e.g., health and integrity of biotic communities, biodiversity).

As states develop their Watershed Protection Approach (WPA), biological assessment and monitoring offer a means of conducting comprehensive evaluations of ecological status and improvements from restoration/rehabilitation activities. Biological assessment integrates the condition of the watershed from tributaries to mainstem through the exposure/response of indigenous aquatic communities.

2.3.4 CWA Section 303(d)—The TMDL Process

The technical backbone of the WPA is the TMDL process. A total maximum daily load (TMDL) is a tool used to achieve applicable water quality standards. The TMDL process quantifies the loading capacity of a waterbody for a given stressor and ultimately provides a quantitative scheme for allocating loadings (or external inputs) among pollutant sources (USEPA 1994a). In doing so, the TMDL quantifies the relationships among sources, stressors, recommended controls, and water quality conditions. For example, a TMDL might mathematically show how a specified percent reduction of a pollutant is necessary to reach the pollutant concentration reflected in a water quality standard.

Section 303(d) of the CWA requires each state to establish, in accordance with its priority rankings, the total maximum daily load for each waterbody or reach identified by the state as failing to meet, or not expected to meet, water quality standards after imposition of technology-based controls. In addition, TMDLs are vital elements of a growing number of state programs. For example, as more permits incorporate water quality-based effluent limits, TMDLs are becoming an increasingly important component of the point-source control program.

TMDLs are suitable for nonchemical as well as chemical stressors (USEPA 1994a). These include all stressors that contribute to the failure to meet water quality standards, as well as any stressor that presently threatens but does not yet impair water quality. TMDLs are applicable to waterbodies impacted by both point and nonpoint sources. Some stressors, such as sediment deposition or physical alteration of instream habitat, might not clearly fit traditional concepts associated with chemical stressors and loadings. For these nonchemical stressors, it might sometimes be difficult to develop TMDLs because of limitations in the data or in the technical methods for analysis and modeling. In the case of nonpoint source TMDLs, another difficulty arises in that the CWA does not provide well-defined support for regulatory control actions as it does for point source controls, and controls based on another statutory authority might be necessary.

Biological assessments and criteria address the cumulative impacts of all stressors, especially habitat degradation, and chemical contamination, which result in a loss of biological diversity. Biological information can help provide an ecologically based assessment of the status of a waterbody and as such can be used to decide which waterbodies need TMDLs (USEPA 1997c) and aid in the ranking process by targeting waters for TMDL development with a more accurate link between bioassessment and ecological integrity.

Finally, the TMDL process is a geographically-based approach to preparing load and wasteload allocations for sources of stress that might impact waterbody integrity. The geographic nature of this process will be complemented and enhanced if ecological regionalization is applied as part of the bioassessment activities. Specifically, similarities among ecosystems can be grouped into homogeneous classes of streams and rivers that provides a geographic framework for more efficient aquatic resource management.

2.3.5 CWA Section 402—NPDES Permits and Individual Control Strategies

All point sources of wastewater must obtain a National Pollutant Discharge Elimination System (NPDES) permit (or state equivalent), which regulates the facility's discharge of pollutants. The approach to controlling and eliminating water pollution is focused on the pollutants determined to be harmful to receiving waters and on the sources of such pollutants. Authority for issuing NPDES permits is established under Section 402 of the CWA (USEPA 1989).

Point sources are generally divided into two types—industrial and municipal. Nationwide, there are approximately 50,000 industrial sources, which include commercial and manufacturing facilities. Municipal sources, also known as publicly owned treatment works (POTWs), number about 15,700 nationwide. Wastewater from municipal sources results from domestic wastewater discharged to POTWs, as well as the "indirect" discharge of industrial wastes to sewers. In addition, stormwater may be discrete or diffuse, but is also covered by NPDES permitting regulations.

USEPA does not recommend the use of biological survey data as the basis for deriving an effluent limit for an NPDES permit (USEPA 1994d). Unlike chemical-specific water quality analyses, biological data do not measure the concentrations or levels of chemical stressors. Instead, they directly measure the impacts of any and all stressors on the resident aquatic biota. Where appropriate, biological assessment can be used within the NPDES process (USEPA 1994d) to obtain information on the status of a waterbody where point sources might cause, or contribute to, a water quality problem. In conjunction with chemical water quality and whole-effluent toxicity data, biological data can be used to detect previously unmeasured chemical water quality problems and to evaluate the effectiveness of implemented controls.

Some states have already demonstrated the usefulness of biological data to indicate the need for additional or more stringent permit limits (e.g., sole-source discharge into a stream where there is no significant nonpoint source discharge, habitat degradation, or atmospheric deposition) (USEPA 1994d). In these situations, the biological findings triggered additional investigations to establish the cause-and-effect relationship and to determine the appropriate limits. In this manner, biological data support regulatory evaluations and decision making. Biological data can also be useful in monitoring highly variable or diffuse sources of pollution that are treated as point sources such as wet-weather discharges and stormwater runoff (USEPA 1994d). Traditional chemical water quality monitoring is usually only minimally informative for these types of point source pollution, and a biological survey of their impact might be critical to effectively evaluate these discharges and associated treatment measures.

2.3.6 Ecological Risk Assessment

Risk assessment is a scientific process that includes stressor identification, receptor characterization and endpoint selection, stress-response assessment, and risk characterization (USEPA 1992, Suter et al. 1993). Risk management is a decision-making process that involves all the human-health and ecological assessment results, considered with political, legal, economic, and ethical values, to develop and enforce environmental standards, criteria, and regulations (Maughan 1993). Risk assessment can be performed on an on-site basis or can be geographically-based (i.e., watershed or regional scale), and it can be used to assess human health risks or to identify ecological impairments. In early 1997, a report prepared by a Presidential/Congressional Commission on risk enlarged the context of risk to include ecological as well as public health risks (Karr and Chu 1997).

Biological monitoring is the essential foundation of ecological risk assessment because it measures present biological conditions — not just chemical contamination — and provides the means to compare them with the conditions expected in the absence of humans (Karr and Chu 1997). Results of regional bioassessment studies can be used in watershed ecological risk assessments to develop broad scale (geographic) empirical models of biological responses to stressors. Such models can then be used, in combination with exposure information, to predict risk due to stressors or to alternative management actions. Risks to biological resources are characterized, and sources of stress can be prioritized. Watershed risk managers can and should use such results for critical management decisions.

2.3.7 USEPA Water Quality Criteria and Standards

The water quality standards program, as envisioned in Section 303(c) of the Clean Water Act, is a joint effort between the states and USEPA. The states have primary responsibility for setting, reviewing, revising, and enforcing water quality standards. USEPA develops regulations, policies, and guidance to help states implement the program and oversees states' activities to ensure that their adopted standards are consistent with the requirements of the CWA and relevant water quality standards regulations (40 CFR Part 131). USEPA has authority to review and approve or disapprove state standards and, where necessary, to promulgate federal water quality standards.

A water quality standard defines the goals of a waterbody, or a portion thereof, by designating the use or uses to be made of the water, setting criteria necessary to protect those uses, and preventing degradation of water quality through antidegradation provisions. States adopt water quality standards to protect public health or welfare, enhance the quality of water, and protect biological integrity.

Chemical, physical, or biological stressors impact the biological characteristics of an aquatic ecosystem (Gibson et al. 1996). For example, chemical stressors can result in impaired functioning or loss of a sensitive species and a change in community structure. Ultimately, the number and intensity of all stressors within an ecosystem will be evidenced by a change in the condition and function of the biotic community. The interactions among chemical, physical, and biological stressors and their cumulative impacts emphasize the need to directly detect and assess the biota as indicators of actual water resource impairments.

Sections 303 and 304 of the CWA require states to protect biological integrity as part of their water quality standards. This can be accomplished, in part, through the development and use of biological criteria. As part of a state or tribal water quality standards program, biological criteria can provide scientifically sound and detailed descriptions of the designated aquatic life use for a specific waterbody or segment. They fulfill an important assessment function in water quality-based programs by establishing the biological benchmarks for (1) directly measuring the condition of the aquatic biota, (2) determining water quality goals and setting priorities, and (3) evaluating the effectiveness of implemented controls and management actions.

Biological criteria for aquatic systems provide an evaluation benchmark for direct assessment of the condition of the biota that live either part or all of their lives in aquatic systems (Gibson et al. 1996) by describing (in narrative or numeric criteria) the expected biological condition of a minimally impaired aquatic community (USEPA 1990b). They can be used to define ecosystem rehabilitation goals and assessment endpoints. Biological criteria supplement traditional measurements (for example, as backup for hard-to-detect chemical problems) and will be particularly useful in assessing impairment due to nonpoint source pollution and nonchemical (e.g., physical and biological) stressors. Thus, biological criteria fulfill a function missing from USEPA's traditionally chemical-oriented approach to pollution control and abatement (USEPA 1994d).

Biological criteria can also be used to refine the aquatic life use classifications for a state. Each state develops its own designated use classification system based on the generic uses cited in the CWA, including protection and propagation of fish, shellfish, and wildlife. States frequently develop

Rapid Bioassessment Protocols for Use in Streams and Wadeable Rivers: Periphyton, Benthic Macroinvertebrates, and Fish, Second Edition

2-9

subcategories to refine and clarify designated use classes when several surface waters with distinct characteristics fit within the same use class or when waters do not fit well into any single category. As data are collected from biosurveys to develop a biological criteria program, analysis may reveal unique and consistent differences between aquatic communities that inhabit different waters with the same designated use. Therefore, measurable biological attributes can be used to refine aquatic life use or to separate 1 class of aquatic life into 2 or more subclasses. For example, Ohio has established an *exceptional warmwater* use class to include all *unique waters* (i.e., not representative of regional streams and different from their standard warmwater class).

3   ELEMENTS OF BIOMONITORING

3.1   BIOSURVEYS, BIOASSAYS, AND CHEMICAL MONITORING

The water quality-based approach to pollution assessment requires various types of data. Biosurvey techniques, such as the Rapid Bioassessment Protocols (RBPs), are best used for detecting aquatic life impairments and assessing their relative severity. Once an impairment is detected, however, additional ecological data, such as chemical and biological (toxicity) testing is helpful to identify the causative agent, its source, and to implement appropriate mitigation (USEPA 1991c). Integrating information from these data types as well as from habitat assessments, hydrological investigations, and knowledge of land use is helpful to provide a comprehensive diagnostic assessment of impacts from the 5 principal factors (see Karr et al. 1986, Karr 1991, Gibson et al. 1996 for description of water quality, habitat structure, energy source, flow regime, and biotic interaction factors). Following mitigation, biosurveys are important for evaluating the effectiveness of such control measures. Biosurveys may be used within a planning and management framework to prioritize water quality problems for more stringent assessments and to document "environmental recovery" following control action and rehabilitation activities. Some of the advantages of using biosurveys for this type of monitoring are:

> !   Biological communities reflect overall ecological integrity (i.e., chemical, physical, and biological integrity). Therefore, biosurvey results directly assess the status of a waterbody relative to the primary goal of the Clean Water Act (CWA).

> !   Biological communities integrate the effects of different stressors and thus provide a broad measure of their aggregate impact.

> !   Communities integrate the stresses over time and provide an ecological measure of fluctuating environmental conditions.

> !   Routine monitoring of biological communities can be relatively inexpensive, particularly when compared to the cost of assessing toxic pollutants, either chemically or with toxicity tests (Ohio EPA 1987).

> !   The status of biological communities is of direct interest to the public as a measure of a pollution free environment.

> !   Where criteria for specific ambient impacts do not exist (e.g., nonpoint-source impacts that degrade habitat), biological communities may be the only practical means of evaluation.

Biosurvey methods have a long-standing history of use for "before and after" monitoring. However, the intermediate steps in pollution control, i.e., identifying causes and limiting sources, require integrating information of various types—chemical, physical, toxicological, and/or biosurvey data. These data are needed to:

Identify the specific stress agents causing impact: This may be a relatively simple task; but, given the array of potentially important pollutants (and their possible combinations), it is likely to be both difficult and costly. In situations where specific chemical stress agents are either poorly understood or too varied to assess individually, toxicity tests can be used to focus specific chemical investigations or to characterize generic stress agents (e.g., whole effluent or ambient toxicity). For situations where habitat degradation is prevalent, a combination of biosurvey and physical habitat assessment is most useful (Barbour and Stribling 1991).

Identify and limit the specific sources of these agents: Although biosurveys can be used to help locate the likely origins of impact, chemical analyses and/or toxicity tests are helpful to confirm the point sources and develop appropriate discharge limits. Impacts due to factors other than chemical contamination will require different ecological data.

Design appropriate treatment to meet the prescribed limits and monitor compliance: Treatment facilities are designed to remove identified chemical constituents with a specific efficiency. Chemical data are therefore required to evaluate treatment effectiveness. To some degree, a biological endpoint resulting from toxicity testing can also be used to evaluate the effectiveness of prototype treatment schemes and can serve as a design parameter. In most cases, these same parameters are limited in discharge permits and, after controls are in place, are used to monitor for compliance. Where discharges are not controlled through a permit system (e.g., nonpoint-source runoff, combined sewer outfalls, and dams) compliance must be assessed in terms of ambient standards. Improvement of the ecosystem both from restoration or rehabilitation activities are best monitored by biosurvey techniques.

Effective implementation of the water quality-based approach requires that various monitoring techniques be considered within a larger context of water resource management. Both biological and chemical methods play critical roles in a successful pollution control program. They should be considered complementary rather than mutually exclusive approaches that will enhance overall program effectiveness when used appropriately.

3.2   USE OF DIFFERENT ASSEMBLAGES IN BIOSURVEYS

The techniques presented in this document focus on the evaluation of water quality (physicochemical constituents), habitat parameters, and analysis of the periphyton, benthic macroinvertebrate, and fish assemblages. Many State water quality agencies employ trained and experienced benthic biologists, have accumulated considerable background data on macroinvertebrates, and consider benthic surveys a useful assessment tool. However, water quality standards, legislative mandate, and public opinion are more directly related to the status of a waterbody as a fishery resource. For this reason, separate protocols were developed for fish and were incorporated as Chapter 8 in this document. The fish survey protocol is based largely on Karr's Index of Biotic Integrity (IBI) (Karr 1981, Karr et al. 1986, Miller et al. 1988), which uses the structure of the fish assemblage to evaluate water quality. The integration of functional and structural/compositional metrics, which forms the basis for the IBI, is a common element to the rapid bioassessment approaches.

The periphyton assemblage (primarily algae) is also useful for water quality monitoring, but has not been incorporated widely in monitoring programs. They represent the primary producer trophic level, exhibit a different range of sensitivities, and will often indicate effects only indirectly observed in the benthic and fish communities. As in the benthic macroinvertebrate and fish assemblages, integration of structural/compositional and functional characteristics provides the best means of assessing impairment (Rodgers et al. 1979).

In selecting the aquatic assemblage appropriate for a particular biomonitoring situation, the advantages of using each assemblage must be considered along with the objectives of the program. Some of the advantages of using periphyton, benthic macroinvertebrates, and fish in a biomonitoring program are presented in this section. References for this list are Cairns and Dickson (1971), American Public Health Association et al. (1971), Patrick (1973), Rodgers et al. (1979), Weitzel (1979), Karr (1981), USEPA (1983), Hughes et al. (1982), and Plafkin et al. (1989).

3.2.1 Advantages of Using Periphyton

! Algae generally have rapid reproduction rates and very short life cycles, making them valuable indicators of short-term impacts.

! As primary producers, algae are most directly affected by physical and chemical factors.

! Sampling is easy, inexpensive, requires few people, and creates minimal impact to resident biota.

! Relatively standard methods exist for evaluation of functional and non-taxonomic structural (biomass, chlorophyll measurements) characteristics of algal communities.

! Algal assemblages are sensitive to some pollutants which may not visibly affect other aquatic assemblages, or may only affect other organisms at higher concentrations (i.e., herbicides).

3.2.2 Advantages of Using Benthic Macroinvertebrates

! Macroinvertebrate assemblages are good indicators of localized conditions. Because many benthic macroinvertebrates have limited migration patterns or a sessile mode of life, they are particularly well-suited for assessing site-specific impacts (upstream-downstream studies).

! Macroinvertebrates integrate the effects of short-term environmental variations. Most species have a complex life cycle of approximately one year or more. Sensitive life stages will respond quickly to stress; the overall community will respond more slowly.

! Degraded conditions can often be detected by an experienced biologist with only a cursory examination of the benthic macroinvertebrate assemblage. Macro-invertebrates are relatively easy to identify to family; many "intolerant" taxa can be identified to lower taxonomic levels with ease.

! Benthic macroinvertebrate assemblages are made up of species that constitute a broad range of trophic levels and pollution tolerances, thus providing strong information for interpreting cumulative effects.

! Sampling is relatively easy, requires few people and inexpensive gear, and has minimal detrimental effect on the resident biota.

Rapid Bioassessment Protocols for Use in Streams and Wadeable Rivers: Periphyton, Benthic Macroinvertebrates, and Fish, Second Edition

3-3

! Benthic macroinvertebrates serve as a primary food source for fish, including many recreationally and commercially important species.

! Benthic macroinvertebrates are abundant in most streams. Many small streams (1st and 2nd order), which naturally support a diverse macroinvertebrate fauna, only support a limited fish fauna.

! Most state water quality agencies that routinely collect biosurvey data focus on macroinvertebrates (Southerland and Stribling 1995). Many states already have background macroinvertebrate data. Most state water quality agencies have more expertise with invertebrates than fish.

3.2.3 Advantages of Using Fish

! Fish are good indicators of long-term (several years) effects and broad habitat conditions because they are relatively long-lived and mobile (Karr et al. 1986).

! Fish assemblages generally include a range of species that represent a variety of trophic levels (omnivores, herbivores, insectivores, planktivores, piscivores). They tend to integrate effects of lower trophic levels; thus, fish assemblage structure is reflective of integrated environmental health.

! Fish are at the top of the aquatic food web and are consumed by humans, making them important for assessing contamination.

! Fish are relatively easy to collect and identify to the species level. Most specimens can be sorted and identified in the field by experienced fisheries professionals, and subsequently released unharmed.

! Environmental requirements of most fish are comparatively well known. Life history information is extensive for many species, and information on fish distributions is commonly available.

! Aquatic life uses (water quality standards) are typically characterized in terms of fisheries (coldwater, coolwater, warmwater, sport, forage). Monitoring fish provides direct evaluation of "fishability" and "fish propagation", which emphasizes the importance of fish to anglers and commercial fishermen.

! Fish account for nearly half of the endangered vertebrate species and subspecies in the United States (Warren and Burr 1994).

3.3   IMPORTANCE OF HABITAT ASSESSMENT

The procedure for assessing physical habitat quality presented in this document (Chapter 5) is an integral component of the final evaluation of impairment. The matrix used to assess habitat quality is based on key physical characteristics of the waterbody and surrounding land, particularly the catchment of the site under investigation. All of the habitat parameters evaluated are related to overall aquatic life use and are a potential source of limitation to the aquatic biota.

The alteration of the physical structure of the habitat is one of 5 major factors from human activities described by Karr (Karr et al. 1986, Karr 1991) that degrade aquatic resources. Habitat, as structured by instream and surrounding topographical features, is a major determinant of aquatic community potential (Southwood 1977, Plafkin et al. 1989, and Barbour and Stribling 1991). Both the quality and quantity of available habitat affect the structure and composition of resident biological communities. Effects of such features on biological assessment results can be minimized by sampling similar habitats at all stations being compared. However, when all stations are not physically comparable, habitat characterization is particularly important for proper interpretation of biosurvey results.

Where physical habitat quality at a test site is similar to that of a reference, detected impacts can be attributed to water quality factors (i.e., chemical contamination) or other stressors. However, where habitat quality differs substantially from reference conditions, the question of appropriate aquatic life use designation and physical habitat alteration/restoration must be addressed. Final conclusions regarding the presence and degree of biological impairment should thus include an evaluation of habitat quality to determine the extent that habitat may be a limiting factor. The habitat characterization matrix included in the Rapid Bioassessment Protocols provides an effective means of evaluating and documenting habitat quality at each biosurvey station.

3.4 THE REGIONAL REFERENCE CONCEPT

The issue of reference conditions is critical to the interpretation of biological surveys. Barbour et al. (1996a) describe 2 types of reference conditions that are currently used in biological surveys: site-specific and regional reference. The former typically consists of measurements of conditions upstream of a point source discharge or from a "paired" watershed. Regional reference conditions, on the other hand, consist of measurements from a population of relatively unimpaired sites within a relatively homogeneous region and habitat type, and therefore are not site-specific.

The reference condition establishes the basis for making comparisons and for detecting use impairment; it should be applicable to an individual waterbody, such as a stream segment, but also to similar waterbodies on a regional scale (Gibson et al. 1996).

Although both site-specific and ecoregional references represent conditions without the influence of a particular discharge, the 2 types of references may not yield equivalent measurements (Barbour et al. 1996a). While site-specific reference conditions represented by the upstream, downstream, or paired-site approach are desirable, they are limited in their usefulness. Hughes (1995) points out three problems with site-specific reference conditions: (1) because they typically lack any broad study design, site-specific reference conditions possess limited capacity for extrapolation— they have only site-specific value; (2) usually site-specific reference conditions allow limited variance estimates; there are too few sites for robust variance evaluations because each site of concern is typically represented by one-to-three reference sites; the result could be an incorrect assessment if the upstream site has especially good or especially poor habitat or chemical quality; and (3) they involve a substantial assessment effort when considered on a statewide basis.

The advantages of measuring upstream reference conditions are these: (1) if carefully selected, the habitat quality is often similar to that measured downstream of a discharge, thereby reducing complications in interpretation arising from habitat differences, and (2) impairments due to upstream influences from other point and nonpoint sources are already factored into the reference condition (Barbour et al. 1996a). New York DEC has found that an upstream-downstream approach aids in diagnosing cause-and-effect to specific discharges and increase precision (Bode and Novak 1995).

Where feasible, effects should be bracketed by establishing a series or network of sampling stations at points of increasing distance from the impact source(s). These stations will provide a basis for delineating impact and recovery zones. In significantly altered systems (i.e., channelized or heavily urbanized streams), suitable reference sites are usually not available (Gibson et al. 1996). In these cases, historical data or simple ecological models may be necessary to establish reference conditions. See Gibson et al. (1996) for more detail.

Innate regional differences exist in forests, lands with high agricultural potential, wetlands, and waterbodies. These regional differences have been mapped by Bailey (1976), U.S. Department of Agriculture (USDA) Soil Conservation Service (1981), Energy, Mines and Resources Canada (1986), and Omernik (1987). Waterbodies reflect the lands they drain (Omernik 1987, Hunsaker and Levine 1995) and it is assumed that similar lands should produce similar waterbodies. This ecoregional approach provides robust and ecologically-meaningful regional maps that are based on an examination of several mapped land variables. For example, hydrologic unit maps are useful for mapping drainage patterns, but have limited value for explaining the substantial changes that occur in water quality and biota independent of stream size and river basin.

Omernik (1987) provided an ecoregional framework for interpreting spatial patterns in state and national data. The geographical framework is based on regional patterns in land-surface form, soil, potential natural vegetation, and land use, which vary across the country. Geographic patterns of similarity among ecosystems can be grouped into ecoregions or subecoregions. Naturally occurring biotic assemblages, as components of the ecosystem, would be *expected* to differ among ecoregions but be relatively similar within a given ecoregion. The ecoregion concept thus provides a geographic framework for efficient management of aquatic ecosystems and their components (Hughes 1985, Hughes et al. 1986, and Hughes and Larsen 1988). For example, studies in Ohio (Larsen et al. 1986), Arkansas (Rohm et al. 1987), and Oregon (Hughes et al. 1987, Whittier et al. 1988) have shown that distributional patterns of fish communities approximate ecoregional boundaries as defined *a priori* by Omernik (1987). This, in turn, implies that similar water quality standards, criteria, and monitoring strategies are likely to be valid throughout a given ecoregion, but should be tailored to accommodate the innate differences among ecoregions (Ohio EPA 1987).

However, some programs, such as EMAP (Klemm and Lazorchak 1994) and the Maryland Biological Stream Survey (MBSS) (Volstad et al. 1995) have found that a surrogate measure of stream size (catchment size) is useful in partitioning the variability of stream segments for assessment. Hydrologic regime can include flow regulation, water withdrawal, and whether a stream is considered intermittent or perennial. Elevation has been found to be an important classification variable when using the benthic macroinvertebrate assemblage (Barbour et al. 1992, Barbour et al. 1994, Spindler 1996). In addition, descriptors at a smaller scale may be needed to characterize streams within regions or classes. For example, even though a given stream segment is classified within a subecoregion or other type of stream class, it may be wooded (deciduous or coniferous) or open within a perennial or intermittent flow regime, and represent one of several orders of stream size.

Individual descriptors *will not apply to all regional reference streams,* nor will all conditions (i.e., deciduous, coniferous, open) be present in all streams. Those streams or stream segments that represent characteristics atypical for that particular ecoregion should be excluded from the regional aggregate of sites and treated as a special situation. For example, Ohio EPA (1987) considered aquatic systems with unique (i.e., unusual for the ecoregion) natural characteristics to be a separate aquatic life use designation (exceptional warmwater aquatic life use) on a statewide basis.

Although the final rapid bioassessment guidance should be generally applicable to all regions of the United States, each agency will need to evaluate the generic criteria suggested in this document for inclusion into specific programs. To this end, the application of the regional reference concept versus the site-specific control approach will need to be examined. When Rapid Bioassessment Protocols (RBPs) are used to assess impact sources (upstream-downstream studies), regional reference criteria may not be as important if an unimpacted site-specific control station can be sampled. However, when a synoptic ("snapshot") or trend monitoring survey is being conducted in a watershed or river basin, use of regional criteria may be the only means of discerning use impairment or assessing impact. Additional investigation will be needed to: delineate areas (classes of streams)that differ significantly in their innate biological potential; locate reference sites within each stream class that fully support aquatic life uses; develop biological criteria (e.g., define optimal values for the metrics) using data generated from each of the assemblages.

3.5   STATION SITING

Site selection for assessment and monitoring can either be "targeted", i.e., relevant to special studies that focus on potential problems, or "probabilistic", which provides information of the overall status or condition of the watershed, basin, or region. In a probabilistic or random sampling regime, stream characteristics may be highly dissimilar among the sites, but will provide a more accurate assessment of biological condition throughout the area than a targeted design. Selecting sites randomly provides an unbiased assessment of the condition of the waterbody at a scale above the individual site or stream. Thus, an agency can address questions at multiple scales. Studies for 305(b) status and trends assessments are best done with a probabilistic design.

Most studies conducted by state water quality agencies for identification of problems and sensitive waters are done with a targeted design. In this case, sampling sites are selected based on known existing problems, knowledge of upcoming events that will adversely affect the waterbody such as a development or deforestation; or installation of BMPs or habitat restoration that are intended to improve waterbody quality. This method provides assessments of individual sites or stream reaches. Studies for aquatic life use determination and those related to TMDLs can be done with a random (watershed or higher level) or targeted (site-specific) design.

To meaningfully evaluate biological condition in a targeted design, sampling locations must be similar enough to have similar biological expectations, which, in turn, provides a basis for comparison of impairment. If the goal of an assessment is to evaluate the effects of water chemistry degradation, comparable physical habitat should be sampled at all stations, otherwise, the differences in the biology attributable to a degraded habitat will be difficult to separate from those resulting from chemical pollution water quality degradation. Availability of appropriate habitat at each sampling location can be established during preliminary reconnaissance. In evaluations where several stations on a waterbody will be compared, the station with the greatest habitat constraints (in terms of productive habitat availability) should be noted. The station with the least number of productive habitats available will often determine the type of habitat to be sampled at all sample stations.

Locally modified sites, such as small impoundments and bridge areas, should be avoided unless data are needed to assess their effects. Sampling near the mouths of tributaries entering large waterbodies should also be avoided because these areas will have habitat more typical of the larger waterbody (Karr et al. 1986).

For bioassessment activities where the concern is non-chemical stressors, e.g., the effects of habitat degradation or flow alteration, or cumulative impacts, a different approach to station selection is used. Physical habitat differences between sites can be substantial for two reasons: (1) one or a set of sites is

Rapid Bioassessment Protocols for Use in Streams and Wadeable Rivers: Periphyton, Benthic Macroinvertebrates, and Fish, Second Edition

3-7

more degraded (physically) than another, or (2) is unique for the stream class or region due to the essential natural structure resulting from geological characteristics. Because of these situations, the more critical part of the siting process comes from the recognition of the habitat features that are representative of the region or stream class. In basin-wide or watershed studies, sample locations should not be avoided due to habitat degradation or to physical features that are well-represented in the stream class.

3.6   DATA MANAGEMENT AND ANALYSIS

USEPA is developing a biological data management system linked to STORET, which provides a centralized system for storage of biological data and associated analytical tools for data analysis. The field survey file component of STORET provides a means of storing, retrieving, and analyzing biosurvey data, and will process data on the distribution, abundance, and physical condition of aquatic organisms, as well as descriptions of their habitats. Data stored in STORET become part of a comprehensive database that can be used as a reference, to refine analysis techniques or to define ecological requirements for aquatic populations. Data from the Rapid Bioassessment Protocols can be readily managed with the STORET field survey file using header information presented on the field data forms (Appendix A) to identify sampling stations.

Habitat and physical characterization information may also be stored in the field survey file with organism abundance data. Parameters available in the field survey file can be used to store some of the environmental characteristics associated with the sampling event, including physical characteristics, water quality, and habitat assessment. Physical/chemical parameters include stream depth, velocity, and substrate characteristics, as well as many other parameters. STORET also allows storage of other pertinent station or sample information in the comments section.

Entering data into a computer system can provide a substantial time savings. An additional advantage to computerization is analysis documentation, which is an important component for a Quality Assurance/Quality Control (QA/QC) plan. An agency conducting rapid bioassessment programs can choose an existing system within their agency or utilize the STORET system developed as a national database system.

Data collected as part of state bioassessment programs are usually entered, stored and analyzed in easily obtainable spreadsheet programs. This method of data management becomes cumbersome as the database grows in volume. An alternative to spreadsheet programs is a multiuser relational database management system (RDMS). Most relational database software is designed for the Windows operating system and offer menu driven interfaces and ranges of toolbars that provide quick access to many routine database tasks. Automated tools help users quickly create forms for data input and lookup, tables, reports, and complex queries about the data. The USEPA is developing a multiuser relational database management system that can transfer sampling data to STORET. This relational database management system is EDAS (Ecological Data Application System) and allows the user to input, compile, and analyze complex ecological data to make assessments of ecosystem condition. EDAS includes tools to format sampling data so it may be loaded into STORET as a batch file. These batch files are formatted as flat ASCII text and can be loaded (transferred) electronically to STORET. This will eliminate the need to key sample data into STORET.

By using tables and queries as established in EDAS, a user can enter, manipulate, and print data. The metrics used in most bioassessments can be calculated with simple queries that have already been created for the user. New queries may be created so additional metrics can be calculated at the click of the mouse each time data are updated or changed. If an operation on the data is too complex for one of the many default functions then the function can be written in code (e.g., visual basic access) and stored in a module for use in any query. Repetitive steps can be handled with macros. As the user develops the database other database elements such as forms and reports can be added.

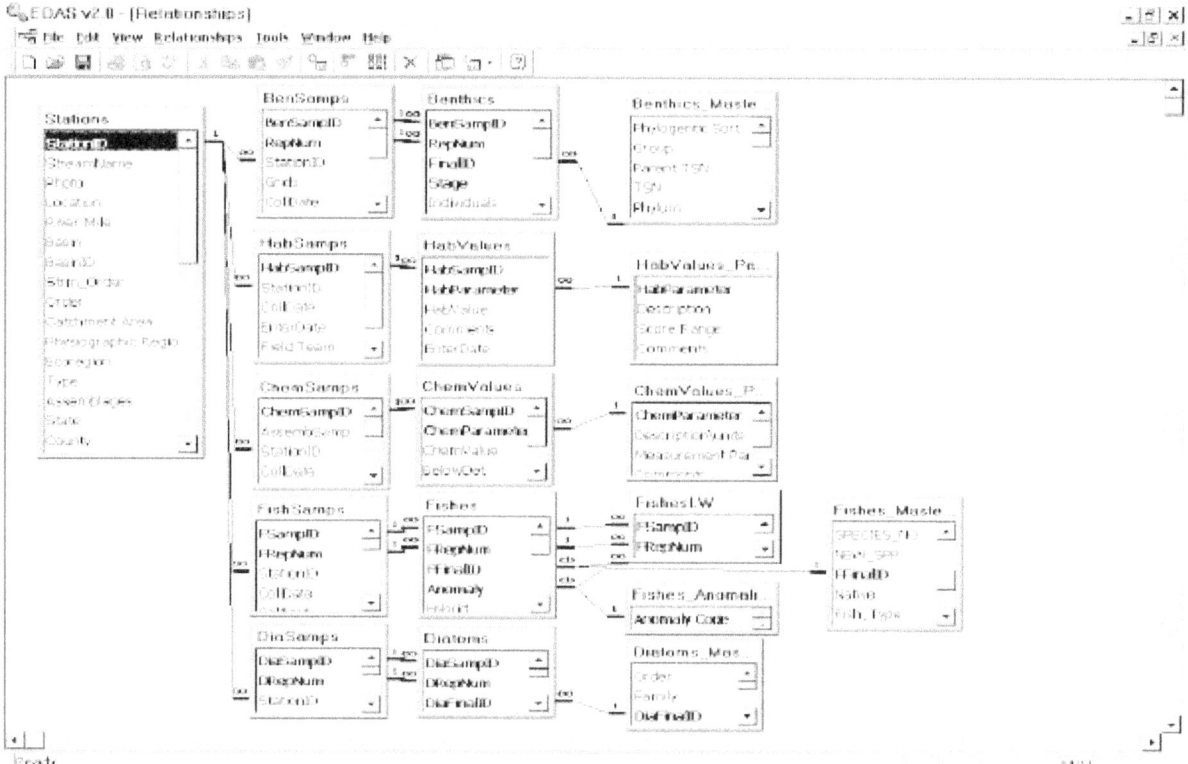

Figure 3-1. Example of the relationship of data tables in a typical relational database.

Table design is the foundation of the relational database, such as EDAS (Figure 3-1), because they function as data containers. Tables are related through the use of a unique identifier or index. In the example database "StationId" links the tables "ChemSamps", "HabSamps", and "BenSamps" to the "Stations" table. The chemical parameters and habitat parameters table act as reference tables and contain descriptive data (e.g., measurement units, detection limits). This method of storing data is more efficient than spreadsheets, because it eliminates a lot of redundant data. Master Taxa tables are created for the biological data to contain all relevant information about each taxon. This information does not have to be repeated each time a taxon is entered into the database.

Input or lookup forms (Figure 3-2) are screens that are designed to aid in entering or retrieving data. Forms are linked to tables so data go to the right cell in the right table. Because of the relationships among the tables, data can be updated across all the tables that are linked to the form. Reports can be generated in a variety of styles, and data can be exported to other databases or spreadsheet programs.

3.7   TECHNICAL ISSUES FOR SAMPLING THE PERIPHYTON ASSEMBLAGE

3.7.1 Seasonality

Stream periphyton have distinct seasonal cycles, with peak abundance and diversity typically occurring in late summer or early fall (Bahls 1993). High flows may scour and sweep away periphyton. For these reasons, the index period for periphyton sampling is usually late summer or early fall, when stream flow is relatively stable (Kentucky DEP 1993, Bahls 1993).

Algae are light limited, and may be sparse in heavily shaded streams. Early spring, before leafout, may be a better sampling index period in shaded streams.

Finally, since algae have short generation times (one to several days), they respond rapidly to environmental changes. Samples of the algal community are "snapshots" in time, and do not integrate environmental effects over entire seasons or years.

3.7.2 Sampling Methodology

Artificial substrates (periphytometers) have long been used in algal investigations, typically using glass slides as the substrate, but also with glass rods, plastic plates, ceramic tiles and other substances. However, many agencies are sampling periphyton from natural substrates to characterize

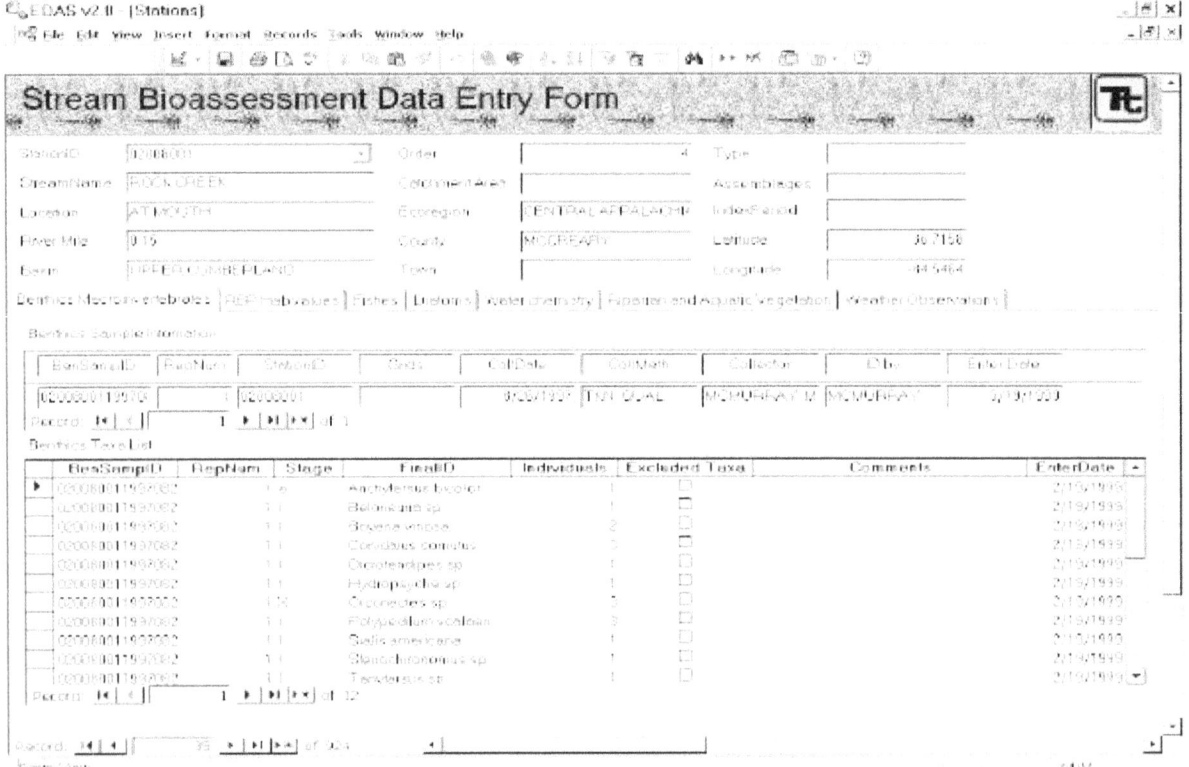

Figure 3-2. Example input or lookup form in a typical relational database.

the natural community. Advantages of artificial and natural substrates are summarized below (Cairns 1982, Bahls 1993).

Advantages of Artificial Substrates:

! Artificial substrates allow sample collection in locations that are typically difficult to sample effectively (e.g., bedrock, boulder, or shifting substrates; deep or high velocity water).

! As a "passive" sample collection device, artificial substrates permit standardized sampling by eliminating subjectivity in sample collection technique. Direct sampling of natural substrate requires similar effort and degree of efficiency for the collection of each sample. Use of artificial substrates requires standardization of setting and retrieval; however, colonization provides the actual sampling mechanism.

! Confounding effects of habitat differences are minimized by providing a standardized microhabitat. Microhabitat standardization may promote selectivity for specific organisms if the artificial substrate provides a different microhabitat than that naturally available at a site.

! Sampling variability is decreased due to a reduction in microhabitat patchiness, improving the potential for spatial and temporal similarity among samples.

! Sample collection using artificial substrates may require less skill and training than direct sampling of natural substrates.

Disadvantages of Artificial Substrates:

! Artificial substrates require a return trip; this may be a significant consideration in large states or those with limited technical resources.

! Artificial substrates are prone to loss, natural damage or vandalism.

! The material of the substrate will influence the composition and structure of the community; solid artificial substrates will favor attached forms over motile forms and compromise the usefulness of the siltation index.

! Orientation and length of exposure of the substrate will influence the composition and structure of the community.

3.8 TECHNICAL ISSUES FOR SAMPLING THE BENTHIC MACROINVERTEBRATE ASSEMBLAGE

3.8.1 Seasonality for Benthic Collections (adapted from Gibson et al. 1996)

The ideal sampling procedure is to survey the biological community with each change of season, then select the appropriate sampling periods that accommodate seasonal variation. Such indexing makes the best use of the biological data. However, resident assemblages integrate stress effects over the course of the year, and their seasonal cycles of abundance and taxa composition are fairly predictable within the limits of interannual variability.

Many programs have found that a single index period provides a strong database that allows all of their management objectives to be addressed. However, if one goal of a program is to understand seasonal variability, then establishing index periods during multiple seasons is necessary. Although a single index period would not likely be adequate for assessing the effects of catastrophic events, such as spill, those assessments should be viewed as special studies requiring sampling of reference sites during the same time period.

Ultimately, selection of the appropriate sampling period should be based on 3 factors that reflect efforts to:

1.   minimize year-to-year variability resulting from natural events,

2.   maximize gear efficiency, and

3.   maximize accessibility of targeted assemblage.

Sampling and comparisons of data from the same seasons (or index periods) as the previous year's sampling provides some correction and minimization of annual variability. The season of the year during which sampling gear is most effective is an important consideration for selecting an index period. For example, low flow or freezing conditions may hamper an agency's ability to sample with its selected gear. Seasons where those conditions are prevalent should be avoided. The targeted assemblage(s) should be accessible and not be inhabiting hard-to-reach portions of the sampling area. For example, if benthos are primarily deep in the substrate in winter, beyond normal sampling depth, that period should be avoided and another index period chosen. If high flows are typical of spring runoff periods, and sampling cannot occur, the index period should be established during typical or low flow periods.

3.8.2   Benthic Sampling Methodology

The benthic RBPs employ direct sampling of natural substrates. Because routine evaluation of a large number of sites is a primary objective of the RBPs, artificial substrates were eliminated from consideration due to time required for both placement and retrieval, and the amount of exposure time required for colonization. However, where conditions are inappropriate for the collection of natural substrate samples, artificial substrates may be an option. The Science Advisory Board (SAB 1993) cautioned that the only appropriate type of artificial substrates to be used for assessment are those that are "introduced substrates", i.e., substrates that are representative of the natural substrate of the stream system, such as rock-filled baskets in cobble- or gravel-bottomed streams. Ohio EPA and Maine DEP, are examples of states that use artificial substrates for their water resource investigations (Davis et al. 1996).

Advantages and disadvantages of artificial substrates (Cairns 1982) relative to the use of natural substrates are presented below.

Advantages of Artificial Substrates:

!   Artificial substrates allow sample collection in locations that are typically difficult to sample effectively (e.g., bedrock, boulder, or shifting substrates; deep or high velocity water).

! As a "passive" sample collection device, artificial substrates permit standardized sampling by eliminating subjectivity in sample collection technique. Direct sampling of natural substrate requires similar effort and degree of efficiency for the collection of each sample. Use of artificial substrates requires standardization of setting and retrieval; however, colonization provides the actual sampling mechanism.

! Confounding effects of habitat differences are minimized by providing a standardized microhabitat. Microhabitat standardization may promote selectivity for specific organisms if the artificial substrate provides a different microhabitat than that naturally available at a site (see second bullet under Disadvantages below). Most artificial substrates, by design, select for the Scraper and Filterer components of the benthic assemblages or for Collectors if accumulation of debris has occured in the substrates.

! Sampling variability is decreased due to a reduction in microhabitat patchiness, improving the potential for spatial and temporal similarity among samples.

! Sample collection using artificial substrates may require less skill and training than direct sampling of natural substrates. Depending on the type of artificial substrate used, properly trained technicians could place and retrieve the substrates. However, an experienced specialist should be responsible for the selection of habitats and sample sites.

Disadvantages of Artificial Substrates:

! Two trips (one to set and one to retrieve) are required for each artificial substrate sample; only one trip is necessary for direct sampling of the natural substrate. Artificial substrates require a long (8-week average) exposure period for colonization. This decreases their utility for certain rapid biological assessments.

! Samples may not be fully representative of the benthic assemblage at a station if the artificial substrate offers different microhabitats than those available in the natural substrate. Artificial substrates often selectively sample certain taxa, misrepresenting relative abundances of these taxa in the natural substrate. Artificial substrate samples would thus indicate colonization potential rather than the resident community structure. This could be advantageous if a study is designed to isolate water quality effects from substrate and other microhabitat effects. Where habitat quality is a limiting factor, artificial substrates could be used to discriminate between physical and chemical effects and assess a site's potential to support aquatic life on the basis of water quality alone.

! Sampler loss or perturbation commonly occurs due to sedimentation, extremely high or low flows, or vandalism during the relatively long (at least several weeks) exposure period required for colonization.

! Depending on the configuration of the artificial substrate used, transport and storage can be difficult. The number of artificial substrate samplers required for sample collection increases such inconvenience.

3.9 TECHNICAL ISSUES FOR THE SURVEY OF THE FISH ASSEMBLAGE

3.9.1 Seasonality for Fish Collections

Seasonal changes in the relative abundances of the fish community primarily occur during reproductive periods and (for some species) the spring and fall migratory periods. However, because larval fish sampling is not recommended in this protocol, reproductive period changes in relative abundance are not of primary importance.

Generally, the preferred sampling season is mid to late summer, when stream and river flows are moderate to low, and less variable than during other seasons. Although some fish species are capable of extensive migration, fish populations and individual fish tend to remain in the same area during summer (Funk 1957, Gerking 1959, Cairns and Kaesler 1971). The Ohio Environmental Protection Agency (1987) stated that few fishes in perennial streams migrate long distances. Hill and Grossman (1987) found that the three dominant fish species in a North Carolina stream had home ranges of 13 to 19 meters over a period of 18 months. Ross et al. (1985) and Matthews (1986) found that stream fish assemblages were stable and persistent for 10 years, recovering rapidly from droughts and floods indicating that substantial population fluctuations are not likely to occur in response to purely natural environmental phenomena. However, comparison of data collected during different seasons is discouraged, as are data collected during or immediately after major flow changes.

3.9.2 Fish Sampling Methodology

Although various gear types are routinely used to sample fish, electrofishing equipment and seines are the most commonly used collection methods in fresh water habitats. Each method has advantages and disadvantages (Hendricks et al. 1980, Nielsen and Johnson 1983). However, electrofishing is recommended for most fish field surveys because of its greater applicability and efficiency. Local conditions may require consideration of seining as an optional collection method. Advantages and disadvantages of each gear type are presented below.

3.9.2.1 Advantages and Disadvantages of Electrofishing

Advantages of Electrofishing:

!   Electrofishing allows greater standardization of catch per unit of effort.

!   Electrofishing requires less time and a reduced level of effort than some sampling methods (e.g., use of ichthyocides) (Hendricks et al. 1980).

!   Electrofishing is less selective than seining (although it is selective towards size and species) (Hendricks et al. 1980). (See second bullet under Disadvantages below).

!   If properly used, adverse effects on fish are minimized.

!   Electrofishing is appropriate in a variety of habitats.

Disadvantages of Electrofishing:

! Sampling efficiency is affected by turbidity and conductivity.

! Although less selective than seining, electrofishing is size and species selective. Effects of electrofishing increase with body size. Species specific behavioral and anatomical differences also determine vulnerability to electroshocking (Reynolds 1983).

! Electrofishing is a hazardous operation that can injure field personnel if proper safety procedures are ignored.

3.9.2.2 Advantages and Disadvantages of Seining

Advantages of Seining:

! Seines are relatively inexpensive.

! Seines are lightweight and are easily transported and stored.

! Seine repair and maintenance are minimal and can be accomplished onsite.

! Seine use is not restricted by water quality parameters.

! Effects on the fish population are minimal because fish are collected alive and are generally unharmed.

Disadvantages of Seining:

! Previous experience and skill, knowledge of fish habitats and behavior, and sampling effort are probably more important in seining than in the use of any other gear (Hendricks et al. 1980).

! Sample effort and results for seining are more variable than sampling with electrofishing.

! Use of seines is generally restricted to slower water with smooth bottoms, and is most effective in small streams or pools with little cover.

! Standardization of unit of effort to ensure data comparability is difficult.

3.10   SAMPLING REPRESENTATIVE HABITAT

Effort should be made when sampling to avoid regionally unique natural habitat. Samples from such situations, when compared to those from sites lacking the unique habitat, will appear different, i.e., assess as in either better or worse condition, than those not having the unique habitat. This is due to the usually high habitat specificity that different taxa have to their range of habitat conditions; unique habitat will have unique taxa. Thus, all RBP sampling is focused on sampling of representative habitat.

Composite sampling is the norm for RBP investigations to characterize the reach, rather than individual small replicates. However, a major source of variance can result from taking too few samples for a composite. Therefore, each of the protocols (i.e., for periphyton, benthos, fish) advocate compositing several samples or efforts throughout the stream reach. Replication is strongly encouraged for precision evaluation of the methods.

When sampling wadeable streams, rivers, or waterbodies with complex habitats, a complete inventory of the entire reach is not necessary for bioassessment. However, the sampling area should be representative of the reach, incorporating riffles, runs, and pools if these habitats are typical of the stream in question. Midchannel and wetland areas of large rivers, which are difficult to sample effectively, may be avoided. Sampling effort may be concentrated in near-shore habitats where most species will be collected. Although some deep water or wetland species may be undersampled, the data should be adequate for the objective of bioassessment.

4 PERFORMANCE-BASED METHODS SYSTEM (PBMS)

Determining the performance characteristics of individual methods enables agencies to share data to a certain extent by providing an estimate of the level of confidence in assessments from one method to the next. The purpose of this chapter is to provide a framework for measuring the performance characteristics of various methods. The contents of this chapter are taken liberally from Diamond et al. 1996, which is a refinement of the PBMS approach developed for ITFM (1995b). This chapter is best assimilated if the reader is familiar with data analysis for bioassessment. Therefore, the reader may wish to review Chapter 9 on data analysis before reading this PBMS material. Specific quality assurance aspects of the methods are included in the assemblage chapters.

Regardless of the type of data being collected, field methods share one important feature in common—they cannot tell whether the information collected is an accurate portrayal of the system of interest (Intergovernmental Task Force on Monitoring Water Quality [ITFM] 1995a). Properties of a given field sample can be known, but research questions typically relate to much larger spatial and temporal scales. It is possible to know, with some accuracy, properties or characteristics of a given sample taken from the field; but typically, research questions relate to much larger spatial and temporal scales. To grapple with this problem, environmental scientists and statisticians have long recognized that field methods must strive to obtain information that is representative of the field conditions at the time of sampling.

An accurate assessment of stream biological data is difficult because natural variability cannot be controlled (Resh and Jackson 1993). Unlike analytical assessments conducted in the laboratory, in which accuracy can be verified in a number of ways, the accuracy of macroinvertebrate assessments in the field cannot be objectively verified. For example, it isn't possible to "spike" a stream with a known species assemblage and then determine the accuracy of a bioassessment method. This problem is not theoretical. Different techniques may yield conflicting interpretations at the same sites, underscoring the question of accuracy in bioassessment. Depending on which methods are chosen, the actual structure and condition of the assemblage present, or the trends in status of the assemblage over time may be misinterpreted. Even with considerable convergence in methods used in the U.S. by states and other agencies (Southerland and Stribling 1995, Davis et al. 1996), direct sharing of data among agencies may cause problems because of the uncertainty associated with unfamiliar methods, misapplication of familiar methods, or varied data analyses and interpretation (Diamond et al. 1996).

4.1 APPROACHES FOR ACQUIRING COMPARABLE BIOASSESSMENT DATA

Water quality management programs have different reasons for doing bioassessments which may not require the same level or type of effort in sample collection, taxonomic identification, and data analysis (Gurtz and Muir 1994). However, different methods of sampling and analysis may yield comparable data for certain objectives despite differences in effort. There are 2 general approaches for acquiring comparable bioassessment data among programs or among states. The first is for everyone to use the same method on every study. Most water resource agencies in the U.S. have developed standard operating procedures (SOPs). These SOPs would be adhered to throughout statewide or regional areas

Rapid Bioassessment Protocols for Use in Streams and Wadeable Rivers: Periphyton, Benthic Macroinvertebrates, and Fish, Second Edition

4-1

to provide comparable assessments within each program. The Rapid Bioassessment Protocols (RBPs) developed by Plafkin et al. (1989) and refined in this document are attempts to provide a framework for agencies to develop SOPs. However, the use of a single method, even for a particular type of habitat, is probably not likely among different agencies, no matter how exemplary (Diamond et al. 1996).

The second approach to acquiring comparable data from different organizations, is to encourage the documentation of performance characteristics (e.g., precision, sensitivity) for all methods and to use those characteristics to determine comparability of different methods (ITFM 1995b). This documentation is known as a performance-based method system (PBMS) which, in the context of biological assessments, is defined as a system that permits the use of any method (to sample and analyze stream assemblages) that meets established requirements for data quality (Diamond et al. 1996). Data quality objectives (DQOs) are qualitative and quantitative expressions that define requirements for data precision, bias, method sensitivity, and range of conditions over which a method yields satisfactory data (Klemm et al. 1990). The determination of DQOs for a given study or agency program is central to all data collection and to a PBMS, particularly, because these objectives establish not only the necessary quality of a given method (Klemm et al. 1990) but also the types of methods that are likely to provide satisfactory information.

In practice, DQO's are developed in 3 stages: (1) determine what information is needed and why and how that information will be used; (2) determine methodological and practical constraints and technical specifications to achieve the information desired; and (3) compare different available methods and choose the one that best meets the desired specifications within identified practical and technical limitations (USEPA 1984, 1986, Klemm et al. 1990, USEPA 1995a, 1997c). It is difficult to make an informed decision regarding which methods to use if data quality characteristics are unavailable. The successful introduction of the PBMS concept in laboratory chemistry, and more recently in laboratory toxicity testing (USEPA 1990c, American Society of Testing and Materials [ASTM] 1995), recommends adapting such a system for biological monitoring and assessment.

If different methods are similar with respect to the quality of data each produces, then results of an assessment from those methods may be used interchangeably or together. As an example, a method for sample sorting and organism identification, through repeated examination using trained personnel, could be used to determine that the proportion of missed organisms is less than 10% of the organisms present in a given sample and that taxonomic identifications (to the genus level) have an accuracy rate of at least 90% (as determined by samples verified by recognized experts). A study could require the above percentages of missed organisms and taxonomic accuracy as DQOs to ensure the collection of satisfactory data (Ettinger 1984, Clifford and Casey 1992, Cuffney et al. 1993a). In a PBMS approach, any laboratory sorting and identification method that documented the attainment of these DQOs would yield comparable data and the results would therefore be satisfactory for the study.

For the PBMS approach to be useful, 4 basic assumptions must be met (ITFM 1995b):

1.   DQOs must be set that realistically define and measure the quality of the data needed; reference (validated) methods must be made available to meet those DQOs;

2.   to be considered satisfactory, an alternative method must be as good or better than the reference method in terms of its resulting data quality characteristics;

3.   there must be proof that the method yields reproducible results that are sensitive enough for the program; and

4.   the method must be effective over the prescribed range of conditions in which it is to be used. For bioassessments, the above assumptions imply that a given method for sample collection and analysis produces data of known quality, including precision, the range of habitats over which the collection method yields a specified precision, and the magnitude of difference in data among sites with different levels or types of impairment (Diamond et al. 1996).

Thus, for multimetric assessment methods, such as RBPs, the precision of the total multimetric score is of interest as well as the individual metrics that make up the score (Diamond et al. 1996). Several performance characteristics must be characterized for a given method to utilize a PBMS approach. These characteristics include method precision, bias, performance range, interferences, and sensitivity (detection limit). These characteristics, as well as method accuracy, are typically demonstrated in analytical chemistry systems through the use of blanks, standards, spikes, blind samples, performance evaluation samples, and other

> **PERFORMANCE CHARACTERISTICS**
>
> * **Precision**
> * **Bias**
> * **Performance range**
> * **Interferences**
> * **Sensitivity**

techniques to compare different methods and eventually derive a reference method for a given analyte. Many of these performance characteristics are applicable to biological laboratory and field methods and other prelaboratory procedures as well (Table 4-1). It is known that a given collection method is not equally accurate over all ecological conditions even within a general aquatic system classification (e.g., streams, lakes, estuaries). Therefore, assuming a given method is a "reference method" on the basis of regulatory or programmatic reasons does not allow for possible translation or sharing of data derived from different methods because the performance characteristics of different methods have not been quantified. One can evaluate performance characteristics of methods in 2 ways: (1) with respect to the collection method itself and, (2) with respect to the overall assessment process. Method performance is characterized using quantifiable data (metrics, scores) derived from data collection and analysis. Assessment performance, on the other hand, is a step removed from the actual data collected. Interpretive criteria (which may be based on a variety of approaches) are used to rank sites and thus, PBMS in this case is concerned with performance characteristics of the ranking procedures as well as the methods that lead to the assessment.

Table 4-1. Progression of a generic bioassessment field and laboratory method with associated examples of performance characteristics.

Step	Procedure	Examples of Performance Characteristics
1	Sampling device	*Precision*—repeatability in a habitat. *Bias*—exclusion of certain taxa (mesh size). *Performance range*—different efficiency in various habitat types or substrates. *Interferences*—matrix or physical limitations (current velocity, water depth).
2	Sampling method	*Precision*—variable metrics or measures among replicate samples at a site. *Bias*—exclusion of certain taxa (mesh size) or habitats. *Performance range*—limitations in certain habitats or substrates. *Interferences*—high river flows, training of personnel.

Table 4-1. Progression of a generic bioassessment field and laboratory method with associated examples of performance characteristics. (Continued)

Step	Procedure	Examples of Performance Characteristics
3	Field sample processing (subsampling, sample transfer, preservation)	*Precision*—variable metrics among splits of subsamples. *Bias*— efficiency of locating small organisms. *Performance range*—sample preservation and holding time. *Interferences*—Weather conditions. **Additional characteristics:** *Accuracy*—of sample transfer process and labeling.
4	Laboratory sample processing (sieving, sorting)	*Precision*—split samples. *Bias*—sorting certain taxonomic groups or organism size. *Performance range*—sorting method depending on sample matrix (detritus, mud). *Interferences*—distractions; equipment. **Additional characteristics:** *Accuracy*—sorting method; lab equipment.
5	Taxonomic enumeration	*Precision*—split samples. *Bias*—counts and identifications for certain taxonomic groups. *Performance range*—dependent on taxonomic group and (or) density. *Interferences*—appropriateness of taxonomic keys. *Sensitivity*— level of taxonomy related to type of stressor **Additional characteristics:** *Accuracy*—identification and counts.

Data quality and performance characteristics of methods for analytical chemistry are typically validated through the use of quality control samples including blanks, calibration standards, and samples spiked with a known quantity of the analyte of interest. Table 4-2 summarizes some performance characteristics used in analytical chemistry and how these might be translated to biological methods.

The collection of high-quality data, particularly for bioassessments, depends on having adequately trained people. One way to document satisfactory training is to have newly trained personnel use the method and then compare their results with those previously considered acceptable. Although field crews and laboratory personnel in many organizations are trained in this way (Cuffney et al. 1993b), the results are rarely documented or quantified. As a result, an organization cannot assure either itself or other potential data users that different personnel performing the same method at the same site yield comparable results and that data quality specifications of the method (e.g., precision of metrics or scores) are consistently met. Some of this information is published for certain bioassessment sampling methods, but is defined qualitatively (see Elliott and Tullett 1978, Peckarsky 1984, Resh et al. 1990, Merritt et al. 1996 for examples), not quantitatively. Quantitative information needs to be more available so that the quality of data obtained by different methods is documented.

Table 4-2. Translation of some performance characteristics, derived for laboratory analytical systems, to biological laboratory systems (taken from Diamond et al. 1996).

Performance Characteristics	Analytical Chemical Methods	Biological Methods
Precision	Replicate samples	Multiple taxonomists identifying 1 sample; split sample for sorting, identification, enumeration; replicate samples within sites; duplicate reaches
Bias	Matrix-spiked samples; standard reference materials; performance evaluation samples	Taxonomic reference samples; "spiked" organism samples
Performance range	Standard reference materials at various concentrations; evaluation of spiked samples by using different matrices	Efficiency of field sorting procedures under different sample conditions (mud, detritus, sand, low light)
Interferences	Occurrence of chemical reactions involved in procedure; spiked samples; procedural blanks; contamination	Excessive detrital material or mud in sample; identification of young life stages; taxonomic uncertainty
Sensitivity	Standards; instrument calibration	Organism-spiked samples; standard level of identification
Accuracy	Performance standards; procedural blanks	Confirmation of identification, percentage of "missed" specimens

It is imperative that the specific range of environmental conditions (or performance range) is quantitatively defined for a sampling method (Diamond et al. 1996). As an example, the performance range for macroinvertebrate sampling is usually addressed qualitatively by characterizing factors such as stream size, hydrogeomorphic reach classification, and general habitat features (riffle vs. pool, shallow vs. deep water, rocky vs. silt substrate; Merritt et al. 1996). In a PBMS framework, different methods could be classified based on the ability of the method to achieve specified levels of performance characteristics such as data precision and sensitivity to impairment over a range of appropriate habitats. Thus, the precision of individual metrics or scores obtained by different sampling methods can be directly and quantitatively compared for different types of habitats.

4.2 ADVANTAGES OF A PBMS APPROACH FOR CHARACTERIZING BIOASSESSMENT METHODS

Two fundamental requirements for a biological assessment are: (1) that the sample taken and analyzed is representative of the site or the assemblage of interest and, (2) that the data obtained are an accurate reflection of the sample. The latter requirement is ensured using proper quality control (QC) in the laboratory including the types of performance characteristics summarized in Table 4-2. The first requirement is met through appropriate field sampling procedures, including random selection of sampling locations within the habitat type(s) of interest, choice of sampling device, and sample preservation methods. The degree to which a sample is representative of the environment depends on the type of sampling method used (including subsampling) and the ecological endpoint being measured. For example, many benthic samples may be needed from a stream to obtain 95% confidence intervals that are within 50% of the mean value for macroinvertebrate density, whereas fewer benthic samples may be needed to determine the dominant species in a given habitat type at a particular time (Needham and Usinger 1956, Resh 1979, Plafkin et al. 1989).

Rapid Bioassessment Protocols for Use in Streams and Wadeable Rivers: Periphyton, Benthic Macroinvertebrates, and Fish, Second Edition

4-5

Several questions have been raised concerning the appropriateness or "accuracy" of methods such as RBPs, which take few samples from a site and base their measures or scores on subsamples. Subsampling methods have been debated relevant to the "accuracy" of data derived from different methods (Courtemanch 1996, Barbour and Gerritsen 1996, Vinson and Hawkins 1996). Using a PBMS framework, the question is not which subsampling method is more "accurate" or precise but rather what accuracy and precision level can a method achieve, and do those performance characteristics meet the DQOs of the program? Looking at bioassessment methods in this way, (including subsampling and taxonomic identification), forces the researcher or program manager to quantitatively define beforehand the quality control characteristics necessary to make the type of interpretive assessments required by the study or program.

Once the objectives and data quality characteristics are defined for a given study, a method is chosen that meets those objectives. Depending on the data quality characteristics desired, several different methods for collecting and sorting macroinvertebrates may be suitable. Once data precision and "accuracy" are quantified for measures derived from a given bioassessment method, the method's sensitivity (the degree of change in measures or endpoints between a test site and a control or reference site that can be detected as a difference) and reliability (the degree to which an objectively defined impaired site is identified as such) can be quantified and compared with other methods. A method may be modified (e.g., more replicates or larger samples taken) to improve the precision and "accuracy" of the method and meet more stringent data requirements. Thus, a PBMS framework has the advantage of forcing scientists to focus on the ever-important issue: what type of sampling program and data quality are needed to answer the question at hand?

A second advantage of a PBMS framework is that data users and resource managers could potentially increase the amount of available information by combining data based on known comparable methods. The 305(b) process of the National Water Quality Inventory, (USEPA 1997c) is a good example of an environmental program that would benefit from a PBMS framework. This program is designed to determine status and trends of surface water quality in the U.S. A PBMS framework would make explicit the quality and comparability of data derived from different bioassessment methods, would allow more effective sharing of information collected by different states, and would improve the existing national database. Only those methods that met certain DQOs would be used. Such a decision might encourage other organizations to meet those minimum data requirements, thus increasing the amount of usable information that can be shared. For example, the RBPs used by many state agencies for water resources (Southerland and Stribling 1995) could be modified for field and laboratory procedures and still meet similar data quality objectives. The overall design steps of the RBPs, and criteria for determining useful metrics or community measures, would be relatively constant across regions and states to ensure similar quality and comparability of data.

4.3   QUANTIFYING PERFORMANCE CHARACTERISTICS

The following suggested sampling approach (Figure 4-1) need only be performed once for a particular method and by a given agency or research team; it need not be performed for each bioassessment study. Once data quality characteristics for the method are established, limited quality control (QC) sampling and analysis should supplement the required sampling for each bioassessment study to ensure that data quality characteristics of the method are met (USEPA 1995a). The additional effort and expense of such QC are negligible in relation to the potential environmental cost of producing data of poor or unknown quality.

The first step is to define precision of the collection method, also known as "measurement error". This is accomplished by replicate sampling within sites (see Hannaford and Resh 1995). The samples collected are processed and analyzed separately and their metrics compared to obtain a more realistic measure of the method precision and consistency. Repeated samples within sites estimate the precision of the entire method, comprising variability due to several sources including small-scale spatial variability within a site; operator consistency and bias; and laboratory consistency. Finally, it is desirable to sample a range of site classes (stream size, habitat type) over which the method is likely to be used. This kind of sampling, processing, and analysis should reveal potential biases.

Once the precision of the method is known, one can determine the actual variability associated with sampling "replicate" reference sites within an ecoregion or habitat type. This is known as sampling error, referring to the sample (of sites) drawn from a subpopulation (sites in a region). The degree of assemblage similarity observed among "replicate" reference streams, along with the precision of the collection method itself, will determine the overall precision, accuracy, and sensitivity of the bioassessment approach as a whole. This kind of checking has been done, at least in part, by several states (Bode and Novak 1995; Yoder and Rankin 1995a; Hornig et al. 1995;

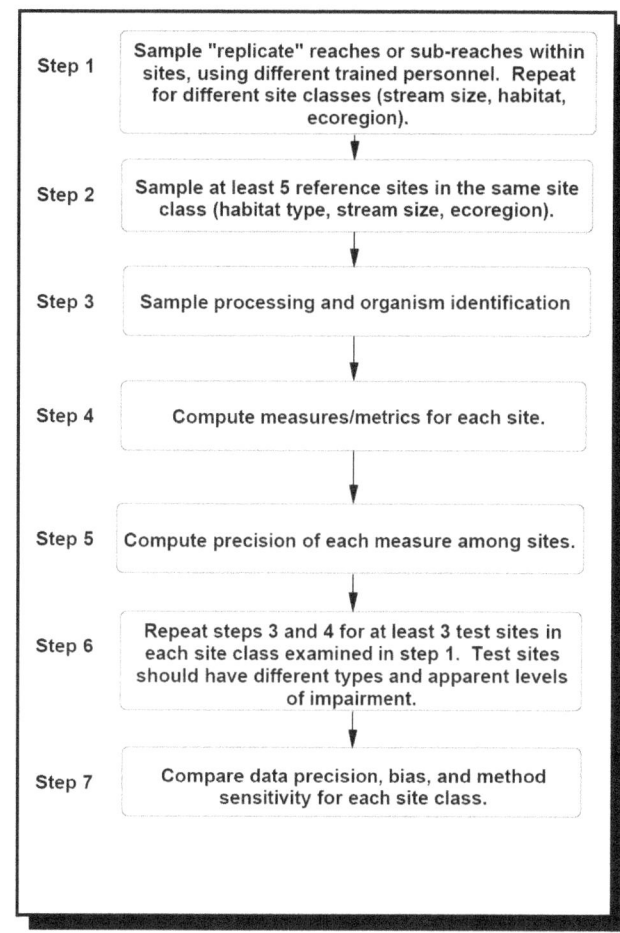

Figure 4-1. Flow chart summarizing the steps necessary to quantify performance characteristics of a bioassessment method (modified from Diamond et al. 1996).

Barbour et al. 1996b), some USEPA programs (Gibson et al. 1996), and the U.S. Geological Survey (USGS) National Water Quality Assessment Program (Cuffney et al. 1993b, Gurtz 1994). Evaluation of metric or score variability among replicate reference sites can result in improved data precision and choices of stream classification. For example, the Arizona Department of Environmental Quality (DEQ) determined that macroinvertebrate assemblage structure varied substantially within ecoregions resulting in large metric variability among reference sites and poor classification (Spindler 1996). Using detrended correspondence and cluster analysis, the state agency determined that discrimination of sites by elevation and watershed area, corresponding to montane upland, desert lowland, and transition zones, resulted in much lower variability among reference sites and a better classification scheme to measure sensitivity to impairment.

If multiple reference sites are sampled in different site classes (where the sampling method is judged to be appropriate), several important method performance characteristics can be quantified, including: (1) precision for a given metric or assessment score across replicate reference sites within a site class; (2) relative precision of a given metric or score among reference sites in different classes; (3) range of classes over which a given method yields similar precision and "accuracy"; (4) potential interferences to a given method that are related to specific class characteristics and qualities; and (5) bias of a given metric, method, or both, owing to differences in classes (Diamond et al. 1996).

A study by Barbour et al. (1996b) for Florida streams, illustrates the importance of documenting method performance characteristics using multiple reference sites in different site classes. Using the same method at all sites, fewer taxa were observed in reference sites from the Florida Peninsula (one site class) compared to the Florida Panhandle (another site class), resulting in much lower reference values for taxa richness metrics in the Peninsula. Although metric precision was similar among reference sites in each site class, method sensitivity (i.e., the ability of a metric to discern a difference between reference and stressed sites) was poorer in the Peninsula for taxa richness. Thus, bioassessment "accuracy" may be more uncertain for the Florida Peninsula; that is, the probability of committing a Type II error (concluding a test site is no different from reference — therefore minimally impaired — when, in fact, it is) may be greater in the Peninsula region. In the context of a PBMS, the state agency can recognize and document differences in method performance characteristics between site classes and incorporate them into their DQOs. The state in this case can also use the method performance results to identify those site classes for which the biological indicator (index, metric, or other measurement endpoint) may not be naturally sensitive to impairment; i.e., the fauna is naturally species-poor and thus less likely to reflect impacts from stressors. If the state agency desires greater sensitivity than the current method provides, it may have to develop and test different region-specific methods and perhaps different indicators.

In the last step of the process, a method is used over a range of impaired conditions so as to determine the method's sensitivity or ability to detect impairment. As discussed earlier, sites with known levels of impairment or analogous standards by which to create a calibration curve for a given bioassessment method are lacking. In lieu of this limitation, sampling sites are chosen that have known stresses (e.g., urban runoff, toxic pollutants, livestock intrusion, sedimentation, pesticides). Because different sites may or may not have the same level of impairment within a site class (i.e., they are not replicate sites), precision of a method in impaired sites may best be examined by taking and analyzing multiple samples from the same site or adjacent reaches (Hannaford and Resh 1995).

The quantification of performance characteristics is a compromise between statistical power and cost while maintaining biological relevance. Given the often wide variation of natural geomorphic conditions and landscape ecology, even within supposedly "uniform" site classes (Corkum 1989, Hughes 1995), it is desirable to examine 10 or more reference sites (Yoder and Rankin 1995a, Gibson et al. 1996). More site classes in the evaluation process would improve documentation of the performance range and bias for a given method. Using the sampling design suggested in Figure 4-1, data from at least 30 sites (reference and test sites combined), sampled within a brief time period (so as to minimize seasonal changes in the target assemblage), are needed to define performance characteristics. An alternative approach might be to use bootstrap resampling of fewer sites to evaluate the nature of variation of these samples (Fore et al. 1996).

A range of "known" stressed sites within a site class is sampled to test the performance characteristics of a given method. It is important that stressed sites meet the following criteria: (1) they belong to the same site class as the reference sites examined; (2) they clearly have been receiving some chemical, physical, or biological stress(es) for some time (months at least); and (3) impairment is not obvious without sampling; i.e., impairment is not severe.

The first criterion is necessary to reduce potential interferences owing to class differences between the test and reference sites. Thus, the condition of the reference site will have high probability of serving as a true blank as discussed earlier. For example, it is clearly inappropriate to use high gradient mountain streams as references for assessing plains streams.

Chapter 4: Performance-Based Methods System (PBMS)

The second criterion, which is the documented presence of potential stresses, is necessary to ensure the likelihood that the test site is truly impaired (Resh and Jackson 1993). A potential test site might include a body of water that receives toxic chemicals from a point-source discharge or from nonpoint sources, or a water body that has been colonized by introduced or exotic "pest" species (for example, zebra mussel or grass carp). Stresses at the test site should be measured quantitatively to document potential cause(s) of impairment.

The third criterion, that the site is not obviously impaired, provides a reasonable test of method sensitivity or "detection limit." Severe impairment (e.g., a site that is dominated by 1 or 2 invertebrate species, or a site apparently devoid of aquatic life) generally requires little biological sampling for detection.

4.4 RECOMMENDED PROCESS FOR DOCUMENTATION OF METHOD COMPARABILITY

Although a comparison of methods at the same reference and test sites at the same time is preferable (same seasons and similar conditions), it is not essential. The critical requirement when comparing different sampling methods is that performance characteristics for each method are derived using similar habitat conditions and site classes at similar times/seasons (Diamond et al. 1996). This approach is most useful when examining the numeric scores upon which the eventual assessment is based. Thus, for a method such as RBP that sums the values of several metrics to derive a single score for a site, the framework described in Figure 4-1 should use the site scores. If one were interested in how a particular multimetric scoring system behaves, or one wishes to compare the same metric across methods, then individual metrics could be examined using the framework in Figure 4-1. For multivariate assessment methods that do not compute metric scores, one could instead examine a measure of community similarity or other variable that the researcher uses in multivariate analyses (Norris 1995).

Method comparability is based on 2 factors: (1) the relative magnitude of the coefficients of variation in measurements within and among site classes, and (2) the relative percent differences in measurements between reference and test sites. It is important to emphasize that comparability is not based on the measurements themselves, because different methods may produce different numeric scores or metrics and some sampling methods may explicitly ignore certain taxonomic groups, which will influence the metrics examined. Instead, detection of a systematic relationship among indices or the same measures among methods is advised. If 2 methods are otherwise comparable based on similar performance characteristics, then results of the 2 methods can be numerically related to each other. This outcome is a clear benefit of examining method comparability using a PBMS framework.

Figure 4-1 summarizes a suggested test design, and Table 4-3 summarizes recommended analyses for documenting both the performance characteristics of a given method, and the degree of data comparability between 2 or more methods. The process outlined in Figure 4-1 is not one that is implemented with every study. Rather, the process should be performed at least once to document the limitations and range of applicability of the methods, and should be cited with subsequent uses of the method(s).

The following performance characteristics are quantified for each bioassessment method and compared: (1) the within-class coefficient of variation for a given metric score or index by examining reference-site data for each site class separately (e.g., CV_{Alr} and CV_{Blr}; Fig. 4-1); (2) difference or bias in precision related to site class for a given metric or index (by comparing reference site coefficient of variation from each class: CV_{A1r}/CV_{B1r}; Table 4-3); and (3) estimates of method sensitivity or discriminatory power, by comparing test site data with reference site data

Table 4-3. Suggested arithmetic expressions for deriving performance characteristics that can be compared between 2 or more methods. In all cases, $\bar{x}$ = mean value, X = test site value, s = standard deviation. Subscripts are as follows: capital letter refers to site class (A or B); numeral refers to method 1 or 2; and lower case letter refers to reference (r) or test site (t) (modified from Diamond et al. 1996).

Performance Characteristic	Parameters for Quantifying Method Comparability	Desired Outcome
Relative *precision* of metric or index *within* a site class	CV_{A1r} and CV_{A2r} ; CV_{B1r} and CV_{B2r}	Low values
Relative *precision* of metric or index *between* sites (population of samples at a site) or site classes (population of sites)	$\dfrac{CV_{A1r}}{CV_{B1r}}$; $\dfrac{CV_{A2r}}{CV_{B2r}}$	High ratio
Relative *sensitivity* or "detection limit" of metric or index *within* a site class. Comparison of those values between methods reveals the most sensitive method	$\dfrac{\bar{x}_{A1r}-X_{A1t}}{s_{A1r}}$; $\dfrac{\bar{x}_{A2r}-X_{A2t}}{s_{A2r}}$ $\dfrac{\bar{x}_{B1r}-X_{B1t}}{s_{B1r}}$; $\dfrac{\bar{x}_{B2r}-X_{B2t}}{s_{B2r}}$	High ratio
Relative *sensitivity* of metric or index *between* site classes	$\dfrac{\bar{x}_{A1r}-X_{A1t}}{s_{A1r}}$; $\dfrac{\bar{x}_{B1r}-X_{B1t}}{s_{B1r}}$ $\dfrac{\bar{x}_{A2r}-X_{A2t}}{s_{A2r}}$; $\dfrac{\bar{x}_{B2r}-X_{B2t}}{s_{B2r}}$	High ratio

within each site class as a function of reference site variability (Table 4-3), e.g.,

$$\frac{\bar{x}_{A1r}-X_{A1t}}{s_{A1r}}$$

A method that yields a smaller difference between test and reference sites in relation to the reference site variability measured (Table 4-3) would indicate less discriminatory power or sensitivity; that is, the test site is erroneously perceived to be similar to or better than the reference condition and not impaired (Type II error).

Relatively few methods may be able to consistently meet the above data quality criterion and also maintain high sensitivity to impairment because both characteristics require a method that produces relatively precise, accurate data. For example, if the agency's intent is to screen many sites so as to prioritize "hot spots" or significant impairment in need of corrective action, then a method that is inexpensive, quick, and tends to show impairment when significant impairment is actually present

Rapid Bioassessment Protocols for Use in Streams and Wadeable Rivers: Periphyton, Benthic Macroinvertebrates, and Fish, Second Edition

4-11

(such as some volunteer monitoring methods) (Barbour et al. 1996a) can meet prescribed DQOs with less cost and effort. In this case, the data requirements dictate high priority for method sensitivity or discriminatory power (detection if impaired sites), understanding that there is likely also to be a high Type I error rate (misidentification of unimpaired sites).

Relative accuracy of each method is addressed to the extent that the test sites chosen are likely to be truly impaired on the basis of independent factors such as the presence of chemical stresses or suboptimal habitat. A method with relatively low precision (high variance) among reference sites compared with another method may suggest lower method accuracy. Note that a method having lower precision may still be satisfactory for some programs if it has other advantages, such as high ability to detect impaired sites with less cost and effort to perform.

Once performance characteristics are defined for each method, data comparability can be determined. If 2 methods are similarly precise, sensitive, and biased over the habitat types sampled, then the different methods should produce comparable data. Interpretive judgements could then be made concerning the quality of aquatic life using data produced by either or both methods combined. Alternatively, the comparison may show that 2 methods are comparable in their performance characteristics in certain habitats or regions and not others. If this is so, results of the 2 methods can be combined for the type for the types of habitats in which data comparability was demonstrated, but not for other regions or habitat types.

In practice, comparability of bioassessment methods would be judged relative to a reference method that has already been fully characterized (using the framework summarized in Figure 4-1) and which produces data with the quality needed by a certain program or agency. The qualities of this reference method are then defined as method performance criteria. If an alternative method yields less precision among reference sites within the same site class than the reference method (e.g., $CV_{A1r} > CV_{A2r}$ in Table 4-3), then the alternative method probably is not comparable to the reference method. A program or study could require that alternative methods are acceptable only if they are as precise as the reference method. A similar process would be accomplished for other performance characteristics that a program or agency deems important based on the type of data required by the program or study.

4.5 CASE EXAMPLE DEFINING METHOD PERFORMANCE CHARACTERISTICS

Florida Department of Environmental Protection (DEP) has developed a statewide network for monitoring and assessing the state's surface waters using macroinvertebrate data. Florida DEP has rigorously examined performance characteristics of their collection and assessment methods to provide better overall quality assurance of their biomonitoring program and to provide defensible and appropriate assessments of the state's surface waters (Barbour et al. 1996b, c). Much of the method characterization process developed for Florida DEP is easily communicated in the context of a PBMS approach.

In addition to characterizing data quality and method performance based on ecoregional site classes, Florida DEP also characterized their methods based on season (summer vs. winter sampling index periods), and size of subsample analyzed (100, 200, or 300-organism subsample). In addition, analyses were performed on the individual component metrics which composed the Florida stream condition index (SCI). For the sake of brevity, the characterization process and results for the SCI in the summer index period and the Peninsula and Northeast bioregions are summarized. The same process was used for other bioregions in the state and in the winter index period.

Performance Criteria Characteristics of Florida SCI (see Figure 4-1 for process)

Characterize Measurement Error (Method Precision Within a Site)—A total of 7 sites in the Peninsula bioregion were subjected to multiple sampling (adjacent reaches). The DEP observed a mean SCI = 28.4 and a CV (within a stream) = 6.8%. These data suggest low measurement error associated with the method and the index score. Given this degree of precision in the reference condition SCI score, power analysis indicated that 80% of the time, a test site with an SCI 5 points less (based on only a single sample at the test site) than the reference criterion, could be distinguished as impaired with 95% confidence. This analysis also indicated that if duplicate samples were taken at the test site, a difference of 3 points in the SCI score between the test site and the reference criterion could be distinguished as impaired with 95% confidence.

Characterize Sampling Error (Method Precision on a Population of Reference Sites)—A total of 56 reference sites were sampled in the Peninsula bioregion (Step 1, Figure 4-1). The SCI score could range from a minimum of 7 to a theoretical maximum of 31 based on the component metric scores. However, in the Peninsula, reference site SCI scores generally ranged between 21 and 31. A mean SCI score of 27.6 was observed with a CV of 12.0%.

Determine Method and Index Sensitivity—Distribution of SCI scores of the 56 reference sites showed that the 5th percentile was a score of 20. Thus, 95% of Peninsula reference sites had a score >20. Accuracy of the method, using known stressed sites, indicated that approximately 80% of the test sites had SCI scores ≤ 20 (Fig. 4-2). In other words, a stressed site would be assessed as impaired 80% of the time using the collection method in the Peninsula bioregion in the summer, and an impairment criterion of the 5th percentile of reference sites. The criterion could also be raised to, say, the 25th percentile of reference sites, which would increase accuracy of correctly classifying stressed sites to approximately 90%, but would decrease accuracy of correctly assessing unimpaired sites to 75%.

Determination of Method Bias and Relative Sensitivity in Different Site Classes—A comparative analysis of precision, sensitivity, and ultimately bias, can be performed for the Florida DEP method and the SCI index outlined in Table 4-3. For example, the mean SCI score in the Panhandle bioregion, during the same summer index period, was 26.3 with a CV = 12.8% based on 16 reference sites. Comparing this CV to the one reported for the Peninsula in the previous step, it is apparent that the precision of this method in the Panhandle was similar to that observed in the Peninsula bioregion.

The 5th percentile of the Panhandle reference sites was an SCI score of 17, such that actual sensitivity of the method in the Panhandle was slightly lower than in the Peninsula bioregion (Figure 4-2). An impaired site would be assessed as such only 50% of the time in the Panhandle bioregion in the summer as opposed to 80% of the time in the Peninsula bioregion during the same index period. Part of the difference in accuracy of the method among the 2 bioregions can be attributed to differences in sample size. Data from only 4 "known" impaired sites were available in the Panhandle bioregion while the Peninsula bioregion had data from 12 impaired sites. The above analyses show, however, that there may be differences in method performance between the 2 regions (probably attributable to large habitat differences between the regions) which should be further explored using data from additional "known" stressed sites, if available.

Rapid Bioassessment Protocols for Use in Streams and Wadeable Rivers: Periphyton, Benthic Macroinvertebrates, and Fish, Second Edition

4-13

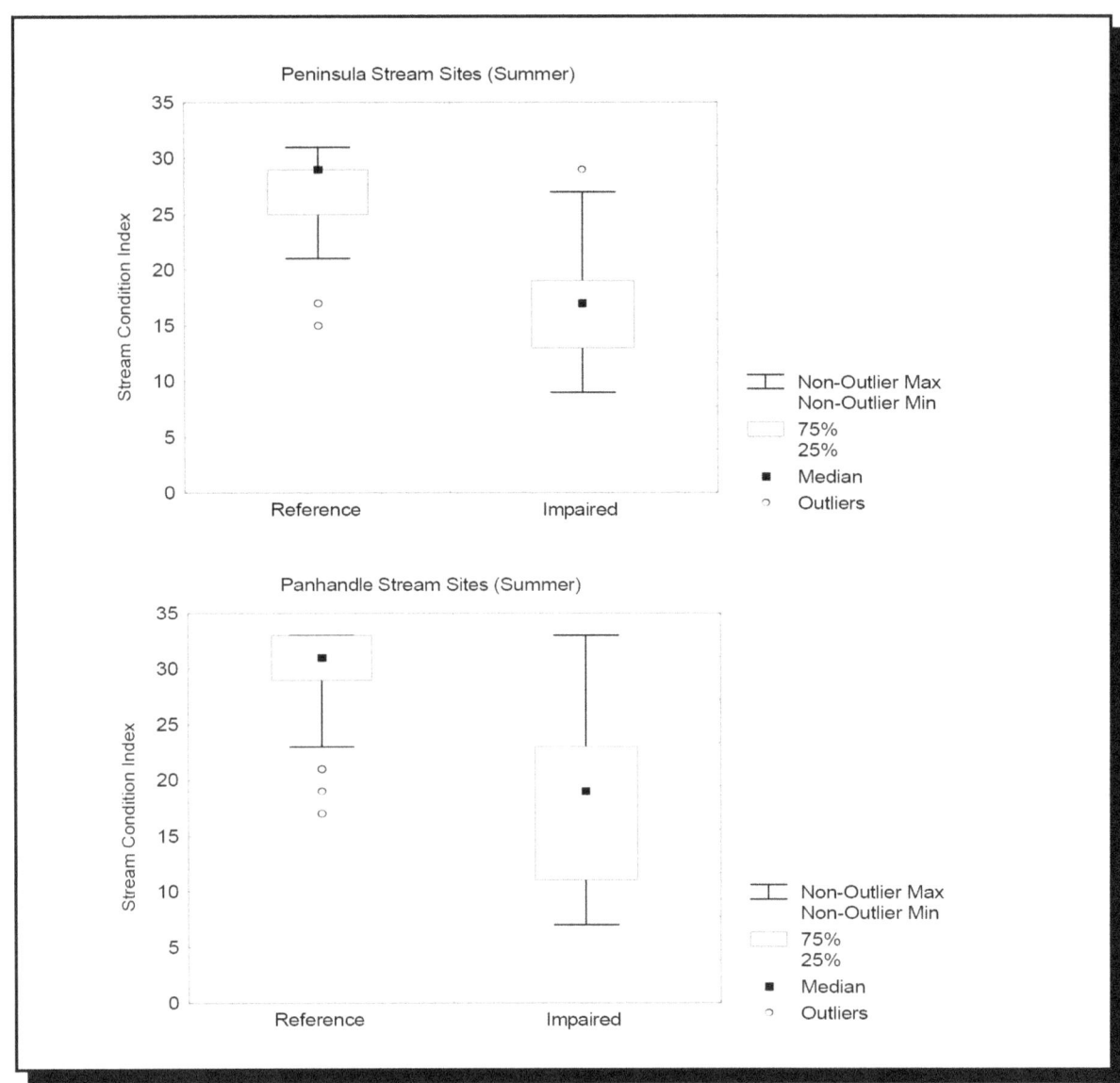

Figure 4-2. Comparison of the discriminatory ability of the SCI between Florida's Peninsula and Panhandle Bioregions. Percentiles used (not x̄, sd) to depict relationship.

4.6 APPLICATION OF THE PBMS

The PBMS approach is intended to provide information regarding the confidence of an assessment, given a particular method. By having some measure of confidence in the endpoint and the subsequent decision pertinent to the condition of the water resource, assessment and monitoring programs are greatly strengthened. Three primary questions can be identified that enable agencies to ascertain the value and scientific validity of using information derived from different methods. Use of PBMS is necessary for these questions to be answered.

Question 1 — How rigorous must a method be to accurately detect impairment?

The analyses of Ohio EPA (1992) reveal that the power and ability of a bioassessment technique to accurately portray biological community performance and ecological integrity, and to discriminate even finer levels of aquatic life use impairments, are directly related to the data dimensions (i.e., ecological complexity, environmental accuracy, discriminatory power) produced by each (Barbour et al. 1996b). For example, a technique that includes the identification of macroinvertebrate taxa to genus and species will produce a higher attainment of data dimensions than a technique that is limited to family-level taxonomy. In general, this leads to a greater discrimination of the biological condition of sites.

Some states use one method for screening assessments and a second method for more intensive and confirmatory assessments. Florida DEP uses a BioRecon (see description in Chapter 7) to conduct statewide screening for their watershed-based monitoring. A more rigorous method based on a multihabitat sampling (see Chapter 7) is used for targeted surveys related to identified or suspected problem areas. North Carolina Water Quality Division (WQD) has a rapid EPT index (cumulative number of species of Ephemeroptera, Plecoptera, Trichoptera) to conduct screening assessments. Their more intensive method is used to monitor biological condition on a broader basis.

Use of various methods having differing levels of rigor can be examined with estimates of precision and sensitivity. These performance characteristics will help agencies make informed decisions of how resulting data can be used in assessing condition.

Question 2 — How can data derived from different methods be compared to locate additional reference sites?

Many agencies are increasingly confronted with the issue of locating appropriate reference sites from which to develop impairment/unimpairment thresholds. In some instances, sites outside of jurisdictional boundaries are needed to refine the reference condition. As watershed-based monitoring becomes implemented throughout the U.S., jurisdictional boundaries may become impediments to effective monitoring. County governments, tribal associations, local environmental interest groups, and state water resource agencies are all examples of entities that would benefit from collaborative efforts to identify common reference sites.

In most instances, all of the various agencies conducting monitoring and assessment will be using different methods. A knowledge of the precision and sensitivity of the methods will allow for an agency to decide whether the characterization of a site as reference or minimally impaired by a second agency or other entity fits the necessary criteria to be included as an additional reference site.

Question 3 — How can data from different methods be combined or integrated for increasing a database for assessment?

The question of combining data for a comprehensive assessment is most often asked by states and tribes that want to increase the spatial coverage of an assessment beyond their own limited datasets. From a national or regional perspective, the ability to combine datasets is desirable to make judgements on the condition of the water resource at a higher geographical scale. Ideally, each dataset will have been collected with the same methods.

This question is the most difficult to answer even with a knowledge of the precision and sensitivity. Widely divergent methodologies having highly divergent performance characteristics are not likely to be appropriate for combining under any circumstances. The risk of committing error in judgement of biological condition from a combined dataset of this sort would be too high.

Divergent methodologies with similar or nearly identical performance characteristics are plausible candidates for combining data at metric or index levels. However, a calibration of the methods is

Rapid Bioassessment Protocols for Use in Streams and Wadeable Rivers: Periphyton, Benthic Macroinvertebrates, and Fish, Second Edition

4-15

necessary to ensure that extrapolations of data from one method to the other is scientifically valid. The best fit for a calibrated model is a 1:1 ratio for each metric and index. Realistically, the calibration will be on a less-than-perfect relationship; extrapolations may be via range of values rather than absolute numbers. Thus, combining datasets from dissimilar methods may be valuable for characterizing severe impairment or sites of excellent condition. However, sites with slight to moderate impairment might not be detected with a high level of confidence.

For example, a 6-state collaborative study was conducted on Mid-Atlantic coastal plain streams to determine whether a combined reference condition could be established (Maxted et al. in review). In this study, a single method was applied to all sites in the coastal plain in all 6 states (New Jersey, Delaware, Maryland, Virginia, North Carolina, and South Carolina). The results indicated that two Bioregions exist for the coastal plain ecoregion—the northern portion, including coastal plain streams in New Jersey, Delaware, and Maryland; and the southern portion that includes Virginia, North Carolina, and South Carolina. In most situations, agencies have databases from well-established methods that differ in specific ways. The ability to combine unlike datasets has historically been a problem for scientific investigations. The usual practice has been to aggregate the data to the least common denominator and discard data that do not fit the criteria.

5 HABITAT ASSESSMENT AND PHYSICOCHEMICAL PARAMETERS

An evaluation of habitat quality is critical to any assessment of ecological integrity and should be performed at each site at the time of the biological sampling. In general, habitat and biological diversity in rivers are closely linked (Raven et al. 1998). In the truest sense, "habitat" incorporates all aspects of physical and chemical constituents along with the biotic interactions. In these protocols, the definition of "habitat" is narrowed to the quality of the instream and riparian habitat that influences the structure and function of the aquatic community in a stream. The presence of an altered habitat structure is considered one of the major stressors of aquatic systems (Karr et al. 1986). The presence of a degraded habitat can sometimes obscure investigations on the effects of toxicity and/or pollution. The assessments performed by many water resource agencies include a general description of the site, a physical characterization and water quality assessment, and a visual assessment of instream and riparian habitat quality. Some states (e.g., Idaho DEQ and Illinois EPA) include quantitative measurements of physical parameters in their habitat assessment. Together these data provide an integrated picture of several of the factors influencing the biological condition of a stream system. These assessments are not as comprehensive as needed to adequately identify all causes of impact. However, additional investigation into hydrological modification of water courses and drainage patterns can be conducted, once impairment is noted.

The habitat quality evaluation can be accomplished by characterizing selected physicochemical parameters in conjunction with a systematic assessment of physical structure. Through this approach, key features can be rated or scored to provide a useful assessment of habitat quality.

5.1 PHYSICAL CHARACTERISTICS AND WATER QUALITY

Both physical characteristics and water quality parameters are pertinent to characterization of the stream habitat. An example of the data sheet used to characterize the physical characteristics and water quality of a site is shown in Appendix A. The information required includes measurements of physical characterization and water quality made routinely to supplement biological surveys.

Physical characterization includes documentation of general land use, description of the stream origin and type, summary of the riparian vegetation features, and measurements of instream parameters such as width, depth, flow, and substrate. The water quality discussed in these protocols are *in situ* measurements of standard parameters that can be taken with a water quality instrument. These are generally instantaneous measurements taken at the time of the survey. Measurements of certain parameters, such as temperature, dissolved oxygen, and turbidity, can be taken over a diurnal cycle and will require instrumentation that can be left in place for extended periods or collects water samples at periodic intervals for measurement. In addition, water samples may be desired to be collected for selected chemical analysis. These chemical samples are transported to an analytical laboratory for processing. The combination of this information (physical characterization and water quality) will provide insight as to the ability of the stream to support a healthy aquatic community, and to the presence of chemical and non-chemical stressors to the stream ecosystem. Information requested in this section (Appendix A-1, Form 1) is standard to many aquatic studies and allows for some comparison among sites. Additionally, conditions that may significantly affect aquatic biota are documented.

5.1.1 Header Information (Station Identifier)

The header information is identical on all data sheets and requires sufficient information to identify the station and location where the survey was conducted, date and time of survey, and the investigators responsible for the quality and integrity of the data. The stream name and river basin identify the watershed and tributary; the location of the station is described in the narrative to help identify access to the station for repeat visits. The rivermile (if applicable) and latitude/longitude are specific locational data for the station. The station number is a code assigned by the agency that will associate the sample and survey data with the station. The STORET number is assigned to each datapoint for inclusion in USEPA's STORET system. The stream class is a designation of the grouping of homogeneous characteristics from which assessments will be made. For instance, Ohio EPA uses ecoregions and size of stream, Florida DEP uses bioregions (aggregations of subecoregions), and Arizona DEQ uses elevation as a means to identify stream classes. Listing the agency and investigators assigns responsibility to the data collected from the station at a specific date and time. The reason for the survey is sometimes useful to an agency that conducts surveys for various programs and purposes.

5.1.2 Weather Conditions

Note the present weather conditions on the day of the survey and those immediately preceding the day of the survey. This information is important to interpret the effects of storm events on the sampling effort.

5.1.3 Site Location/Map

To complete this phase of the bioassessment, a photograph may be helpful in identifying station location and documenting habitat conditions. Any observations or data not requested but deemed important by the field observer should be recorded. A hand-drawn map is useful to illustrate major landmarks or features of the channel morphology or orientation, vegetative zones, buildings, etc. that might be used to aid in data interpretation.

5.1.4 Stream Characterization

Stream Subsystem: In regions where the perennial nature of streams is important, or where the tidal influence of streams will alter the structure and function of communities, this parameter should be noted.

Stream Type: Communities inhabiting coldwater streams are markedly different from those in warmwater streams, many states have established temperature criteria that differentiate these 2 stream types.

Stream Origin: Note the origination of the stream under study, if it is known. Examples are glacial, montane, swamp, and bog. As the size of the stream or river increases, a mixture of origins of tributaries is likely.

Chapter 5: Habitat Assessment and Physicochemical Parameters

5.1.5 Watershed Features

Collecting this information usually requires some effort initially for a station. However, subsequent surveys will most likely not require an in-depth research of this information.

Predominant Surrounding Land Use Type: Document the prevalent land-use type in the catchment of the station (noting any other land uses in the area which, although not predominant, may potentially affect water quality). Land use maps should be consulted to accurately document this information.

Local Watershed Nonpoint Source Pollution: This item refers to problems and potential problems in the watershed. Nonpoint source pollution is defined as diffuse agricultural and urban runoff. Other compromising factors in a watershed that may affect water quality include feedlots, constructed wetlands, septic systems, dams and impoundments, mine seepage, etc.

Local Watershed Erosion: The existing or potential detachment of soil within the local watershed (the portion of the watershed or catchment that directly affects the stream reach or station under study) and its movement into the stream is noted. Erosion can be rated through visual observation of watershed and stream characteristics (note any turbidity observed during water quality assessment below).

5.1.6 Riparian Vegetation

An acceptable riparian zone includes a buffer strip of a minimum of 18 m (Barton et al. 1985) from the stream on either side. The acceptable width of the riparian zone may also be variable depending on the size of the stream. Streams over 4 m in width may require larger riparian zones. The vegetation within the riparian zone is documented here as the dominant type and species, if known.

5.1.7 Instream Features

Instream features are measured or evaluated in the sampling reach and catchment as appropriate.

Estimated Reach Length: Measure or estimate the length of the sampling reach. This information is important if reaches of variable length are surveyed and assessed.

Estimated Stream Width (in meters, m): Estimate the distance from bank to bank at a transect representative of the stream width in the reach. If variable widths, use an average to find that which is representative for the given reach.

Sampling Reach Area (m²): Multiply the sampling reach length by the stream width to obtain a calculated surface area.

Estimated Stream Depth (m): Estimate the vertical distance from water surface to stream bottom at a representative depth (use instream habitat feature that is most common in reach) to obtain average depth.

Rapid Bioassessment Protocols for Use in Streams and Wadeable Rivers: Periphyton, Benthic Macroinvertebrates, and Fish, Second Edition

5-3

Velocity: Measure the surface velocity in the thalweg of a representative run area. If measurement is not done, estimate the velocity as slow, moderate, or fast.

Canopy Cover: Note the general proportion of open to shaded area which best describes the amount of cover at the sampling reach or station. A densiometer may be used in place of visual estimation.

High Water Mark (m): Estimate the vertical distance from the bankfull margin of the stream bank to the peak overflow level, as indicated by debris hanging in riparian or floodplain vegetation, and deposition of silt or soil. In instances where bank overflow is rare, a high water mark may not be evident.

Proportion of Reach Represented by Stream Morphological Types: The proportion represented by riffles, runs, and pools should be noted to describe the morphological heterogeneity of the reach.

Channelized: Indicate whether or not the area around the sampling reach or station is channelized (e.g., straightening of stream, bridge abutments and road crossings, diversions, etc.).

Dam Present: Indicate the presence or absence of a dam upstream in the catchment or downstream of the sampling reach or station. If a dam is present, include specific information relating to alteration of flow.

5.1.8 Large Woody Debris

Large Woody Debris (LWD) density, defined and measured as described below, has been used in regional surveys (Shields et al. 1995) and intensive studies of degraded and restored streams (Shields et al. 1998). The method was developed for sand or sand-and-gravel bed streams in the Southeastern U.S. that are wadeable at baseflow, with water widths between 1 and 30 m (Cooper and Testa 1999).

Cooper and Testa's (1999) procedure involves measurements based on visual estimates taken by a wading observer. Only woody debris actually in contact with stream water is counted. Each woody debris formation with a surface area in the plane of the water surface >0.25 m^2 is recorded. The estimated length and width of each formation is recorded on a form or marked directly onto a stream reach drawing. Estimates are made to the nearest 0.5 m , and formations with length or width less than 0.5 m are not counted. Recorded length is maximum width in the direction perpendicular to the length. Maximum actual length and width of a limb, log, or accumulation are not considered.

If only a portion of the log/limb is in contact with the water, only that portion in contact is measured. Root wads and logs/limbs in the water margin are counted if they contact the water, and are arbitrarily given a width of 0.5 m Lone individual limbs and logs are included in the determination if their diameter is 10 cm or larger (Keller and Swanson 1979, Ward and Aumen 1986). Accumulations of smaller limbs and logs are included if the formation total length or width is 0.5 m or larger. Standing trees and stumps within the stream are also recorded if their length and width exceed 0.5 m.

The length and width of each LWD formation are then multiplied, and the resulting products are summed to give the aquatic habitat area directly influenced. This area is then divided by the water

Chapter 5: Habitat Assessment and Physicochemical Parameters

surface area (km^2) within the sampled reach (obtained by multiplying the average water surface width by reach length) to obtain LWD density. Density values of 10^3 to 10^4 m^2/km^2 have been reported for channelized and incised streams and on the order of 10^5 m^2/km^2 for non-incised streams (Shields et al. 1995 and 1998). This density is not an expression of the volume of LWD, but rather a measure of LWD influence on velocity, depth, and cover.

5.1.9 Aquatic Vegetation

The general type and relative dominance of aquatic plants are documented in this section. Only an estimation of the extent of aquatic vegetation is made. Besides being an ecological assemblage that responds to perturbation, aquatic vegetation provides refugia and food for aquatic fauna. List the species of aquatic vegetation, if known.

5.1.10 Water Quality

Temperature (°C), Conductivity or "Specific Conductance" (µohms), Dissolved Oxygen (µg/L), pH, Turbidity: Measure and record values for each of the water quality parameters indicated, using the appropriate calibrated water quality instrument(s). Note the type of instrument and unit number used.

Water Odors: Note those odors described (or include any other odors not listed) that are associated with the water in the sampling area.

Water Surface Oils: Note the term that best describes the relative amount of any oils present on the water surface.

Turbidity: If turbidity is not measured directly, note the term which, based upon visual observation, best describes the amount of material suspended in the water column.

5.1.11 Sediment/Substrate

Sediment Odors: Disturb sediment in pool or other depositional areas and note any odors described (or include any other odors not listed) which are associated with sediment in the sampling reach.

Sediment Oils: Note the term which best describes the relative amount of any sediment oils observed in the sampling area.

Sediment Deposits: Note those deposits described (or include any other deposits not listed) that are present in the sampling reach. Also indicate whether the undersides of rocks not deeply embedded are black (which generally indicates low dissolved oxygen or anaerobic conditions).

Inorganic Substrate Components: Visually estimate the relative proportion of each of the 7 substrate/particle types listed that are present over the sampling reach.

Organic Substrate Components: Indicate relative abundance of each of the 3 substrate types listed.

Rapid Bioassessment Protocols for Use in Streams and Wadeable Rivers: Periphyton, Benthic Macroinvertebrates, and Fish, Second Edition

5-5

5.2   A VISUAL-BASED HABITAT ASSESSMENT

Biological potential is limited by the quality of the physical habitat, forming the template within which biological communities develop (Southwood 1977). Thus, habitat assessment is defined as the evaluation of the structure of the surrounding physical habitat that influences the quality of the water resource and the condition of the resident aquatic community (Barbour et al. 1996a). For streams, an encompassing approach to assessing structure of the habitat includes an evaluation of the variety and quality of the substrate, channel morphology, bank structure, and riparian vegetation. Habitat parameters pertinent to the assessment of habitat quality include those that characterize the stream "micro scale" habitat (e.g., estimation of embeddeddness), the "macro scale" features (e.g., channel morphology), and the riparian and bank structure features that are most often influential in affecting the other parameters.

Rosgen (1985, 1994) presented a stream and river classification system that is founded on the premise that dynamically-stable stream channels have a morphology that provides appropriate distribution of flow energy during storm events. Further, he identifies 8 major variables that affect the stability of channel morphology, but are not mutually independent: channel width, channel depth, flow velocity, discharge, channel slope, roughness of channel materials, sediment load and sediment particle size distribution. When streams have one of these characteristics altered, some of their capability to dissipate energy properly is lost (Leopold et al. 1964, Rosgen 1985) and will result in accelerated rates of channel erosion. Some of the habitat structural components that function to dissipate flow energy are:

> **EQUIPMENT/SUPPLIES NEEDED FOR HABITAT ASSESSMENT AND PHYSICAL/WATER QUALITY CHARACTERIZATION**
>
> - Physical Characterization and Water Quality Field Data Sheet[*]
> - Habitat Assessment Field Data Sheet[*]
> - clipboard
> - pencils or waterproof pens
> - 35 mm camera (may be digital)
> - video camera (optional)
> - upstream/downstream "arrows" or signs for photographing and documenting sampling reaches
> - Flow or velocity meter
> - *In situ* water quality meters
> - Global Positioning System (GPS) Unit
>
> [*] It is helpful to copy field sheets onto water-resistant paper for use in wet weather conditions

- ! sinuosity

- ! roughness of bed and bank materials

- ! presence of point bars (slope is an important characteristic)

- ! vegetative conditions of stream banks and the riparian zone

- ! condition of the floodplain (accessibility from bank, overflow, and size are important characteristics).

Measurement of these parameters or characteristics serve to stratify and place streams into distinct classifications. However, none of these habitat classification techniques attempt to differentiate the quality of the habitat and the ability of the habitat to support the optimal biological condition of the region. Much of our understanding of habitat relationships in streams has emerged from comparative studies that describe statistical relationships between habitat variables and abundance of biota (Hawkins et al. 1993). However, in response to the need to incorporate broader scale habitat assessments in water resource programs, 2 types of approaches for evaluating habitat structure have been developed. In the first, the Environmental Monitoring and Assessment Program (EMAP) of the USEPA and the National Water-Quality Assessment Program (NAWQA) of the USGS developed techniques that incorporate measurements of various features of the instream, channel, and bank morphology (Meader et al. 1993, Klemm and Lazorchak 1994). These techniques provide a relatively comprehensive characterization of the physical structure of the stream sampling reach and its surrounding floodplain. The second type was a more rapid and qualitative habitat assessment approach that was developed to describe the overall quality of the physical habitat (Ball 1982, Ohio EPA 1987, Plafkin et al. 1989, Barbour and Stribling 1991, 1994, Rankin 1991, 1995). In this document, the more rapid visual-based approach is described. A cursory overview of the more quantitative approaches to characterizing the physical structure of the habitat is provided.

The habitat assessment matrix developed for the Rapid Bioassessment Protocols (RBPs) in Plafkin et al. (1989) were originally based on the Stream Classification Guidelines for Wisconsin developed by Ball (1982) and *"Methods of Evaluating Stream, Riparian, and Biotic Conditions"* developed by Platts et al. (1983). Barbour and Stribling (1991, 1994) modified the habitat assessment approach originally developed for the RBPs to include additional assessment parameters for high gradient streams and a more appropriate parameter set for low gradient streams (Appendix A-1, Forms 2,3). All parameters are evaluated and rated on a numerical scale of 0 to 20 (highest) for each sampling reach. The ratings are then totaled and compared to a reference condition to provide a final habitat ranking. Scores increase as habitat quality increases. To ensure consistency in the evaluation procedure, descriptions of the physical parameters and relative criteria are included in the rating form.

The Environmental Agency of Great Britain (Environment Agency of England and Wales, Scottish Environment Protection Agency, and Environment and Heritage Service of Northern Ireland) have developed a River Habitat Survey (RHS) for characterizing the quality of their streams and rivers (Raven et al. 1998). The approach used in Great Britain is similar to the visual-based habitat assessment used in the US in that scores are assigned to ranges of conditions of various habitat parameters.

A biologist who is well versed in the ecology and zoogeography of the region can generally recognize optimal habitat structure as it relates to the biological community. The ability to accurately assess the quality of the physical habitat structure using a visual-based approach depends on several factors:

!   the parameters selected to represent the various features of habitat structure need to be relevant and clearly defined

!   a continuum of conditions for each parameter must exist that can be characterized from the optimum for the region or stream type under study to the poorest situation reflecting substantial alteration due to anthropogenic activities

!   the judgement criteria for the attributes of each parameter should minimize
         subjectivity through either quantitative measurements or specific categorical
         choices

!   the investigators are experienced in or adequately trained for stream assessments
         in the region under study (Hannaford et al. 1997)

!   adequate documentation and ongoing training is maintained to evaluate and correct
         errors resulting in outliers and aberrant assessments.

Habitat evaluations are first made on instream habitat, followed by channel morphology, bank structural features, and riparian vegetation. Generally, a single, comprehensive assessment is made that incorporates features of the entire sampling reach as well as selected features of the catchment. Additional assessments may be made on neighboring reaches to provide a broader evaluation of habitat quality for the stream ecosystem. The actual habitat assessment process involves rating the 10 parameters as optimal, suboptimal, marginal, or poor based on the criteria included on the Habitat Assessment Field Data Sheets (Appendix A-1, Forms 2,3). Some state programs, such as Florida Department of Environmental Protection (DEP) (1996) and Mid-Atlantic Coastal Streams Workgroup (MACS) (1996) have adapted this approach using somewhat fewer and different parameters.

Reference conditions are used to scale the assessment to the "best attainable" situation. This approach is critical to the assessment because stream characteristics will vary dramatically across different regions (Barbour and Stribling 1991). The ratio between the score for the test station and the score for the reference condition provides a percent comparability measure for each station. The station of interest is then classified on the basis of its similarity to expected conditions (reference condition), and its apparent potential to support an acceptable level of biological health. Use of a percent comparability evaluation allows for regional and stream-size differences which affect flow or velocity, substrate, and channel morphology. Some regions are characterized by streams having a low channel gradient, such as coastal plains or prairie regions.

Other habitat assessment approaches or a more rigorously quantitative approach to measuring the habitat parameters may be used (See Klemm and Lazorchak 1994, Kaufmann and Robison 1997, Meader et al. 1993). However, holistic and rapid assessment of a wide variety of habitat attributes along with other types of data is critical if physical measurements are to be used to best advantage in interpreting biological data. A more detailed discussion of the relationship between habitat quality and biological condition is presented in Chapter 10.

A generic habitat assessment approach based on visual observation can be separated into 2 basic approaches—one designed for high-gradient streams and one designed for low-gradient streams. High-gradient or riffle/run prevalent streams are those in moderate to high gradient landscapes. Natural high-gradient streams have substrates primarily composed of coarse sediment particles (i.e., gravel or larger) or frequent coarse particulate aggregations along stream reaches. Low-gradient or glide/pool prevalent streams are those in low to moderate gradient landscapes. Natural low-gradient streams have substrates of fine sediment or infrequent aggregations of more coarse (gravel or larger) sediment particles along stream reaches. The entire sampling reach is evaluated for each parameter. Descriptions of each parameter and its relevance to instream biota are presented in the following discussion. Parameters that are used only for high-gradient prevalent streams are marked with an "a"; those for low-gradient dominant streams, a "b". If a parameter is used for both stream types, it is not marked with a letter. A brief set of decision criteria is given for each parameter corresponding to each of the 4 categories reflecting a continuum of conditions on the field sheet (optimal, suboptimal, marginal, and poor). Refer to Appendix A-1, Forms 2 and 3, for a complete field assessment guide.

PROCEDURE FOR PERFORMING HABITAT ASSESSMENT

1. Select the reach to be assessed. The habitat assessment is performed on the same 100 m reach (or other reach designation [e.g., 40 x stream wetted width]) from which the biological sampling is conducted. Some parameters require an observation of a broader section of the catchment than just the sampling reach.

2. Complete the station identification section of each field data sheet and habitat assessment form.

3. It is best for the investigators to obtain a close look at the habitat features to make an adequate assessment. If the physical and water quality characterization and habitat assessment are done before the biological sampling, care must be taken to avoid disturbing the sampling habitat.

4. Complete the **Physical Characterization and Water Quality Field Data Sheet**. Sketch a map of the sampling reach on the back of this form.

5. Complete the **Habitat Assessment Field Data Sheet**, in a team of 2 or more biologists, if possible, to come to a consensus on determination of quality. Those parameters to be evaluated on a scale greater than a sampling reach require traversing the stream corridor to the extent deemed necessary to assess the habitat feature. As a general rule-of-thumb, use 2 lengths of the sampling reach to assess these parameters.

QUALITY ASSURANCE PROCEDURES

1. Each biologist is to be trained in the visual-based habitat assessment technique for the applicable region or state.

2. The judgment criteria for each habitat parameter are calibrated for the stream classes under study. Some text modifications may be needed on a regional basis.

3. Periodic checks of assessment results are completed using pictures of the sampling reach and discussions among the biologists in the agency.

Parameters to be evaluated in sampling reach:

1   EPIFAUNAL SUBSTRATE/AVAILABLE COVER

high and low gradient streams

Includes the relative quantity and variety of natural structures in the stream, such as cobble (riffles), large rocks, fallen trees, logs and branches, and undercut banks, available as refugia, feeding, or sites for spawning and nursery functions of aquatic macrofauna. A wide variety and/or abundance of submerged structures in the stream provides macroinvertebrates and fish with a large number of niches, thus increasing habitat diversity. As variety and abundance of cover decreases, habitat structure becomes monotonous, diversity decreases, and the potential for recovery following disturbance decreases. Riffles and runs are critical for maintaining a variety and abundance of insects in most high-gradient streams and serving as spawning and feeding refugia for certain fish. The extent and quality of the riffle is an important factor in the support of a healthy biological condition in high-gradient streams. Riffles and runs offer a diversity of habitat through variety of particle size, and, in many small high-gradient streams, will provide the most stable habitat. Snags and submerged logs are among the most productive habitat structure for macroinvertebrate colonization and fish refugia in low-gradient streams. However, "new fall" will not yet be suitable for colonization.

Selected References

Wesche et al. 1985, Pearsons et al. 1992, Gorman 1988, Rankin 1991, Barbour and Stribling 1991, Plafkin et al. 1989, Platts et al. 1983, Osborne et al. 1991, Benke et al. 1984, Wallace et al. 1996, Ball 1982, MacDonald et al. 1991, Reice 1980, Clements 1987, Hawkins et al. 1982, Beechie and Sibley 1997.

Habitat Parameter	Condition Category			
	Optimal	Suboptimal	Marginal	Poor
1. Epifaunal Substrate/ Available Cover (high and low gradient)	Greater than 70% (50% for low gradient streams) of substrate favorable for epifaunal colonization and fish cover; mix of snags, submerged logs, undercut banks, cobble or other stable habitat and at stage to allow full colonization potential (i.e., logs/snags that are not new fall and not transient).	40-70% (30-50% for low gradient streams) mix of stable habitat; well-suited for full colonization potential; adequate habitat for maintenance of populations; presence of additional substrate in the form of newfall, but not yet prepared for colonization (may rate at high end of scale).	20-40% (10-30% for low gradient streams) mix of stable habitat; habitat availability less than desirable; substrate frequently disturbed or removed.	Less than 20% (10% for low gradient streams) stable habitat; lack of habitat is obvious; substrate unstable or lacking.
SCORE	20 19 18 17 16	15 14 13 12 11	10 9 8 7 6	5 4 3 2 1 0

1a. Epifaunal Substrate/Available Cover—High Gradient

Poor Range

Optimal Range

1b. Epifaunal Substrate/Available Cover—Low Gradient

Optimal Range   *(Mary Kay Corazalla, U. of Minn.)*   Poor Range

2a EMBEDDEDNESS

high gradient streams   Refers to the extent to which rocks (gravel, cobble, and boulders) and snags are covered or sunken into the silt, sand, or mud of the stream bottom. Generally, as rocks become embedded, the surface area available to macroinvertebrates and fish (shelter, spawning, and egg incubation) is decreased. Embeddedness is a result of large-scale sediment movement and deposition, and is a parameter evaluated in the riffles and runs of high-gradient streams. The rating of this parameter may be variable depending on where the observations are taken. To avoid confusion with sediment deposition (another habitat parameter), observations of embeddedness should be taken in the upstream and central portions of riffles and cobble substrate areas.

Selected References   Ball 1982, Osborne et al. 1991, Barbour and Stribling 1991, Platts et al. 1983, MacDonald et al. 1991, Rankin 1991, Reice 1980, Clements 1987, Benke et al. 1984, Hawkins et al. 1982, Burton and Harvey 1990.

Habitat Parameter	Condition Category			
	Optimal	Suboptimal	Marginal	Poor
2.a Embeddedness (high gradient)	Gravel, cobble, and boulder particles are 0-25% surrounded by fine sediment. Layering of cobble provides diversity of niche space.	Gravel, cobble, and boulder particles are 25-50% surrounded by fine sediment.	Gravel, cobble, and boulder particles are 50-75% surrounded by fine sediment.	Gravel, cobble, and boulder particles are more than 75% surrounded by fine sediment.
SCORE	20 19 18 17 16	15 14 13 12 11	10 9 8 7 6	5 4 3 2 1 0

2a. Embeddedness—High Gradient

Optimal Range   *(William Taft, MI DNR)*

Poor Range   *(William Taft, MI DNR)*

2b POOL SUBSTRATE CHARACTERIZATION

low gradient
streams
Evaluates the type and condition of bottom substrates found in pools. Firmer sediment types (e.g., gravel, sand) and rooted aquatic plants support a wider variety of organisms than a pool substrate dominated by mud or bedrock and no plants. In addition, a stream that has a uniform substrate in its pools will support far fewer types of organisms than a stream that has a variety of substrate types.

Selected
References
Beschta and Platts 1986, U.S. EPA 1983.

Habitat Parameter	Condition Category			
	Optimal	Suboptimal	Marginal	Poor
2b. Pool Substrate Characterization (low gradient)	Mixture of substrate materials, with gravel and firm sand prevalent; root mats and submerged vegetation common.	Mixture of soft sand, mud, or clay; mud may be dominant; some root mats and submerged vegetation present.	All mud or clay or sand bottom; little or no root mat; no submerged vegetation.	Hard-pan clay or bedrock; no root mat or submerged vegetation.
SCORE	20 19 18 17 16	15 14 13 12 11	10 9 8 7 6	5 4 3 2 1 0

2b. Pool Substrate Characterization—Low Gradient

Optimal Range
(Mary Kay Corazalla, U. of Minn.)

Poor Range

Chapter 5: Habitat Assessment and Physicochemical Parameters

3a VELOCITY/DEPTH COMBINATIONS

high gradient streams

Patterns of velocity and depth are included for high-gradient streams under this parameter as an important feature of habitat diversity. The best streams in most high-gradient regions will have all 4 patterns present: (1) slow-deep, (2) slow-shallow, (3) fast-deep, and (4) fast-shallow. The general guidelines are 0.5 m depth to separate shallow from deep, and 0.3 m/sec to separate fast from slow. The occurrence of these 4 patterns relates to the stream's ability to provide and maintain a stable aquatic environment.

Selected References

Ball 1982, Brown and Brussock 1991, Gore and Judy 1981, Oswood and Barber 1982.

Habitat Parameter	Condition Category			
	Optimal	Suboptimal	Marginal	Poor
3a. Velocity/ Depth Regimes (high gradient)	All 4 velocity/depth regimes present (slow-deep, slow-shallow, fast-deep, fast-shallow). (slow is <0.3 m/s, deep is >0.5 m)	Only 3 of the 4 regimes present (if fast-shallow is missing, score lower than if missing other regimes).	Only 2 of the 4 habitat regimes present (if fast-shallow or slow-shallow are missing, score low).	Dominated by 1 velocity/depth regime (usually slow-deep).
SCORE	20 19 18 17 16	15 14 13 12 11	10 9 8 7 6	5 4 3 2 1 0

3a.   Velocity/Depth Regimes—High Gradient

Optimal Range   *(Mary Kay Corazalla, U. of Minn.)*
(arrows emphasize different velocity/depth regimes)

Poor Range   *(William Taft, MI DNR)*

3b POOL VARIABILITY

low gradient streams

Rates the overall mixture of pool types found in streams, according to size and depth. The 4 basic types of pools are large-shallow, large-deep, small-shallow, and small-deep. A stream with many pool types will support a wide variety of aquatic species. Rivers with low sinuosity (few bends) and monotonous pool characteristics do not have sufficient quantities and types of habitat to support a diverse aquatic community. General guidelines are any pool dimension (i.e., length, width, oblique) greater than half the cross-section of the stream for separating large from small and 1 m depth separating shallow and deep.

Selected References

Beschta and Platts 1986, USEPA 1983.

Habitat Parameter	Condition Category			
	Optimal	Suboptimal	Marginal	Poor
3b. Pool Variability (low gradient)	Even mix of large-shallow, large-deep, small-shallow, small-deep pools present.	Majority of pools large-deep; very few shallow.	Shallow pools much more prevalent than deep pools.	Majority of pools small-shallow or pools absent.
SCORE	20 19 18 17 16	15 14 13 12 11	10 9 8 7 6	5 4 3 2 1 0

3b. Pool Variability—Low Gradient

Optimal Range   *(Peggy Morgan, FL DEP)*

Poor Range   *(William Taft, MI DNR)*

4 SEDIMENT DEPOSITION

high and low gradient streams   Measures the amount of sediment that has accumulated in pools and the changes that have occurred to the stream bottom as a result of deposition. Deposition occurs from large-scale movement of sediment. Sediment deposition may cause the formation of islands, point bars (areas of increased deposition usually at the beginning of a meander that increase in size as the channel is diverted toward the outer bank) or shoals, or result in the filling of runs and pools. Usually deposition is evident in areas that are obstructed by natural or manmade debris and areas where the stream flow decreases, such as bends. High levels of sediment deposition are symptoms of an unstable and continually changing environment that becomes unsuitable for many organisms.

Selected References   MacDonald et al. 1991, Platts et al. 1983, Ball 1982, Armour et al. 1991, Barbour and Stribling 1991, Rosgen 1985.

Habitat Parameter	Condition Category			
	Optimal	Suboptimal	Marginal	Poor
4. Sediment Deposition **(high and low gradient)**	Little or no enlargement of islands or point bars and less than 5% (<20% for low-gradient streams) of the bottom affected by sediment deposition.	Some new increase in bar formation, mostly from gravel, sand or fine sediment; 5-30% (20-50% for low-gradient) of the bottom affected; slight deposition in pools.	Moderate deposition of new gravel, sand or fine sediment on old and new bars; 30-50% (50-80% for low-gradient) of the bottom affected; sediment deposits at obstructions, constrictions, and bends; moderate deposition of pools prevalent.	Heavy deposits of fine material, increased bar development; more than 50% (80% for low-gradient) of the bottom changing frequently; pools almost absent due to substantial sediment deposition.
SCORE	20 19 18 17 16	15 14 13 12 11	10 9 8 7 6	5 4 3 2 1 0

4a.   Sediment Deposition—High Gradient

Optimal Range

Poor Range
(arrow pointing to sediment deposition)

4b.   Sediment Deposition—Low Gradient

Optimal Range

Poor Range
(arrows pointing to sediment deposition)

5 CHANNEL FLOW STATUS

high and low
gradient streams
The degree to which the channel is filled with water. The flow status will change as the channel enlarges (e.g., aggrading stream beds with actively widening channels) or as flow decreases as a result of dams and other obstructions, diversions for irrigation, or drought. When water does not cover much of the streambed, the amount of suitable substrate for aquatic organisms is limited. In high-gradient streams, riffles and cobble substrate are exposed; in low-gradient streams, the decrease in water level exposes logs and snags, thereby reducing the areas of good habitat. Channel flow is especially useful for interpreting biological condition under abnormal or lowered flow conditions. This parameter becomes important when more than one biological index period is used for surveys or the timing of sampling is inconsistent among sites or annual periodicity.

Selected
References
Rankin 1991, Rosgen 1985, Hupp and Simon 1986, MacDonald et al. 1991, Ball 1982, Hicks et al. 1991.

Habitat Parameter	Condition Category			
	Optimal	Suboptimal	Marginal	Poor
5. Channel Flow Status (high and low gradient)	Water reaches base of both lower banks, and minimal amount of channel substrate is exposed.	Water fills >75% of the available channel; or <25% of channel substrate is exposed.	Water fills 25-75% of the available channel, and/or riffle substrates are mostly exposed.	Very little water in channel and mostly present as standing pools.
SCORE	20　19　18　17　16	15　14　13　12　11	10　9　8　7　6	5　4　3　2　1　0

5a.   Channel Flow Status—High Gradient

Poor Range
(arrow showing that water is not reaching both banks; leaving much of channel uncovered)

Optimal Range

5b.   Channel Flow Status—Low Gradient

Optimal Range

Poor Range   *(James Stahl, IN DEM)*

Parameters to be evaluated broader than sampling reach:

6 CHANNEL ALTERATION

high and low gradient streams

Is a measure of large-scale changes in the shape of the stream channel. Many streams in urban and agricultural areas have been straightened, deepened, or diverted into concrete channels, often for flood control or irrigation purposes. Such streams have far fewer natural habitats for fish, macroinvertebrates, and plants than do naturally meandering streams. Channel alteration is present when artificial embankments, riprap, and other forms of artificial bank stabilization or structures are present; when the stream is very straight for significant distances; when dams and bridges are present; and when other such changes have occurred. Scouring is often associated with channel alteration.

Selected References

Barbour and Stribling 1991, Simon 1989a, b, Simon and Hupp 1987, Hupp and Simon 1986, Hupp 1992, Rosgen 1985, Rankin 1991, MacDonald et al. 1991.

Habitat Parameter	Condition Category			
	Optimal	Suboptimal	Marginal	Poor
6. Channel Alteration (high and low gradient)	Channelization or dredging absent or minimal; stream with normal pattern.	Some channelization present, usually in areas of bridge abutments; evidence of past channelization, i.e., dredging, (greater than past 20 yr) may be present, but recent channelization is not present.	Channelization may be extensive; embankments or shoring structures present on both banks; and 40 to 80% of stream reach channelized and disrupted.	Banks shored with gabion or cement; over 80% of the stream reach channelized and disrupted. Instream habitat greatly altered or removed entirely.
SCORE	20 19 18 17 16	15 14 13 12 11	10 9 8 7 6	5 4 3 2 1 0

Rapid Bioassessment Protocols for Use in Streams and Wadeable Rivers: Periphyton, Benthic Macroinvertebrates, and Fish, Second Edition

5-21

6a.   Channel Alteration—High Gradient

Optimal Range

Poor Range
(arrows emphasizing large-scale channel
alterations)

6b.   Channel Alteration—Low Gradient

Optimal Range

Poor Range   *(John Maxted, DE DNREC)*

7a FREQUENCY OF RIFFLES (OR BENDS)

high gradient streams

Is a way to measure the sequence of riffles and thus the heterogeneity occurring in a stream. Riffles are a source of high-quality habitat and diverse fauna, therefore, an increased frequency of occurrence greatly enhances the diversity of the stream community. For high gradient streams where distinct riffles are uncommon, a run/bend ratio can be used as a measure of meandering or sinuosity (see 7b). A high degree of sinuosity provides for diverse habitat and fauna, and the stream is better able to handle surges when the stream fluctuates as a result of storms. The absorption of this energy by bends protects the stream from excessive erosion and flooding and provides refugia for benthic invertebrates and fish during storm events. To gain an appreciation of this parameter in some streams, a longer segment or reach than that designated for sampling should be incorporated into the evaluation. In some situations, this parameter may be rated from viewing accurate topographical maps. The "sequencing" pattern of the stream morphology is important in rating this parameter. In headwaters, riffles are usually continuous and the presence of cascades or boulders provides a form of sinuosity and enhances the structure of the stream. A stable channel is one that does not exhibit progressive changes in slope, shape, or dimensions, although short-term variations may occur during floods (Gordon et al. 1992).

Selected References

Hupp and Simon 1991, Brussock and Brown 1991, Platts et al. 1983, Rankin 1991, Rosgen 1985, 1994, 1996, Osborne and Hendricks 1983, Hughes and Omernik 1983, Cushman 1985, Bain and Boltz 1989, Gislason 1985, Hawkins et al. 1982, Statzner et al. 1988.

Habitat Parameter	Condition Category			
	Optimal	Suboptimal	Marginal	Poor
7a. Frequency of Riffles (or bends) **(high gradient)**	Occurrence of riffles relatively frequent; ratio of distance between riffles divided by width of the stream <7:1 (generally 5 to 7); variety of habitat is key. In streams where riffles are continuous, placement of boulders or other large, natural obstruction is important.	Occurrence of riffles infrequent; distance between riffles divided by the width of the stream is between 7 to 15.	Occasional riffle or bend; bottom contours provide some habitat; distance between riffles divided by the width of the stream is between 15 to 25.	Generally all flat water or shallow riffles; poor habitat; distance between riffles divided by the width of the stream is a ratio of >25.
SCORE	20 19 18 17 16	15 14 13 12 11	10 9 8 7 6	5 4 3 2 1 0

Rapid Bioassessment Protocols for Use in Streams and Wadeable Rivers: Periphyton, Benthic Macroinvertebrates, and Fish, Second Edition

5-23

7a.   Frequency of Riffles (or bends)—High Gradient

Poor Range

Optimal Range
(arrows showing frequency of riffles and bends)

7b   CHANNEL SINUOSITY

low gradient streams

Evaluates the meandering or sinuosity of the stream. A high degree of sinuosity provides for diverse habitat and fauna, and the stream is better able to handle surges when the stream fluctuates as a result of storms. The absorption of this energy by bends protects the stream from excessive erosion and flooding and provides refugia for benthic invertebrates and fish during storm events. To gain an appreciation of this parameter in low gradient streams, a longer segment or reach than that designated for sampling may be incorporated into the evaluation. In some situations, this parameter may be rated from viewing accurate topographical maps. The "sequencing" pattern of the stream morphology is important in rating this parameter. In "oxbow" streams of coastal areas and deltas, meanders are highly exaggerated and transient. Natural conditions in these streams are shifting channels and bends, and alteration is usually in the form of flow regulation and diversion. A stable channel is one that does not exhibit progressive changes in slope, shape, or dimensions, although short-term variations may occur during floods (Gordon et al. 1992).

Selected References

Hupp and Simon 1991, Brussock and Brown 1991, Platts et al. 1983, Rankin 1991, Rosgen 1985, 1994, 1996, Osborne and Hendricks 1983, Hughes and Omernik 1983, Cushman 1985, Bain and Boltz 1989, Gislason 1985, Hawkins et al. 1982, Statzner et al. 1988.

Habitat Parameter	Condition Category			
	Optimal	Suboptimal	Marginal	Poor
7b. Channel Sinuosity (low gradient)	The bends in the stream increase the stream length 3 to 4 times longer than if it was in a straight line. (Note - channel braiding is considered normal in coastal plains and other low-lying areas. This parameter is not easily rated in these areas.)	The bends in the stream increase the stream length 2 to 3 times longer than if it was in a straight line.	The bends in the stream increase the stream length 1 to 2 times longer than if it was in a straight line.	Channel straight; waterway has been channelized for a long distance.
SCORE	20 19 18 17 16	15 14 13 12 11	10 9 8 7 6	5 4 3 2 1 0

7b.   Channel Sinuosity—Low Gradient

Optimal Range

Poor Range

Rapid Bioassessment Protocols for Use in Streams and Wadeable Rivers: Periphyton, Benthic Macroinvertebrates, and Fish, Second Edition

5-25

8 BANK STABILITY (condition of banks)

high and low gradient streams

Measures whether the stream banks are eroded (or have the potential for erosion). Steep banks are more likely to collapse and suffer from erosion than are gently sloping banks, and are therefore considered to be unstable. Signs of erosion include crumbling, unvegetated banks, exposed tree roots, and exposed soil. Eroded banks indicate a problem of sediment movement and deposition, and suggest a scarcity of cover and organic input to streams. Each bank is evaluated separately and the cumulative score (right and left) is used for this parameter.

Selected References

Ball 1982, MacDonald et al. 1991, Armour et al. 1991, Barbour and Stribling 1991, Hupp and Simon 1986, 1991, Simon 1989a, Hupp 1992, Hicks et al. 1991, Osborne et al. 1991, Rosgen 1994, 1996.

Habitat Parameter	Condition Category											
	Optimal			Suboptimal			Marginal			Poor		
8. Bank Stability (score each bank) **Note: determine left or right side by facing downstream** **(high and low gradient)**	Banks stable; evidence of erosion or bank failure absent or minimal; little potential for future problems. <5% of bank affected.			Moderately stable; infrequent, small areas of erosion mostly healed over. 5-30% of bank in reach has areas of erosion.			Moderately unstable; 30-60% of bank in reach has areas of erosion; high erosion potential during floods.			Unstable; many eroded areas; "raw" areas frequent along straight sections and bends; obvious bank sloughing; 60-100% of bank has erosional scars.		
SCORE ___ (LB)	Left Bank	10	9	8	7	6	5	4	3	2	1	0
SCORE ___ (RB)	Right Bank	10	9	8	7	6	5	4	3	2	1	0

8a.   Bank Stability (condition of banks)—High Gradient

Optimal Range
(arrow pointing to stable streambanks)

Poor Range   *(MD Save Our Streams)*
(arrow highlighting unstable streambanks)

8b.   Bank Stability (condition of banks)—Low Gradient

Optimal Range   *(Peggy Morgan, FL DEP)*

Poor Range
(arrow highlighting unstable streambanks)

9 BANK VEGETATIVE PROTECTION

high and low gradient streams

Measures the amount of vegetative protection afforded to the stream bank and the near-stream portion of the riparian zone. The root systems of plants growing on stream banks help hold soil in place, thereby reducing the amount of erosion that is likely to occur. This parameter supplies information on the ability of the bank to resist erosion as well as some additional information on the uptake of nutrients by the plants, the control of instream scouring, and stream shading. Banks that have full, natural plant growth are better for fish and macroinvertebrates than are banks without vegetative protection or those shored up with concrete or riprap. This parameter is made more effective by defining the native vegetation for the region and stream type (i.e., shrubs, trees, etc.). In some regions, the introduction of exotics has virtually replaced all native vegetation. The value of exotic vegetation to the quality of the habitat structure and contribution to the stream ecosystem must be considered in this parameter. In areas of high grazing pressure from livestock or where residential and urban development activities disrupt the riparian zone, the growth of a natural plant community is impeded and can extend to the bank vegetative protection zone. Each bank is evaluated separately and the cumulative score (right and left) is used for this parameter.

Selected References

Platts et al. 1983, Hupp and Simon 1986, 1991, Simon and Hupp 1987, Ball 1982, Osborne et al. 1991, Rankin 1991, Barbour and Stribling 1991, MacDonald et al. 1991, Armour et al. 1991, Myers and Swanson 1991, Bauer and Burton 1993.

Habitat Parameter	Condition Category										
	Optimal			Suboptimal			Marginal			Poor	
9. Vegetative Protection (score each bank) **Note: determine left or right side by facing downstream.** **(high and low gradient)**	More than 90% of the streambank surfaces and immediate riparian zones covered by native vegetation, including trees, understory shrubs, or nonwoody macrophytes; vegetative disruption through grazing or mowing minimal or not evident; almost all plants allowed to grow naturally.			70-90% of the streambank surfaces covered by native vegetation, but one class of plants is not well-represented; disruption evident but not affecting full plant growth potential to any great extent; more than one-half of the potential plant stubble height remaining.			50-70% of the streambank surfaces covered by vegetation; disruption obvious; patches of bare soil or closely cropped vegetation common; less than one-half of the potential plant stubble height remaining.			Less than 50% of the streambank surfaces covered by vegetation; disruption of streambank vegetation is very high; vegetation has been removed to 5 centimeters or less in average stubble height.	
SCORE ___ (LB)	Left Bank	10	9	8	7	6	5	4	3	2	1	0
SCORE ___ (RB)	Right Bank	10	9	8	7	6	5	4	3	2	1	0

9a. Bank Vegetative Protection—High Gradient

Optimal Range
(arrow pointing to streambank with high level of vegetative cover)

Poor Range
(arrow pointing to streambank with almost no vegetative cover)

9b. Bank Vegetative Protection—Low Gradient

Optimal Range   *(Peggy Morgan, FL DEP)*

Poor Range   *(MD Save Our Streams)*
(arrow pointing to channelized streambank with no vegetative cover)

10 RIPARIAN VEGETATIVE ZONE WIDTH

high and low
gradient streams

Measures the width of natural vegetation from the edge of the stream bank out through the riparian zone. The vegetative zone serves as a buffer to pollutants entering a stream from runoff, controls erosion, and provides habitat and nutrient input into the stream. A relatively undisturbed riparian zone supports a robust stream system; narrow riparian zones occur when roads, parking lots, fields, lawns, bare soil, rocks, or buildings are near the stream bank. Residential developments, urban centers, golf courses, and rangeland are the common causes of anthropogenic degradation of the riparian zone. Conversely, the presence of "old field" (i.e., a previously developed field not currently in use), paths, and walkways in an otherwise undisturbed riparian zone may be judged to be inconsequential to altering the riparian zone and may be given relatively high scores. For variable size streams, the specified width of a desirable riparian zone may also be variable and may be best determined by some multiple of stream width (e.g., 4 x wetted stream width). Each bank is evaluated separately and the cumulative score (right and left) is used for this parameter.

Selected
References

Barton et al. 1985, Naiman et al. 1993, Hupp 1992, Gregory et al. 1991, Platts et al. 1983, Rankin 1991, Barbour and Stribling 1991, Bauer and Burton 1993.

Habitat Parameter	Condition Category											
	Optimal			**Suboptimal**			**Marginal**			**Poor**		
10. Riparian Vegetative Zone Width (score each bank riparian zone) **(high and low gradient)**	Width of riparian zone >18 meters; human activities (i.e., parking lots, roadbeds, clear-cuts, lawns, or crops) have not impacted zone.			Width of riparian zone 12-18 meters; human activities have impacted zone only minimally.			Width of riparian zone 6-12 meters; human activities have impacted zone a great deal.			Width of riparian zone <6 meters; little or no riparian vegetation due to human activities.		
SCORE ___ (LB)	Left Bank	10	9	8	7	6	5	4	3	2	1	0
SCORE ___ (RB)	Right Bank	10	9	8	7	6	5	4	3	2	1	0

10a. Riparian Vegetative Zone Width—High Gradient

Optimal Range
(arrow pointing out an undisturbed riparian zone)

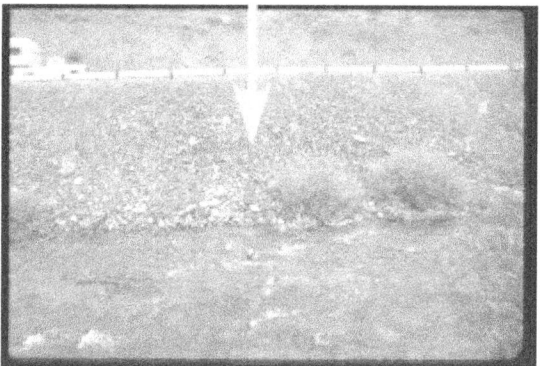

Poor Range
(arrow pointing out lack of riparian zone)

10b. Riparian Vegetative Zone Width—Low Gradient

Optimal Range
(arrow emphasizing an undisturbed riparian zone)

Poor Range   *(MD Save Our Streams)*
(arrow emphasizing lack of riparian zone)

Rapid Bioassessment Protocols for Use in Streams and Wadeable Rivers: Periphyton, Benthic Macroinvertebrates, and Fish, Second Edition

5-31

5.3   ADDITIONS OF QUANTITATIVE MEASURES TO THE HABITAT ASSESSMENT

Kaufmann (1993) identified 7 general physical habitat attributes important in influencing stream ecology. These include:

- !   channel dimensions

- !   channel gradient

- !   channel substrate size and type

- !   habitat complexity and cover

- !   riparian vegetation cover and structure

- !   anthropogenic alterations

- !   channel-riparian interaction.

All of these attributes vary naturally, as do biological characteristics; thus expectations differ even in the absence of anthropogenic disturbances. Within a given physiographic-climatic region, stream drainage area and overall stream gradient are likely to be strong natural determinants of many aspects of stream habitat, because of their influence on discharge, flood stage, and stream power (the product of discharge times gradient). In addition, all of these attributes may be directly or indirectly altered by anthropogenic activities.

In Section 5.2, an approach is described whereby habitat quality is interpreted directly in the field by biologists while sampling the stream reach. This Level 1 approach is observational and requires only one person (although a team approach is recommended) and takes about 15 to 20 minutes per stream reach. This approach more quickly yields a habitat quality assessment. However, it depends upon the knowledge and experience of the field biologist to make the proper interpretation of observed of both the natural expectations (potentials) and the biological consequences (quality) that can be attributed to the observed physical attributes. Hannaford et al. (1997) found that training in habitat assessment was necessary to reduce the subjectivity in a visual-based approach. The authors also stated that training on different types of streams may be necessary to adequately prepare investigators.

The second conceptual approach described here confines observations to habitat characteristics themselves (whether they are quantitative or qualitative), then later ascribing quality scoring to these measurements as part of the data analysis process. Typically, this second type of habitat assessment approach employs more quantitative data collection, as exemplified by field methods described by Kaufmann and Robison (1997) for EMAP, Simonson et al. (1994), Meador et al. (1993) for NAWQA, and others cited by Gurtz and Muir (1994). These field approaches typically define a reach length proportional to stream width and employ transect measurements that are systematically spaced (Simonson et al. 1994, Kaufmann and Robison 1997) or spaced by judgement to be representative (Meador et al. 1993). They usually include measurement of substrate, channel and bank dimensions, riparian canopy cover, discharge, gradient, sinuosity, in-channel cover features, and counts of large woody debris and riparian human disturbances. They may employ systematic visual estimates of substrate embeddedness, fish cover features, habitat types, and riparian vegetation structure. The time commitment in the field to these more quantitative habitat assessment methods is usually 1.5 to 3 hours with a crew of two people. Because of the greater amount of data collected, they also require more time for data summarization, analysis, and interpretation. On the other hand, the more quantitative methods and less ambiguous field parameters result in considerably greater precision. The USEPA applied both quantitative and visual-based (RBPs) methods in a stream survey undertaken over 4 years in the mid-Atlantic region of the Appalachian Mountains. An earlier version of the RBP techniques were applied on 301 streams with repeat visits to 29 streams; signal-to-noise ratios varied from 0.1 to 3.0 for the twelve RBP metrics and averaged (1.1 for the RBP total habitat quality score). The quantitative methods produced a higher level of precision; signal-to-noise ratios were typically between 10 and 50, and sometimes in excess of 100 for quantitative measurements of channel morphology, substrate, and canopy densiometer measurements made on a random subset of 186 streams with 27 repeat visits in the same survey. Similarly, semi-quantitative estimates of fish cover and riparian human disturbance estimates obtained from multiple, systematic visual observations of otherwise measurable features had signal:noise ratios from 5 to 50. Many riparian vegetation cover and structure metrics were moderately precise (signal:noise ranging from 2 to 30). Commonly used flow dependent measures (e.g., riffle/pool and width/depth ratios), and some visual riparian cover estimates were less precise, with signal:noise ratios more in the range of those observed for metrics of the EPA's RBP habitat score (<2).

The USEPA's EMAP habitat assessment field methods are presented as an option for a second level (II) of habitat assessment. These methods have been applied in numerous streams throughout the Mid-Atlantic region, the Midwest, Colorado, California, and the Pacific Northwest. Table 5-1 is a summary of these field methods; more detail is presented in the field manual by Kaufmann and Robison (1997).

Table 5-1. Components of EMAP physical habitat protocol.

Component	Description
1. Thalweg Profile	Measure maximum depth, classify habitat, determine presence of soft/small sediment at 10-15 equally spaced intervals between each of 11 channel cross-sections (100-150 along entire reach). Measure wetted width at 11 channel cross-sections and mid-way between cross-sections (21 measurements).
2. Woody Debris	Between each of the channel cross sections, tally large woody debris numbers within and above the bankfull channel according to size classes.
3. Channel and Riparian Cross-Sections	At 11 cross-section stations placed at equal intervals along reach length: • **Measure**: channel cross section dimensions, bank height, undercut, angle (with rod and clinometer); gradient (clinometer), sinuosity (compass backsite), riparian canopy cover (densiometer). • **Visually Estimate***: substrate size class and embeddedness; areal cover class and type (e.g., woody) of riparian vegetation in Canopy, Mid-Layer and Ground Cover; areal cover class of fish concealment features, aquatic macrophytes and filamentous algae. • **Observe & Record***: human disturbances and their proximity to the channel.
4. Discharge	In medium and large streams (defines later) measure water depth and velocity @ 0.6 depth (with electromagnetic or impeller-type flow meter) at 15 to 20 equally spaced intervals across one carefully chosen channel cross-section. In very small streams, measure discharge with a portable weir or time the filling of a bucket.

* Substrate size class and embeddedness are estimated, and depth is measured for 55 particles taken at 5 equally-spaced points on each of 11 cross-sections. The cross-section is defined by laying the surveyor's rod or tape to span the wetted channel. Woody

Rapid Bioassessment Protocols for Use in Streams and Wadeable Rivers: Periphyton, Benthic Macroinvertebrates, and Fish, Second Edition

5-33

debris is tallied over the distance between each cross-section and the next cross-section upstream. Riparian vegetation and human disturbances are observed 5 m upstream and 5 m downstream from the cross section station. They extend shoreward 10 m from left and right banks. Fish cover types, aquatic macrophytes, and algae are observed within channel 5 m upstream and 5 m downstream from the cross section stations. These boundaries for visual observations are estimated by eye.

Table 5-2 lists the physical habitat metrics that can be derived from applying these field methods. Once these habitat metrics are calculated from the available physical habitat data, an assessment would be obtained from comparing these metric values to those of known reference sites. A strong deviation from the reference expectations would indicate a habitat alteration of the particular parameter. The close connectivity of the various attributes would most likely result in an impact on multiple metrics if habitat alteration was occurring. The actual process for interpreting a habitat assessment using this approach is still under development.

Table 5-2. Example of habitat metrics that can be calculated from the EMAP physical habitat data.

Channel mean width and depth
Channel volume and Residual Pool volume
Mean channel slope and sinuosity
Channel incision, bankfull dimensions, and bank characteristics
Substrate mean diameter, % fines, % embeddedness
Substrate stability
Fish concealment features (areal cover of various types, e.g., undercut banks, brush)
Large woody debris (volume and number of pieces per 100 m)
Channel habitat types (e.g., % of reach composed of pools, riffles, etc.)
Canopy cover
Riparian vegetation structure and complexity
Riparian disturbance measure (proximity-weighted tally of human disturbances)

PERIPHYTON PROTOCOLS

By R. Jan Stevenson, University of Louisville, and Loren L. Bahls, University of Montana

Benthic algae (periphyton or phytobenthos) are primary producers and an important foundation of many stream food webs. These organisms also stabilize substrata and serve as habitat for many other organisms. Because benthic algal assemblages are attached to substrate, their characteristics are affected by physical, chemical, and biological disturbances that occur in the stream reach during the time in which the assemblage developed.

Diatoms in particular are useful ecological indicators because they are found in abundance in most lotic ecosystems. Diatoms and many other algae can be identified to species by experienced algologists. The great numbers of species provide multiple, sensitive indicators of environmental change and the specific conditions of their habitat. Diatom species are differentially adapted to a wide range of ecological conditions.

Periphyton indices of biotic integrity have been developed and tested in several regions (Kentucky Department of Environmental Protection 1993, Hill 1997). Since the ecological tolerances for many species are known (see section 6.1.4), changes in community composition can be used to diagnose the environmental stressors affecting ecological health, as well as to assess biotic integrity (Stevenson 1998, Stevenson and Pan 1999).

Periphyton protocols may be used by themselves, but they are most effective when used with one or more of the other assemblages and protocols. They should be used with habitat and benthic macroinvertebrate assessments particularly because of the close relation between periphyton and these elements of stream ecosystems.

Presently, few states have developed protocols for periphyton assessment. Montana, Kentucky, and Oklahoma have developed periphyton bioassessment programs. Others states are exploring the possibility of developing periphyton programs. Algae have been widely used to monitor water quality in rivers of Europe, where many different approaches have been used for sampling and data analysis (see reviews in Whitton and Rott 1996, Whitton et al. 1991). The protocols presented here are a composite of the techniques used in Kentucky, Montana, and Oklahoma (Bahls 1993, Kentucky Department of Environmental Protection 1993, Oklahoma Conservation Commission 1993).

Two Rapid Bioassessment Protocols for periphyton are presented. These protocols are meant to provide examples of methods that can be used. Other methods are available and should be considered based on the objectives of the assessment program, resources available for study, numbers of streams sampled, hypothesized stressors, and the physical habitat of the streams studied. Examples of other methods are presented in textboxes throughout the chapter.

The first protocol (6.1) is a standard approach in which species composition and/or biomass of a sampled assemblage is assessed in the laboratory. The second protocol (6.2) is a field-based rapid survey of periphyton biomass and coarse-level taxonomic composition (e.g., diatoms, filamentous greens, blue-green algae) and requires little taxonomic expertise. The two protocols can be used together. The first protocol has the advantage of providing much more accuracy in assessing biotic integrity and in diagnosing causes of impairment than the second protocol, but it requires more effort than the second protocol. Additionally, the first protocol provides the option of sampling the natural substrate of the stream or placing artificial substrates for colonization.

6.1   STANDARD LABORATORY-BASED APPROACH

6.1.1   Field Sampling Procedures: Natural Substrates

Periphyton samples should be collected during periods of stable stream flow. High flows can scour the stream bed, flushing the periphyton downstream. Recolonization of substrates will be faster after less severe floods and in streams with nutrient enrichment. Peterson and Stevenson (1990) recommend a three-week delay following high, bottom-scouring stream flows to allow for recolonization and succession to a mature periphyton community. However, recovery after high discharge can be as rapid as 7 days if severe scouring of substrata did not occur (Stevenson 1990).

Two sampling approaches are described for natural substrate sampling. Multihabitat sampling best characterizes the benthic algae in the reach, but results may not be sensitive to subtle water quality changes because of habitat variability between reaches. Species composition of assemblages from a single habitat should reflect water quality differences among streams more precisely than multi-habitat sampling, but impacts in other habitats in the reach may be missed.

The length of stream sampled depends upon the objectives of the project, budget, and expected results. Multihabitat sampling should be conducted at the reach scale (30-40 stream widths) to ensure sampling the diversity of habitats that occur in the stream. Ideally, single habitat sampling should also be conducted at the reach scale. A shorter length of stream can probably be sampled for single habitat samples than multihabitat samples because the chosen single habitat (e.g., riffles) is usually common within the study streams.

6.1.1.1 Multihabitat Sampling

The following procedures for multihabitat sampling of algae have been adapted from the Kentucky and Montana protocols (Kentucky DEP 1993, Bahls 1993). These procedures are recommended when subsequent laboratory assessments of species composition of algal assemblages will be performed.

1. Establish the reach for multihabitat sampling as per the macroinvertebrate protocols (Chapter 7). In most cases, the reach required for periphyton sampling will be the same size as the reach required for

---

FIELD EQUIPMENT FOR PERIPHYTON SAMPLING--NATURAL SUBSTRATES

- stainless steel teaspoon, toothbrush, or similar brushing and scraping tools
- section of PVC pipe (3" diameter or larger) fitted with a rubber collar at one end
- field notebook or field forms*; pens and pencils
- white plastic or enamel pan
- petri dish and spatula (for collecting soft sediment)
- forceps, suction bulb, and disposable pipettes
- squeeze bottle with distilled water
- sample containers (125 ml wide-mouth jars)
- sample container labels
- preservative [Lugol's solution, 4% buffered formalin, "M3" fixative, or 2% glutaraldehyde (APHA 1995)]
- first aid kit
- cooler with ice

* During wet weather conditions, waterproof paper is useful or copies of field forms can be stored in a metal storage box (attached to a clip-board).

---

macroinvertebrate or fish sampling (30-40 stream widths) so that as many algal habitats can be sampled as is practical.

2. Before sampling, complete the physical/chemical field sheet (see Chapter 5; Appendix A-1, Form 1) and the periphyton field data sheet (Appendix A-2, Form 1). Visual estimates or quantitative transect-based assessments can be used to determine the percent coverage of each substrate type and the estimated relative abundance of macrophytes, macroscopic filamentous algae, diatoms and other microscopic algal accumulations (periphyton), and other biota (see section 6.2).

3. Collect algae from all available substrates and habitats. The objective is to collect a single composite sample that is representative of the periphyton assemblage present in the reach. Sample all substrates (Table 6-1) and habitats (riffles, runs, shallow pools, nearshore areas) roughly in proportion to their areal coverage in the reach. Within a stream reach, light, depth, substrate, and current velocity can affect species composition of periphyton assemblages. Changes in species composition of algae among habitats are often evident as changes in color and texture of the periphyton. Small amounts (about 5 mL or less) of subsample from each habitat are usually sufficient. Pick specimens of macroalgae by hand in proportion to their relative abundance in the reach. Combine all samples into a common container.

Table 6-1. Summary of collection techniques for periphyton from wadeable streams (adapted from Kentucky DEP 1993, Bahls 1993).

Substrate Type	Collection Technique
Removable substrates (hard): gravel, pebbles, cobble, and woody debris	Remove representative substrates from water; brush or scrape representative area of algae from surface and rinse into sample jar.
Removable substrates (soft): mosses, macroalgae, vascular plants, root masses	Place a portion of the plant in a sample container with some water. Shake it vigorously and rub it gently to remove algae. Remove plant from sample container.
Large substrates (not removable): boulders, bedrock, logs, trees, roots	Place PVC pipe with a neoprene collar at one end on the substrate so that the collar is sealed against the substrate. Dislodge algae in the pipe with a toothbrush, nail brush, or scraper. Remove algae from pipe with pipette.
Loose sediments: sand, silt, fine particulate organic matter, clay	Invert petri dish over sediments. Trap sediments in petri dish by inserting spatula under dish. Remove sediments from stream and rinse into sampling container. Algal samples from depositional habitats can also be collected with spoons, forceps, or pipette.

4. Place all samples into a single water-tight, unbreakable, wide-mouth container. A composite sample measuring four ounces (ca. 125 ml) is sufficient (Bahls 1993). Add recommended amount of Lugol's (IKI) solution, "M3" fixative, buffered 4% formalin, 2% glutaraldehyde, or other preservative (APHA 1995).

Rapid Bioassessment Protocols for Use in Streams and Wadeable Rivers: Periphyton, Benthic Macroinvertebrates, and Fish, Second Edition

6-3

5. Place a permanent label on the outside of the sample container with the following information: waterbody name, location, station number, date, name of collector, and type of preservative. Record this information and relevant ecological information in a field notebook or on the periphyton field data sheet (Appendix A-2, Form 1). Place another label with the same information inside the sample container. (Caution! Lugol's solution and other iodine-based preservatives will turn paper labels black.)

6. After sampling, review the recorded information on all labels and forms for accuracy and completeness.

7. Examine all brushing and scraping tools for residues. Rub them clean and rinse them in distilled water before sampling the next site and before putting them away.

8. Transport samples back to the laboratory in a cooler with ice (keep them cold and dark) and store preserved samples in the dark until they are processed. Be sure to stow samples in a way so that transport and shifting does not allow samples to leak. When preserved, check preservative every few weeks and replenish as necessary until taxonomic evaluation is completed.

9. Log in all incoming samples (Appendix A-2, Form 2). At a minimum, record sample identification code, date, stream name, sampling location, collector's name, sampling method, and area sampled (if it was determined).

6.1.1.2 Single Habitat Sampling

Variability due to differences in habitat between streams may be reduced by collecting periphyton from a single substrate/habitat combination that characterizes the study reach (Rosen 1995). For comparability of results, the same substrate/habitat combination should be sampled in all reference and test streams. Single habitat sampling should be used when biomass of periphyton will be assessed.

1. Define the sampling reach. The area sampled for single habitat sampling can be smaller than the area used for multihabitat sampling. Valuable results have been achieved in past projects by sampling just one riffle or pool.

2. Before sampling, complete the physical/chemical field sheet (see Chapter 5; Appendix A-1, Form 1) and the periphyton field data sheet

CHLOROPHYLL *a* SUBSAMPLING (OPTIONAL)

1. Chlorophyll *a* subsamples should be taken as soon as possible (< 12 hours after sampling). Generally, if chlorophyll subsamples can not be taken in the lab on the day of collection, subsample in the field.

2. Homogenize samples. In the field, shake vigorously. In the lab, use a tissue homogenizer.

3. Record the initial volume of sample on the periphyton sample log form.

4. Stir the sample on a magnetic stirrer and subsample. When subsampling, take at least two aliquots from the sample for each chlorophyll sample (two aliquots provides a more representative subsample than one). Record the subsample volume for chlorophyll *a* on the periphyton sample log form.

5. Concentrate the chlorophyll subsample on a glass fiber filter (e.g., Whatman® GFC or equivalent).

6. Fold the filter and wrap with aluminum to exclude light.

7. Store the filter in a cold cooler (not in water) and eventually in a freezer.

(Appendix A-2, Form 1). Complete habitat assessments as in multihabitat sampling so that the relative importance of the habitats sampled can be characterized.

3. The recommended substrate/habitat combination is cobble obtained from riffles and runs with current velocities of 10-50 cm/sec. Samples from this habitat are often easier to analyze than from slow current habitats because they contain less silt. These habitats are common in many streams. In low gradient streams where riffles are rare, algae on snags or in depositional habitats can be collected. Shifting sand is not recommended as a targeted substrate because the species composition on sand is limited due to the small size and unstable nature of the substratum. Phytoplankton should be considered as an alternative to periphyton in large, low gradient streams.

4. Collect several subsamples from the same substrate/habitat combination and composite them into a single container. Three or more subsamples should be collected from each reach or study stream.

5. The area sampled should always be determined if biomass (e.g., chlorophyll) per unit area is to be measured.

6. If you plan to assay samples for chlorophyll *a*, do not preserve samples until they have been subsampled (see textbox entitled "Chlorophyll *a* Subsampling").

7. Store, transport, process, and log in samples as in steps 4-9 in section 6.1.1.1.

6.1.2 Field Sampling Procedures: Artificial Substrates

Most monitoring groups prefer sampling natural substrates whenever possible to reduce field time and improve ecological applicability of information. However, periphyton can also be sampled by collecting from artificial substrates that are placed in aquatic habitats and colonized over a period of time. This procedure is particularly useful in non-wadeable streams, rivers with no riffle areas, wetlands, or the littoral zones of lentic habitats. Both natural and artificial substrates are useful in monitoring and assessing waterbody conditions, and have corresponding advantages and disadvantages (Stevenson and Lowe 1986, Aloi 1990). The methods summarized here are a composite of those specified by Kentucky (Kentucky DEP 1993), Florida (Florida DEP 1996), and Oklahoma (Oklahoma CC 1993).
Although glass microslides are preferred, a variety of artificial substrates have been used with success (see #2 below and textbox on p 6-6).

QUALITY CONTROL (QC) IN THE FIELD

1. Sample labels must be accurately and thoroughly completed, including the sample identification code, date, stream name, sampling location, and collector's name. The outside and any inside labels of the container should contain the same information. Chain of custody and sample log forms must include the same information as the sample container labels. **Caution!** Lugol's solution and iodine-based preservatives will turn paper labels black.

2. After sampling has been completed at a given site, all brushes, suction and scraping devices that have come in contact with the sample should be rubbed clean and rinsed thoroughly in distilled water. The equipment should be examined again prior to use at the next sampling site, and rinsed again if necessary.

3. After sampling, review the recorded information on all labels and forms for accuracy and completeness.

4. Collect and analyze one replicate sample from 10% of the sites to evaluate precision or repeatability of sampling technique, collection team, sample analysis, and taxonomy.

Rapid Bioassessment Protocols for Use in Streams and Wadeable Rivers: Periphyton, Benthic Macroinvertebrates, and Fish, Second Edition

6-5

1. Microslides should be thoroughly cleaned before placing in periphytometers (e.g., Patrick et al. 1954). Rinse slides in acetone and clean with Kimwipes®.

2. Place surface (floating) or benthic (bottom) periphytometers fitted with glass slides, glass rods, clay tiles, plexiglass plates or similar substrates in the study area. Allow 2 to 4 weeks for periphyton recruitment and colonization.

3. Replicate a minimum of 3 periphytometers at each site to account for spatial variability. The total number should depend upon the study design and hypotheses tested. Samples can either be composited or analyzed individually.

4. Attach periphytometers to rebars pounded into the stream bottom or to other stable structures. Periphytometers should be hidden from view to minimize disturbance or vandalism. Avoid the main channel of floatable, recreational streams. Each periphytometer should be oriented with the shield directed upstream.

5. If flooding or a similar scouring event occurs during incubation, allow waterbody to equilibrate and reset periphytometers with clean slides.

6. After the incubation period (2-4 weeks), collect substrates. Remove algae using rubber spatulas, toothbrushes and razor blades. You can tell when all algae have been removed from substrates by a change from smooth, mucilaginous feel (even when no visible algae are present) to a non-slimy or rough texture.

7. Store, transport, process, and log in samples as in steps 4-9 in section 6.1.1.1.

8. One advantage of using artificial substrates is that containers (e.g., whirl-pack bags or sample jars) can be purchased that will hold the substrates so

---

FIELD EQUIPMENT/SUPPLIES NEEDED FOR PERIPHYTON SAMPLING-- ARTIFICIAL SUBSTRATES

- periphytometer (frame to hold artificial substrata)
- microslides or other suitable substratum (e.g., clay tiles, sanded Plexiglass® plates, or wooden or acrylic dowels)
- sledge hammer and rebars
- toothbrush, razor blade, or other scraping tools
- water bottle with distilled water
- white plastic or enamel pan
- aluminium foil
- sample containers
- sample container labels
- field notebook (waterproof)
- preservative [Lugol's solution, 4% buffered formalin, "M^3" fixative, or 2% glutaraldehyde (APHA 1995)]
- cooler with ice

---

that substrates need not be scraped in the field. Different substrates can be designated for microscopic analysis and chlorophyll assay. Then algae and substrates can be placed in sampling containers and preserved for later processing and microscopic analysis or placed in a cooler on ice for later chlorophyll *a* analysis. Laboratory sample processing is preferred; so if travel and holding time are less than 12 hours, it is not necessary to split samples before returning to the lab.

6.1.3 Assessing Relative Abundances of Algal Taxa: Both "Soft" (Non-Diatom) Algae and Diatoms

The Methods summarized here are a modified version of those used by Kentucky (Kentucky DEP 1993), Florida (Florida DEP 1996), and Montana (Bahls 1993). For more detail or for alternative methods, see Standard Methods for the Examination of Water and Wastewater (APHA 1995).

Many algae are readily identifiable to species level by trained personnel who have a good library of literature on algal taxonomy (see section 6.3). All algae can not be identified to species because: the growth forms of some algal species are morphologically indistinguishable with the light microscope (e.g., zoospores of many green algae); the species has not been described previously; or the species is not in the laboratory's literature. Consistency in identifications within a laboratory and program is very important, because most bioassessment are based on contrasts between reference and test sites. Accuracy of identifications becomes most important when using autecological information from other studies. Quality assurance techniques are designed to ensure "internal consistency" and also improve comparisons with information in other algal assessment and monitoring programs.

6.1.3.1 "Soft" (Non-Diatom) Algae Relative Abundance and Taxa Richness

1. Homogenize algal samples with a tissue homogenizer or blender.

2. Thoroughly mix the homogenized sample and pipette into a Palmer counting cell (see textbox for alternative methods). Algal suspensions that produce between 10 and 20 cells in a field provide good densities for counting and identifying cells. Lower densities slow counting. Dilute samples if cells overlap too much for counting.

3. Fill in the top portion of the benchsheet for "soft" algae (Appendix A-2, Form 3) with enough information from the sample label and other sources to uniquely identify the sample.

4. Identify and count 300 algal "cell units" to the lowest possible taxonomic level at 400X magnification with the use of the references in Section 6.3.

    ! Distinguishing cells of coenocytic algae (e.g., *Vaucheria*) and small filaments of blue-green algae is a problem in cell counts. "Cell units" can be defined for these algae as 10mm sections of the thallus or filament.
    ! For diatoms, only count live diatoms and do not identify to lower taxonomic levels if a subsequent count of cleaned diatoms is to be undertaken (See section 6.1.3.2).
    ! Record numbers of cells or cell units observed for each taxon on a benchsheet.
    ! Make taxonomic notes and drawings on benchsheets of important specimens.

5. Optional - To better determine non-diatom taxa richness, continue counting until you have not observed any new taxa for 100 cell units or about three minutes of observation.

6.1.3.2 Diatom Relative Abundances and Taxa Richness

1. Subsample at least 5-10 mL of concentrated preserved sample while vigorously shaking the sample (or using magnetic stirrer). Oxidize (clean) samples for diatom analysis (APHA 1995, see textbox entitled "Oxidation Methods for Cleaning Diatoms").

2. Mount diatoms in Naphrax® or another high refractive index medium to make permanent slides. Label slides with same information as on the sample container label.

3. Fill in the top portion of the bench sheet for diatom counts (Appendix A-2, Form 4) with enough information from the sample label to uniquely identify the sample.

4. Identify and count diatom valves to the lowest possible taxonomic level, which should be species and perhaps variety level, under oil immersion at 1000X magnification with the use of the

---

references in Section 6.3. At minimum, count 600 valves (300 cells) and at least until 10 valves of 10 species have been observed. Be careful to distinguish and count both valves of intact frustules. The 10 valves of 10 species rule ensures relatively precise estimates of relative abundances of the dominant taxa when one or two taxa are highly dominant. Six hundred valve counts were chosen to conform with methods used in other national bioassessment programs (Porter et al. 1993). Record numbers of valves observed for each taxon on the bench sheet. Make taxonomic notes and drawings on benchsheets and record stage coordinates of important specimens.

5. Optional - To estimate total diatom taxa richness, continue counting until you have not observed any new species for 100 specimens or about three minutes of observation.

6.1.3.3 Calculating Species Relative Abundances and Taxa Richness

1. Relative abundances of "soft" algae are determined by dividing the number of cells (cell units) counted for each taxon by the total number of cells counted (e.g., 300). Enter this information on Appendix A-2, Form 3.

2. Relative abundances of diatoms have to be corrected for the number of live diatoms observed in the count of all algae. Therefore, determine the relative abundances of diatom species in the algal assemblage by dividing the number of valves counted for each species by the total number of valves counted (e.g., 600); then multiply the relative abundance of each diatom taxon in the diatom count by the relative abundance of live diatoms in the count of all algae. Enter this information on Appendix A-2, Form 4. Some analysts prefer to treat diatom and soft algal species composition separately. In this case, determine the relative abundances of diatom species in the algal assemblage by dividing the number of valves counted for each species by the total number of valves counted (e.g., 600).

3. Total taxa richness can be estimated by adding the number of "soft" algal taxa and diatom taxa.

6.1.3.4 Alternative Preparation Techniques

Palmer counting cells are excellent for identifying and counting soft-algae in most species assemblages. When samples have many very small blue-green algae or a few, relatively important large cells, other slide preparation techniques may be useful to increase magnification and sample size, respectively. Because accurate diatom identification is not possible in Palmer cells, we have recommended counting cleaned diatoms in special mounts. However, if the taxonomy of algae in samples is well known, preparation and counting time can be reduced by mounting algae in syrup. In syrup, both soft algae and diatoms can be identified, but resolution of morphological details of diatoms is not as great as in mounts of diatoms in resins (e.g., Naphrax®).

Assemblages with many small cells: We recommend a simple wet mount procedure when samples contain many small algae so samples can be observed at 1000X. A small volume of water under the coverglass prevents movement of cells when adjusting focus and using oil immersion. These preparations usually last several days if properly sealed (see below).

Wet mounts:
1. Clean coverglasses and place on flat surface.

2. Pipette 1.0 mL of algal suspension onto the coverglass.

3. Dry the algal suspension on the coverglass. For convenience, the evaporation of water can be increased on a slide-warmer or slowed by drying the sample in a vapor chamber (as simple as a cake pan or aluminum foil hood placed over samples).

4. As soon as the algal suspension dries, invert the coverglass into the 0.02 mL of distilled water on a microscope slide.

5. Seal the water under the microscope slide with fingernail polish or polyurethane varnish.

Assemblages with a few large cells:
Sedgewick-Rafter counting chambers, which are large modified microscope slides with 1.0 mL wells, increase sample size. Counts in Sedgewick-Rafter counting cells should be done after counts in Palmer cells or wet mounts so that the relation between sample proportions with the two methods can be determined. While keeping track of the proportion of sample observed, identify and count large algae in transects at 200X or 100X magnification in the counting cell.

Syrup mounts:
1. Prepare Taft's syrup medium (TSM) by mixing 30 mL of clear corn syrup (e.g., Karo's® Corn Syrup) with 7 mL of formaldehyde and 63 mL of distilled water. Dilute a 10 mL proportion of this 100% TSM with 90 mL of distilled water to make 10% TSM.

2. Place 0.2 mL of 10% TSM on coverglass.

OXIDATION (CLEANING) METHODS FOR DIATOMS

Concentrated Acid Oxidation:
1. Place a 5-10 mL subsample of preserved algal sample in a beaker.

2. Under a fume hood, add enough concentrated nitric or sulfuric acid to produce a strong exothermic reaction. Usually equal parts of sample and acid will produce such a reaction.
(Caution! With some preservatives and samples from hard water, adding concentrated acid will produce a violent exothermic reaction. Use a fume hood, safety glasses, and protective clothing. Separate the sample beakers by a few inches to prevent cross-contamination of samples in the event of overflow.)

3. Allow the sample to oxidize overnight.

4. Fill the beaker with distilled water.

5. Wait 1 hour for each centimeter of water depth in the beaker.

6. Siphon off the supernatant and refill the beaker with distilled water. Siphon from the center of the water column to avoid siphoning light algae that have adsorbed onto the sides and surface of the water column.

7. Repeat steps 4 through 6 until all color is removed and the sample becomes clear or has a circumneutral pH.

Hydrogen Peroxide/Potassium Dichromate Oxidation:
1. Prepare samples as in step 1 above, but use 50% H_2O_2 instead of concentrated acid.

2. Allow the sample to oxidize overnight, then add a microspatula of potassium dichromate.
(Caution! This will cause a violent exothermic reaction. Use a fume hood, safety glasses, and protective clothing. Separate the sample beakers by a few inches to prevent cross-contamination in the event of overflow.)

3. When the sample color changes from purple to yellow and boiling stops, fill the beaker with distilled water.

4. Wait 4 hours, siphon off the supernatant, and refill the beaker with distilled water. Siphon from the center of

---

Rapid Bioassessment Protocols for Use in Streams and Wadeable Rivers: Periphyton, Benthic Macroinvertebrates, and Fish, Second Edition

6-9

3. Place 1.0 mL of algal suspension on coverglass. Consider using several dilutions.

4. Let dry for 24 hours. Alternatively, dry on slide warmer on low setting. Do not overdry or cells will plasmolyze.

5. Place another ≈ 1.0 mL of 10% TSM on cover glass and dry (overnight or 4 hours on a slide warmer). Apply 10% TSM quickly to avoid patchy resuspension of the original layer of TSM and algae.

6. Invert coverglass onto microscope slide; place slide on hot plate to warm the slide and syrup. Do not boil, just warm. Press coverglass gently in place with forceps, being careful to keep all syrup under the coverglass. The syrup should spread under coverglass.

7. Remove the slide from the hotplate. Cooling should partially seal the coverglass to the slide.

8. More permanently seal the syrup under slides by painting fingernail polish around the edge of the cover glass and onto the microscope slide.

Note: Preserve color of chloroplasts by keeping samples in dark.

Special Note: If slides get too warm in storage, syrup will loose viscosity and become runny. Algae and medium may then escape containment under coverglass. Store slides in a horizontal position.

6.1.4 Metrics Based on Species Composition

The periphyton metrics presented here are used by several states and environmental assessment programs throughout the US and Europe (e.g., Kentucky DEP 1993, Bahls 1993, Florida DEP 1996, Whitton et al. 1991, Whitton and Kelly 1995). Each of these metrics should be tested for response to human alterations of streams in the region in which they are used (see Chapter 9, Biological Data Analysis). In many cases, diatom and soft algal metrics have been determined separately because changes in small abundant cyanobacteria (blue-green algae) can numerically overwhelm metrics based on relative abundance and because green algae with large cells (e.g., *Cladophora*) may not have appropriate weight. However, attempts should be made to integrate diatoms and soft algae in as many metrics as possible, especially in cases such as species and generic richness when great variability in relative abundance is not an issue.

Many metrics can be calculated based on presence/absence data or on relative abundances of taxa. For example, percent Pollution Tolerant Diatoms can be calculated as the sum of relative abundances of pollution tolerant taxa in an assemblage or as the number of species that are tolerant to pollution in an assemblage. Percent community similarity can be calculated as presented below, which quantifies the percent of organisms in two assemblages that are the same. Alternatively, it can be calculated as the percent of species that are the same by making all relative abundances greater than 0 equal to 1. The following metrics can also be calculated with presence/absence data instead of species relative abundances: % sensitive taxa, % motile taxa, % acidobiontic, % alkalibiontic, % halobiontic, % saprobiontic, % eutrophic, simple autecological indices, and change in inferred ecological conditions. Although we may find that metrics based on species relative abundances are more sensitive to environmental change, metrics based on presence/absence data may be more appropriate when developing metrics with multihabitat samples and proportional sampling of habitats is difficult. In the latter case, presence/absence of species should remain the same, even if relative abundance of taxa differs with biases in multihabitat sampling.

The metrics have been divided into two groups which may be helpful in developing an Index of Biotic Integrity (IBI). Metrics in the first group are less diagnostic than the second group of metrics. Metrics in the first group (species and generic richness, Shannon diversity, etc.) generally characterize biotic integrity ("natural balance in flora and fauna...." as in Karr and Dudley 1981) without specifically diagnosing ecological conditions and causes of impairment. The second group of metrics more specifically diagnoses causes of impaired biotic integrity.

COSTS AND BENEFITS OF SIMPLER ANALYSES

- We recommend that all algae (soft and diatom) be identified and counted. Information may be lost if soft algae are not identified and counted because some impacts may selectively affect soft algae. Most of the species (and thus information) in a sample will be diatoms. Costs of both analyses are not that great.

- Costs can be reduced by only counting diatoms or soft algae. Since diatoms are usually the most species-rich group of algae in samples and most metrics are based on differences in taxonomic composition, we recommend that diatoms be counted. In addition, permanently preserved and readily archived microslides of diatoms can serve as a historic reference of ecological conditions.

- In general, identifying algae to species is recommended for two reasons: (1) to better characterize differences between assemblages that may occur at the species level and (2) because large differences in ecological preferences do exist among algal species within the same genus.

- However, substantial information can be gained by identifying algae just to the genus level. Whereas identifying algae only to genus may loose valuable ecological information, costs of analyses can be reduced, especially for inexperienced analysts.

- If implementing a new program and only an inexperienced analyst is available for the job, identifying diatom genera in assemblages can provide valuable characterizations of biotic integrity and environmental conditions.

- As analysts get more experience counting, the taxonomic level of their analyses should improve. The cost of an experienced analyst counting and identifying algae to species is not much greater than analysis to genus.

Metrics from both groups could be included in an IBI to make a hierarchically diagnostic IBI. Alternatively, an IBI could be constructed from only metrics of biotic integrity so that inference of biotic integrity and diagnosis of impairment are independent (Stevenson and Pan 1999).

Rapid Bioassessment Protocols for Use in Streams and Wadeable Rivers: Periphyton, Benthic Macroinvertebrates, and Fish, Second Edition

6-11

Autecological information about many algal species and genera has been reported in the literature. This information comes in several forms. In some cases, qualitative descriptions of the ecological conditions in which species were observed were reported in early studies of diatoms. Following the development of the saprobic index by Kolkwitz and Marsson (1908), several categorical classification systems (e.g., halobian spectrum, pH spectrum) were developed to describe the ecological preferences and tolerances of species (see Lowe 1974 for a review). Most recently, the ecological optima and tolerances of species for specific environmental conditions have been quantified by using weighted average regression approaches (see ter Braak and van Dam 1989 for a review). We have compiled a list of references for this information in Section 6.4. These references will be valuable for developing many of the metrics below.

Metrics of Biotic Integrity

1.   **Species richness** is an estimate of the number of algal species (diatoms, soft algae, or both) in a sample. High species richness is assumed to indicate high biotic integrity because many species are adapted to the conditions present in the habitat. Species richness is predicted to decrease with increasing pollution because many species are stressed. However, many habitats may be naturally stressed by low nutrients, low light, or other factors. Slight increases in nutrient enrichment can increase species richness in headwater and naturally unproductive, nutrient-poor streams (Bahls et al. 1992).

2.   **Total Number of Genera** (Generic richness) should be highest in reference sites and lowest in impacted sites where sensitive genera become stressed. Total number of genera (diatoms, soft algae, or both) may provide a more robust measure of diversity than species richness, because numerous closely related species are within some genera and may artificially inflate richness estimates.

3.   **Total Number of Divisions** represented by all taxa should be highest in sites with good water quality and high biotic integrity.

4.   **Shannon Diversity (for diatoms)**. The Shannon Index is a function of both the number of species in a sample and the distribution of individuals among those species (Klemm et al. 1990). Because species richness and evenness may vary independently and complexly with water pollution. Stevenson (1984) suggests that changes in species diversity, rather than the diversity value, may be useful indicators of changes in water quality. Species diversity, despite the controversy surrounding it, has historically been used with success as an indicator of organic (sewage) pollution (Wilhm and Dorris 1968, Weber 1973, Cooper and Wilhm 1975). Bahls et al. (1992) uses Shannon diversity because of its sensitivity to water quality changes. Under certain conditions Shannon diversity values may underestimate water quality e.g., when total number of taxa is less than 10. Assessments for low richness samples can be improved by comparing the assemblage Shannon Diversity to the Maximum Shannon Diversity value (David Beeson[1], personal communication).

5.   **Percent Community Similarity (PS_c) of Diatoms**. The percent community similarity (PS_c) index, discussed by Whittaker (1952), was used by Whittaker and Fairbanks (1958) to compare planktonic copepod communities. It was chosen for use in algal bioassessment because it shows community similarities based on relative abundances, and in doing so, gives

---

[1]David Beeson is a phycologist with Schafer & Associates, Inc.

more weight to dominant taxa than rare ones. Percent similarity can be used to compare control and test sites, or average community of a group of control or reference sites with a test site. Percent community similarity values range from 0 (no similarity) to 100%.

The formula for calculating percent community similarity is:

$$PS_c = 100 - .5\Sigma_{i=1}^{s}|a_i - b_i| = \Sigma_{i=1}^{s}\min(a_i, b_i)$$

where:

    a_i = percentage of species i in sample A
    b_i = percentage of species i in sample B

6. **Pollution Tolerance Index for Diatoms.** The pollution tolerance index (PTI) for algae resembles the Hilsenhoff biotic index for macroinvertebrates (Hilsenhoff 1987). Lange-Bertalot (1979) distinguishes three categories of diatoms according to their tolerance to increased pollution, with species assigned a value of 1 for most tolerant taxa (e.g., *Nitzschia palea* or *Gomphonema parvulum*) to 3 for relatively sensitive species. Relative tolerance for taxa can be found in Lange-Bertalot (1979) and in many of the references listed in section 6.4. Thus, Lange-Bertalot's PTI varies from 1 for most polluted to 3 for least polluted waters when using the following equation:

$$PTI = \frac{\Sigma n_i t_i}{N}$$

where:
        n_i = number of cells counted for species i
        t_i = tolerance value of species i
        N = total number of cells counted

In some cases, the range of values for tolerances has been increased, thereby producing a corresponding increase in the range of PTI values.

7. **Percent Sensitive Diatoms.** The percent sensitive diatoms metric is the sum of the relative abundances of all intolerant species. This metric is especially important in smaller-order streams where primary productivity may be naturally low, causing many other metrics to underestimate water quality.

8. **Percent *Achnanthes minutissima*.** This species is a cosmopolitan diatom that has a very broad ecological amplitude. It is an attached diatom and often the first species to pioneer a recently scoured site, sometimes to the exclusion of all other algae. *A. minutissima* is also frequently dominant in streams subjected to acid mine drainage (e.g., Silver Bow Creek, Montana) and to other chemical insults. The percent abundance of *A. minutissima* has been found to be directly proportional to the time that has elapsed since the last scouring flow or episode of toxic pollution. For use in bioassessment, the quartiles of this metric from a

population of sites has been used to establish judgment criteria, e.g., 0-25% = no disturbance, 25-50% = minor disturbance, 50-75% = moderate disturbance, and 75-100% = severe disturbance. Least-impaired streams in Montana may contain up to 50% *A. minutissima* (Bahls, unpublished data).

9.   **Percent live diatoms** was proposed by Hill (1997) as a metric to indicate the health of the diatom assemblage. Low percent live diatoms could be due to heavy sedimentation and/or relatively old algal assemblages with high algal biomass on substrates.

Diagnostic Metrics that Infer Ecological Conditions

The ecological preferences of many diatoms and other algae have been recorded in the literature. Using relative abundances of algal species in the sample and their preferences for specific habitat conditions, metrics can be calculated to indicate the environment stressors in a habitat. These metrics can more specifically infer environmental stressors than the general pollution tolerance index.

10.   **Percent Aberrant Diatoms** is the percent of diatoms in a sample that have anomalies in striae patterns or frustule shape (e.g, long cells that are bent or cells with indentations). This metric has been positively correlated to heavy metal contamination in streams (McFarland et al. 1997).

11.   **Percent Motile Diatoms.** The percent motile diatoms is a siltation index, expressed as the relative abundance of *Navicula* + *Nitzschia* + *Surirella*. It has shown promise in Montana (Bahls et al. 1992). The three genera are able to crawl towards the surface if they are covered by silt; their abundance is thought to reflect the amount and frequency of siltation. Relative abundances of Gyrosigma, Cylindrotheca, and other motile diatoms may also be added to this metric.

12.   **Simple Diagnostic Metrics** can infer the environmental stressor based on the autecology of individual species in the habitats. For example, if acid mine drainage was impairing stream conditions, then we would expect to find more acidobiontic taxa in samples. Calculate a simple diagnostic metric as the sum of the percent relative abundances (range 0-100%) of species that have environmental optima in extreme environmental conditions. For example (see Table 6-2):

   % acidobiontic + % acidophilic
   % alkalibiontic + % alkaliphilic
   % halophilic
   % mesosaprobic + % oligosaprobic + % saprophilic
   % eutrophic

13.   **Inferred Ecological Conditions with Simple Autecological Indices (SAI)** - The ecological preferences for diatoms are commonly recorded in the literature. Using the standard ecological categories compiled by Lowe (1974, Table 6-2), the ecological preferences for different diatom species can be characterized along an environmental (stressor) gradient. For example, pH preferences for many taxa are known. These preferences (Θ_i) can be ranked from 1-5 (e.g., acidobiontic, acidophilic, indifferent, alkaliphilic, alkalibiontic, Table 6-2) and can be used in the following equation to infer environmental conditions (EC) and effect on the periphyton assemblage.

$$SAI_{EC} = \Sigma \, \Theta_i p_i$$

14.   **Inferred Ecological Conditions with Weighted Average Indices** are based on the specific ecological optima (β_i) for algae, which are being reported more and more commonly in recent publications (see Pan and Stevenson 1996). Caution should be exercised, because we do not know how transferable these optima are among regions and habitats. Using the following equation, the ecological conditions (EC) in a habitat can be inferred more accurately by using the optimum environmental conditions (β_i) and relative abundances (ρ_i) for taxa in the habitat (ter Braak and van Dam 1989, Pan et al., 1996) than if only the ecological categorization were used (as above for the SAI). Optimum environmental conditions are those in which the highest relative abundances of a taxon are observed. These can be determined from the literature or from past surveys of taxa and environmental conditions in the study area (see ter Braak and van Dam 1989). In a pH example, the specific pH in a habitat can be inferred if we know the pH optima (H_i) of taxa in the habitat, and use the following general equation:

$$WAI_{EC} = \Sigma \beta_i p_i$$

and modify for inferring pH:

$$WAI_{pH} = \Sigma \, H_i p_i$$

15.   **Impairment of Ecological Conditions** can be inferred with algal assemblages by calculating the deviation (Δ_{EC}) between inferred environmental conditions at a test site and at a reference site.

Compare inferred ecological conditions at the test site to the expected ecological conditions (EC_{ex}) of regional reference sites by using either simple autecological indices (SAI_{EC}) or weighted average indices (WAI_{EC}):

$$\Delta_{EC} = |SAI_{EC} - EC_{ex}|$$

$$\Delta_{EC} = |WAI_{EC} - EC_{ex}|$$

Table 6-2. Environmental definitions of autecological classification systems for algae (as modified or referenced by Lowe 1974). Definitions for classes are given if no subclass is indicated.

Classification System/ Ecological Parameter	Class	Subclass	Conditions of Highest Relative Abundances
pH Spectrum	Acidobiontic		Below 5.5 pH
	Acidophilic		Above 5.5 and below 7 pH
	Indifferent		Around 7 pH
	Alakaliphilic		Above 7 and below 8.5 pH
	Alkalibiontic		Above 8.5 pH
Nutrient Spectrum - based on P and N concentrations	Eutrophic		High nutrient conditions

Classification System/ Ecological Parameter	Class	Subclass	Conditions of Highest Relative Abundances
	Mesotrophic		Moderate nutrient conditions
	Oligotrophic		Low nutrient conditions
	Dystrophic		High humic (DOC) conditions
Halobion Spectrum - based on chloride concentrations or conductivity	Polyhalobous		Salt concentrations > 40,000 mg/L
	Euhalobous		Marine forms: 30,000-40,000 mg/L
	Mesohalobous	Alpha range	Brackish water forms: 10,000-30,000 mg/L
	Mesohalobous	Beta range	Brackish water forms: 500-10,000 mg/L
	Oligohalobous	Halophilous	Freshwater - stimulated by some salt
	Oligohalobous	Indifferent	Freshwater - tolerates some salt
	Oligohalobous	Halophobic	Freshwater - does not tolerate small amounts of salt
Saprobien System - based on organic pollution	Polysaprobic		Characteristic of zone of degradation and putrefication, oxygen usually absent or low in concentration
	Mesosaprobic	Alpha range	Zone of organic load oxidation — N as amino acids
		Beta range	Zone of organic load oxidation — N as ammonia
	Oligosaprobic		Zone in which oxidation of organics complete, but high nutrient concentrations persist
	Saprophilic		Usually in polluted waters, but also in clean waters
	Saproxenous		Usually in clean waters, but also found in polluted waters
	Saprophobic		Only found in unpolluted waters

6.1.5 Determining Periphyton Biomass

Measurement of periphyton biomass is common in many studies and may be especially important in studies that address nutrient enrichment or toxicity. In many cases, however, sampling benthic algae misses peak biomass, which may best indicate nutrient problems and potential for nuisance algal growths (Biggs 1996, Stevenson 1996).

Biomass measurements can be made with samples collected from natural or artificial substrates. To quantify algal biomass (chl a, ash-free dry mass, cell density, biovolume cm^{-2}), the area of the substrate sampled must be determined. Two national stream assessment programs sample and assess area-specific cell density and biovolume (USGS-NAWQA, Porter et al. 1993; and EMAP, Klemm and Lazorchak 1994). These programs estimate algal biomass in habitats and reaches by collecting composite samples separately from riffle and pool habitats.

Periphyton biomass can be estimated with chl *a*, ash-free dry mass (AFDM), cell densities, and biovolume, usually per cm² (Stevenson 1996). Each of these measures estimates a different component of periphyton biomass (see Stevenson 1996 for discussion).

6.1.5.1 Chlorophyll *a*

Chlorophyll *a* ranges from 0.5 to 2% of total algal biomass (APHA 1995), and this ratio varies with taxonomy, light, and nutrients. A detailed description of chlorophyll *a* analysis is beyond the scope of this chapter. Standard methods (APHA 1995, USEPA 1992) are readily available. The analysis is relatively simple and involves:

1. extracting chlorophyll *a* in acetone;

2. measuring chlorophyll concentration in the extract with a spectrophotometer or fluorometer; and

3. calculating chlorophyll density on substrates by determining the proportion of original sample that was assessed for chlorophyll.

LABORATORY EQUIPMENT FOR PERIPHYTON ANALYSIS

- compound microscope with 10X or 15X oculars and 20X, 40X and 100X (oil) objectives
- tally counter (for species proportional count)
- microscope slides and coverglasses
- immersion oil, lens paper and absorbent tissues
- tissue homogenizer or blender
- magnetic stirrer and stir bar
- forceps
- hot plate
- fume hood
- squeeze bottle with distilled water
- oxidation reagents (HNO_3, H_2SO_4, $K_2Cr_2O_7$, H_2O_2)
- 200-500 ml beakers
- safety glasses and protective clothing
- drying oven for AFDM
- muffle furnace for AFDM
- aluminum weighing pans for AFDM
- spectrophotometer or fluorometer for chl *a*
- centrifuge for chl *a*
- graduated test tubes for chl *a*
- acetone for chl *a*
- $MgCO_3$ for chl *a*

6.1.5.2 Ash-Free Dry Mass

Ash-free dry mass is a measurement of the organic matter in samples, and includes biomass of bacteria, fungi, small fauna, and detritus in samples. A detailed description of analysis is beyond the scope of this chapter, but standard methods (APHA 1995, USEPA 1995) are readily available. The analysis is relatively simple and measures the difference in mass of a sample after drying and after incinerating organic matter in the sample. We recommend using AFDM versus dry mass to measure periphyton biomass because silt can account for a substantial proportion of dry mass in some samples. Ash mass in samples can be used to infer the amount of silt or other inorganic matter in samples.

6.1.5.3 Area-Specific Cell Densities and Biovolumes

Cell densities (cells cm⁻²) are determined by dividing the numbers of cells counted by the proportion of sample counted and the area from which samples were collected. Cell biovolumes (mm³ biovolume cm⁻²) are determined by summing the products of cell density and biovolume of each species counted (see Lowe and Pan 1996) and dividing that sum by the proportion of sample counted and the area from which samples were collected.

6.1.5.4 Biomass Metrics

High algal biomass can indicate eutrophication, but high algal biomass can also accumulate in less productive habitats after long periods of stable flow. Low algal biomass may be due to toxic conditions, but could be due to a recent storm event and spate or naturally heavy grazing. Thus, interpretation of biomass results is ambiguous and is the reason that major emphasis has not been placed on quantifying algal biomass for RBP. However, nuisance levels of algal biomass (e.g., > 10 µg chl a cm^{-2}, > 5 mg AFDM cm^{-2}, $> 40\%$ cover by macroalgae; see review by Biggs 1996) do indicate nutrient or organic enrichment. If repeated measurements of biomass can be made, then the mean and maximum benthic chl a could be used to define trophic status of streams. Dodds et al. (1998) have proposed guidelines in which the oligotrophic-mesotrophic boundary is a mean benthic chl a of 2 µg cm^{-2} or a maximum benthic chl a of 7 µg cm^{-2} and the mesotrophic-eutrophic boundary is a mean of 6 µg chl a cm^{-2} and a maximum of 20 µg chl a cm^{-2}.

6.2 FIELD-BASED RAPID PERIPHYTON SURVEY

Semi-quantitative assessments of benthic algal biomass and taxonomic composition can be made rapidly with a viewing bucket marked with a grid and a biomass scoring system. The advantage of using this technique is that it enables rapid assessment of algal biomass over larger spatial scales than substrate sampling and laboratory analysis. Coarse-level taxonomic characterization of communities is also possible with this technique. This technique is a survey of the

---

QUALITY CONTROL IN THE LABORATORY

1. Upon delivery of samples to the laboratory, complete entries on periphyton sample log-in forms (Appendix 2, Form 2).

2. Maintain a voucher collection of all samples and diatom slides. They should be accurately and completely labeled, preserved, and stored in the laboratory for future reference. Specimens on diatom slides should be clearly circled with a diamond or ink marker to facilitate location. A record of the voucher specimens should be maintained. Photographs of specimens improve "in-house" QA.

3. For every QA/QC sample (replicate sample in every 10th stream), assess relative abundances and taxa richness in replicate wet mounts and a replicate diatom slide to assess variation in metrics due to variability in sampling within reaches (habitats), sample preparation, and analytical variability.

4. QA/QC samples should be counted by another taxonomist to assess taxonomic precision and bias, if possible.

5. Common algal taxa should be the same for the two wet mount replicates. The percent community similarity index (Whittaker 1952) (see Section 6.5.1) calculated from proportional counts of the two replicate diatom slides should exceed 75%.

6. If it is not possible to get another taxonomist in the lab to QA/QC samples, an outside taxonomist should be consulted on a periodic basis to spot-check and verify taxonomic identifications in wet mounts and diatom slides. All common genera in the wet mount and all major species on the diatom slide (>3% relative abundance) should be identified similarly by both analysts (synonyms are acceptable). Any differences in identification should be reconciled and bench sheets should be corrected.

7. A library of basic taxonomic literature is an essential aid in the identification of algae and should be maintained and updated as needed in the laboratory (see taxonomic references for periphyton in Section 6.5). Taxonomists should participate in periodic training to ensure accurate identifications

natural substrate and requires no laboratory proc essing, but hand picked samples can be returned to the laboratory to quickly verify identification. It is a technique developed by Stevenson and Rier[2].

1. Fill in top of Rapid Periphyton Survey (RPS) Field Sheet, Appendix A-2, Form 5.

2. Establish at least 3 transects across the habitat being sampled (preferably riffles or runs in the reach in which benthic algal accumulation is readily observed and characterized).

3. Select 3 locations along each transect (e.g., stratified random locations on right, middle, and left bank).

4. Characterize algae in each selected location by immersing the bucket with 50-dot grid (7 x 7 + 1) in the water.
   ! First, characterize macroalgal biomass.

> **FIELD EQUIPMENT FOR RAPID PERIPHYTON SURVEY**
>
> • viewing bucket with 50-dot grid [Make the viewing bucket by cutting a hole in bottom of large (≥ 0.5 m diameter) plastic bucket, but leave a small ridge around the edge. Attach a piece of clear acrylic sheet to the bottom of the bucket with small screws and silicon caulk. The latter makes water tight seal so that no water enters the bucket when it is partially submerged. Periphyton can be clearly viewed by looking down through the bucket when it is partially submerged in the stream. Mark 50 dots in a 7 x 7 grid on the top surface of the acrylic sheet with a waterproof black marker. Add another dot outside the 7 x 7 grid to make the 50 dot grid.]
> • meter stick
> • pencil
> • Rapid Periphyton Survey Field Sheet

- Observe the bottom of the stream through the bottom of the viewing bucket and count the number of dots that occur over macroalgae (e.g., Cladophora or Spirogyra) under which substrates cannot be seen. Record that number and the kind of macroalgae under the dots on RPS field sheet.
- Measure and record the maximum length of the macroalgae.
- If two or more types of macroalgae are present, count the dots, measure, and record information for each type of macroalgae separately.

   ! Second, characterize microalgal cover.
   - While viewing the same area, record the number of dots under which substrata occur that are suitable size for microalgal accumulation (gravel > 2 cm in size).
   - Determine the kind (usually diatoms and blue-green algae) and estimate the thickness (density) of microalgae under each dot using the following thickness scale:
     0 - substrate rough with no visual evidence of microalgae
     0.5 - substrate slimy, but no visual accumulation of microalgae is evident
     1 - a thin layer of microalgae is visually evident
     2 - accumulation of microalgal layer from 0.5-1 mm thick is evident
     3 - accumulation of microalgae layer from 1 mm to 5 mm thick is evident
     4 - accumulation of microalgal layer from 5 mm to 2 cm thick is evident
     5 - accumulation of microalgal layer greater than 2 cm thick is evident
     Mat thickness can be measured with a ruler.
   - Record the number of dots that are over each of the specific thickness ranks separately for diatoms, blue-green algae, or other microalgae.

---

[2]S.T. Rier is a graduate student at the University of Louisville.

Rapid Bioassessment Protocols for Use in Streams and Wadeable Rivers: Periphyton, Benthic Macroinvertebrates, and Fish, Second Edition

6-19

5. Statistically characterize density of algae on substrate by determining:
    ! total number of grid points (dots) evaluated at the site (D_t);
    ! number of grid points (dots) over macroalgae (D_m);
    ! total number of grid points (dots) over suitable substrate for microalgae at the site (d_t);
    ! number of grid points over microalga of different thickness ranks for each type of microalga (d_i);
    ! average percent cover of the habitat by each type of macroalgae (i.e., $100X\ D_m/D_t$);
    ! maximum length of each type of macroalgae;
    ! mean density (i.e., thickness rank) of each type of macroalgae on suitable substrate (i.e., $\Sigma d_i r_i/d_t$); maximum density of each type of microalgae on suitable substrate.

6. QA/QC between observers and calibration between algal biomass (chl *a*, AFDM, cell density and biovolume cm^{-2} and taxonomic composition) can be developed by collecting samples that have specific microalgal rankings and assaying the periphyton.

6.3 TAXONOMIC REFERENCES FOR PERIPHYTON

A great wealth of taxonomic literature is available for algae. Below is a subset of that literature. It is a list of taxonomic references that are useful for most of the United States and are either in English, are important because no English treatment of the group is adequate, or are valuable for the good illustrations.

Camburn, K.E., R.L. Lowe, and D.L. Stoneburner. 1978. The haptobenthic diatom flora of Long Branch Creek, South Carolina. *Nova Hedwigia* 30:149-279.

Collins, G.B. and R.G. Kalinsky. 1977. Studies on Ohio diatoms: I. Diatoms of the Scioto River Basin. *Bull. Ohio Biological Survey.* 5(3):1-45.

Cox, E. J. 1996. *Identification of freshwater diatoms from live material.* Chapman & Hall, London.

Czarnecki, D.B. and D.W. Blinn. 1978. *Diatoms of the Colorado River in Grand Canyon National Park and vicinity.* (Diatoms of Southwestern USA II). Bibliotheca Phycologia 38. J. Cramer. 181 pp.

Dawes, C. J. 1974. *Marine Algae of the West Coast of Florida.* University of Miami Press.

Dillard, G.E. 1989a. Freshwater algae of the Southeastern United States. Part 1. Chlorophyceae: Volvocales, Testrasporales, and Chlorococcales. *Bibliotheca,* 81.

Dillard, G.E. 1989b. Freshwater algae of the Southeastern United States. Part 2. Chlorophyceae: Ulotrichales, Microsporales, Cylindrocapsales, Sphaeropleales, Chaetophorales, Cladophorales, Schizogoniales, Siphonales, and Oedogoniales. *Bibliotheca Phycologica,* 83.

Dillard, G.E. 1990. Freshwater algae of the Southeastern United States. Part 3. Chlorophyceae: Zygnematales: Zygenmataceae, Mesotaeniaceae, and Desmidaceae (Section 1). *Bibliotheca Phycologica,* 85.

Dillard, G.E. 1991. Freshwater algae of the Southeastern United States. Part 4. Chlorophyceae: Zygnemateles: Desmidaceae (Section 2). *Bibliotheca Phycologica,* 89.

Drouet, F. 1968. *Revision of the classification of the oscillatoriaceae.* Monograph 15. Academy of Natural Sciences, Philadelphia. Fulton Press, Lancaster, Pennsylvania.

Hohn, M.H. and J. Hellerman. 1963. The taxonomy and structure of diatom populations from three North American rivers using three sampling methods. *Transaction of the American Microscopal Society* 82:250-329.

Hustedt, F. 1927-1966. Die kieselalgen In Rabenhorst's Kryptogamen-flora von Deutschland Osterreich und der Schweiz VII. Leipzig, West Germany.

Hustedt, F. 1930. *Bacillariophyta (Diatomae).* In Pascher, A. (ed). Die suswasser Flora Mitteleuropas. (The freshwater flora of middle Europe). Gustav Fischer Verlag, Jena, Germany.

Jarrett, G.L. and J.M. King. 1989. The diatom flora (Bacillariphyceae) of Lake Barkley. U.S. Army Corps of Engineers, Nashville Dist. #DACW62-84-C-0085.

Krammer, K. and H. Lange-Bertalot. 1986-1991. Susswasserflora von Mitteleuropa. Band 2. Parts 1-4. Bacillariophyceae. Gustav Fischer Verlag. Stuttgart. New York.

Lange-Bertalot, H. and R. Simonsen. 1978. A taxonomic revision of the Nitzschia lanceolatae Grunow: 2. European and related extra-European freshwater and brackish water taxa. *Bacillaria* 1:11-111.

Lange-Bertalot, H. 1980. New species, combinations and synonyms in the genus Nitzschia. *Bacillaria* 3:41-77.

Patrick, R. and C.W. Reimer. 1966. *The diatoms of the United States, exclusive of Alaska and Hawaii.* Monograph No. 13. Academy of Natural Sciences, Philadelphia, Pennsylvania.

Patrick, R. and C.W. Reimer. 1975. *The Diatoms of the United States.* Vol. 2, Part 1. Monograph No. 13. Academy of Natural Sciences, Philadelphia, Pennsylvania.

Prescott, G.W. 1962. *The algae of the Western Great Lakes area.* Wm. C. Brown Co., Dubuque, Iowa.

Prescott, G.W., H.T. Croasdale, and W.C. Vinyard. 1975. *A Synopsis of North American desmids. Part II. Desmidaceae: Placodermae.* Section 1. Univ. Nebraska Press, Lincoln, Nebraska.

Prescott, G.W., H.T. Croasdale, and W.C. Vinyard. 1977. *A synopsis of North American desmids. Part II. Desmidaceae: Placodermae.* Section 2. Univ. Nebraska Press, Lincoln, Nebraska.

Prescott, G.W., H.T. Croasdale, and W.C. Vinyard. 1981. *A synopsis of North American desmids. Part II. Desmidaceae: Placodermae.* Section 3. Univ. Nebraska Press, Lincoln, Nebraska.

Prescott, G.W. 1978. *How to know the freshwater algae.* 3rd Edition. Wm. C. Brown Co., Dubuque, Iowa.

Simonsen, R. 1987. *Atlas and catalogue of the diatom types of Friedrich Hustedt.* Vol. 1-3. J. Cramer. Berlin, Germany.

Smith, M. 1950. *The Freshwater Algae of the United States*. McGraw-Hill, New York, New York.

Taylor, W. R. 1960. *Marine algae of the eastern tropical and subtropical coasts of the Americas*. University of Michigan Press, Ann Arbor, Michigan.

VanLandingham, S. L. 1982. *Guide to the identification, environmental requirements and pollution tolerance of freshwater blue-green algae (Cyanophyta)*. EPA-600/3-82-073.

Whitford, L.A. and G.J. Schumacher. 1973. *A manual of freshwater algae*. Sparks Press, Raleigh, North Carolina.

Wujek, D.E. and R.F. Rupp. 1980. Diatoms of the Tittabawassee River, Michigan. *Bibliotheca Phycologia* 50:1-100.

6.4 AUTECOLOGICAL REFERENCES FOR PERIPHYTON

Beaver, J. 1981. *Apparent ecological characteristics of some common freshwater diatoms*. Ontario Ministry of the Environment. Rexdale, Ontario, Canada.

Cholnoky, B. J. 1968. Ökologie der Diatomeen in Binnegewässern. Cramer, Lehre.

Fabri, R. and L. Leclercq. 1984. Etude écologique des riviéres du nord du massif Ardennais (Belgique): flore et végétation de diatomeées et physico-chimie des eaux. 1. Station scientifique des Hautes Fagnes, Robertville. 379 pp.

Fjerdingstad, E. 1950. The microflora of the River Molleaa with special reference to the relation of benthic algae to pollution. *Folia Limnologica Scandanavica* 5, 1-123.

Hustedt, F. 1938-39. Systematische und ökologische Untersuchungen über die Diatomeen-Flora von Java, Bali und Sumatra nach dem Material deter Deutschen Limnologischen Sunda-Expedition. Allgemeiner Teil. I. Ubersicht über das Untersuchungsmaterial und Charakterisktik der Diatomeenflora der einzelnen Gebiete. II. Die Diatomeen flora der untersuchten Gesässertypen. III. Die ökologische Faktoren und ihr Einfluss auf die Diatomeenflora. Archiv für Hydrobiologie, Supplement Band, 15:638-790 (1938); 16:1-155 (1938); 16:274-394 (1939).

Hustedt, F. 1957. Die Diatomeenflora des Flusssystems der Weser im Gebiet der Hansestadt Bremen. Abhandlungen naturwissenschaftlichen. Verein zu Bremen, Bd. 34, Heft 3, S. 181-440, 1 Taf.

Lange-Bertalot, H. 1978. Diatomeen-Differentialarten anstelle von Leitformen: ein geeigneteres Kriterium der Gewässerbelastung. *Archiv für Hydrobiologie Supplement 51*, 393-427.

Lange-Bertalot, H. 1979. Pollution tolerance of diatoms as a criterion for water quality estimation. *Nova Hedwigia* 64, 285-304.

LeCointe C., M. Coste, and J. Prygiel. 1993. "OMNIDIA" software for taxonomy, calculation of diatom indices and inventories management. *Hydrobiologia* 269/270: 509-513.

Lowe, R. L. 1974. *Environmental Requirements and Pollution Tolerance of Freshwater Diatoms*. US Environmental Protection Agency, EPA-670/4-74-005. Cincinnati, Ohio, USA.

Palmer, C. M. 1969. A composite rating of algae tolerating organic pollution. *Journal of Phycology* 5, 78-82.

Rott, E., G. Hofmann, K. Pall, P. Pfister, and E. Pipp. 1997. Indikationslisten für Aufwuchsalgen in österreichischen Fliessgewässern. Teil 1: Saprobielle Indikation. Wasserwirtschaftskataster. Bundesminsterium für Land- und Forstwirtschaft. Stubenring 1, 1010 Wein, Austria.

Slàdecek, V. 1973. System of water quality from the biological point of view. *Archiv für Hydrobiologie und Ergebnisse Limnologie* 7, 1-218.

Van Dam, H., Mertenes, A., and Sinkeldam, J. 1994. A coded checklist and ecological indicator values of freshwater diatoms from the Netherlands. *Netherlands Journal of Aquatic Ecology* 28, 117-33.

Vanlandingham, S. L. 1982. *Guide to the identification, environmental requirement and pollution tolerance of freshwater blue-green algae (Cyanophyta)*. U. S. Environmental Protection Agency. EPA-600/3-82-073.

Watanabe, T., Asai, K., Houki, A. Tanaka, S., and Hizuka, T. 1986. Saprophilous and eurysaprobic diatom taxa to organic water pollution and diatom assemblage index (DAIpo). *Diatom* 2:23-73.

7 BENTHIC MACROINVERTEBRATE PROTOCOLS

Rapid bioassessment using the benthic macroinvertebrate assemblage has been the most popular set of protocols among the state water resource agencies since 1989 (Southerland and Stribling 1995). Most of the development of benthic Rapid Bioassessment Protocols (RBPs) has been oriented toward RBP III (described in Plafkin et al. 1989). As states have focused attention on regional specificity, which has included a wide variety of physical characteristics of streams, the methodology of conducting stream surveys of the benthic assemblage has advanced. Some states have preferred to retain more traditional methods such as the Surber or Hess samplers (e.g., Wyoming Department of Environmental Quality [DEQ]) over the kick net in cobble substrate. Other agencies have developed techniques for streams lacking cobble substrate, such as those streams in coastal plains. State water resource agencies composing the Mid-Atlantic Coastal Streams (MACS) Workgroup, i.e., New Jersey Department of Environmental Protection (DEP), Delaware Department of Natural Resources and Environmental Control (DNREC), Maryland Department of Natural Resources (DNR) and Maryland Department of the Environment (MDE), Virginia DEQ, North Carolina Department of Environmental Management (DEM), and South Carolina Department of Health and Environmental Control (DHEC), and a workgroup within the Florida Department of Environmental Protection (DEP) were pioneers in this effort. These 2 groups (MACS and FLDEP) developed a multihabitat sampling procedure using a D-frame dip net. Testing of this procedure by these 2 groups indicates that this technique is scientifically valid for low-gradient streams. Research conducted by the U.S. Environmental Protection

---

STANDARD BENTHIC MACROINVERTEBRATE SAMPLING GEAR TYPES FOR STREAMS
(assumes standard mesh size of 500 μ nytex screen)

- **Kick net:** Dimensions of net are 1 meter (m) x 1 m attached to 2 poles and functions similarly to a fish kick seine. Is most efficient for sampling cobble substrate (i.e., riffles and runs) where velocity of water will transport dislodged organisms into net. Designed to sample 1 m² of substrate at a time and can be used in any depth from a few centimeters to just below 1m (Note -- Depths of 1m or greater will be difficult to sample with any gear).

- **D-frame dip net:** Dimensions of frame are 0.3 m width and 0.3 m height and shaped as a "D" where frame attaches to long pole. Net is cone or bag-shaped for capture of organisms. Can be used in a variety of habitat types and used as a kick net, or for "jabbing", "dipping", or "sweeping".

- **Rectangular dip net:** Dimensions of frame are 0.5 m width and 0.3 m height and attached to a long pole. Net is cone or bag-shaped. Sampling is conducted similarly to the D-frame.

- **Surber:** Dimensions of frame are 0.3 m x 0.3 m, which is horizontally placed on cobble substrate to delineate a 0.09 m² area. A vertical section of the frame has the net attached and captures the dislodged organisms from the sampling area. Is restricted to depths of less than 0.3 m.

- **Hess:** Dimensions of frame are a metal cylinder approximately 0.5 m in diameter and samples an area 0.8 m². Is an advanced design of the Surber and is intended to prevent escape of organisms and contamination from drift. Is restricted to depths of less than 0.5 m.

Rapid Bioassessment Protocols for Use in Streams and Wadeable Rivers: Periphyton, Benthic Macroinvertebrates, and Fish, Second Edition

7-1

Agency (USEPA) for their Environmental Monitoring and Assessment Program (EMAP) program and the United States Geological Survey (USGS) for their National Water Quality Assessment Program (NAWQA) program have indicated that the rectangular dip net is a reasonable compromise between the traditional Surber or Hess samplers and the RBP kick net described the original RBPs.

From the testing and implementation efforts that have been conducted around the country since 1989, refinements have been made to the procedures while maintaining the original concept of the RBPs. Two separate procedures that are oriented toward a "single, most productive" habitat and a multihabitat approach represent the most rigorous benthic RBP and are essentially a replacement of the original RBP III. The primary differences between the original RBP II and III are the decision on field versus lab sorting and level of taxonomy. These differences are not considered sufficient reasons to warrant separate protocols. In addition, a third protocol has been developed as a more standardized biological reconnaissance or screening and replaces RBP I of the original document.

Kicknet

D-frame Dipnet

Rectangular Dipnet

Hess sampler

(Mary Kay Corazalla, Univ. of Minnesota)

7.1 SINGLE HABITAT APPROACH: 1 METER KICK NET

The original RBPs (Plafkin et al. 1989) emphasized the sampling of a single habitat, in particular riffles or runs, as a means to standardize assessments among streams having those habitats. This approach is still valid, because macroinvertebrate diversity and abundance are usually highest in cobble substrate (riffle/run) habitats. Where cobble substrate is the predominant habitat, this sampling approach provides a representative sample of the stream reach. However, some streams naturally lack the cobble substrate. In cases where the cobble substrate represents less than 30% of the sampling reach in reference streams (i.e., those streams that are representative of the region), alternate habitat(s) will need to be sampled (See Section 7.2). The appropriate sampling method should be selected based on the habitat availability of the reference condition and not of potentially impaired streams. For example, methods would not be altered for situations where the extent of cobble substrate in streams influenced by heavy sediment deposition may be substantially reduced from the amount of cobble substrate expected for the region.

7.1.1 Field Sampling Procedures for Single Habitat

1.   A 100 m reach representative of the characteristics of the stream should be selected. Whenever possible, the area should be at least 100 meters upstream from any road or bridge crossing to minimize its effect on stream velocity, depth, and overall habitat quality. There should be no major tributaries discharging to the stream in the study area.

> **FIELD EQUIPMENT/SUPPLIES NEEDED FOR BENTHIC MACROINVERTEBRATE SAMPLING —SINGLE HABITAT APPROACH**
>
> - standard kick-net, 500 μ opening mesh, 1.0 meter width
> - sieve bucket, with 500 μ opening mesh
> - 95% ethanol
> - sample containers, sample container labels
> - forceps
> - pencils, clipboard
> - Benthic Macroinvertebrate Field Data Sheet[*]
> - first aid kit
> - waders (chest-high or hip boots)
> - rubber gloves (arm-length)
> - camera
> - Global Positioning System (GPS) Unit
>
> [*] It is helpful to copy fieldsheets onto water-resistant paper for use in wet weather conditions

2.   Before sampling, complete the physical/chemical field sheet (see Chapter 5; Appendix A-1, Form 1) to document site description, weather conditions, and land use. After sampling, review this information for accuracy and completeness.

3.   Draw a map of the sampling reach. This map should include in-stream attributes (e.g., riffles, falls, fallen trees, pools, bends, etc.) and important structures, plants, and attributes of the bank and near stream areas. Use an arrow to indicate the direction of flow. Indicate the areas that were sampled for macroinvertebrates on the map. Estimate "river mile" for sampling reach for probable use in data management of the water resource agency. If available, use hand-held Global Positioning System (GPS) for latitude and longitude determination taken at the furthest downstream point of the sampling reach.

Rapid Bioassessment Protocols for Use in Streams and Wadeable Rivers: Periphyton, Benthic Macroinvertebrates, and Fish, Second Edition

7-3

4. All riffle and run areas within the 100-m reach are candidates for sampling macroinvertebrates. A composite sample is taken from individual sampling spots in the riffles and runs representing different velocities. Generally, a minimum of 2 m² composited area is sampled for RBP efforts.

5. Sampling begins at the downstream end of the reach and proceeds upstream. Using a 1 m kick net, 2 or 3 kicks are sampled at various velocities in the riffle or series of riffles. A *kick* is a stationary sampling accomplished by positioning the net and disturbing one square meter upstream of the net. Using the toe or heel of the boot, dislodge the upper layer of cobble or gravel and scrape the underlying bed. Larger substrate particles should be picked up and rubbed by hand to remove attached organisms. If different gear is used (e.g., a D-frame or rectangular net), a composite is obtained from numerous kicks (See Section 7.2).

ALTERNATIVES FOR STREAM REACH DESIGNATION

• **Fixed-distance designation**—A standard length of stream, such as a reach, is commonly used to obtain an estimate of natural variability. Conceptually, this approach should provide a mixture of habitats in the reach and provide, at a minimum, duplicate physical and structural elements such as a riffle/pool sequence.

• **Proportional-distance designation**—Alternatively, a standard number of stream "widths" is used to measure the stream distance, e.g., 40 times the stream width is defined by EMAP for sampling (Klemm and Lazorchak 1995). This approach allows variation in the length of the reach based on the size of the stream.

---

6. The jabs or kicks collected from different locations in the cobble substrate will be composited to obtain a single homogeneous sample. After every kick, wash the collected material by running clean stream water through the net 2 to 3 times. If clogging does occur, discard the material in the net and redo that portion of the sample in a different location. Remove large debris after rinsing and inspecting it for organisms; place any organisms found into the sample container. Do not spend time inspecting small debris in the field. [Note — an alternative is to keep the samples from different habitats separated as done in EMAP (Klemm and Lazorchak 1995).]

7. Transfer the sample from the net to sample container(s) and preserve in enough 95 percent ethanol to cover the sample. Forceps may be needed to remove organisms from the dip net. Place a label indicating the sample identification code or lot number, date, stream name, sampling location, and collector name into the sample container. The outside of the container should include the same information and the words "preservative: 95% ethanol". If more than one container is needed for a sample, each container label should contain all the information for the sample and should be numbered (e.g., 1 of 2, 2 of 2, etc.). This information will be recorded in the "Sample Log" at the biological laboratory (Appendix A-3, Form 2).

8. Complete the top portion of the "Benthic Macroinvertebrate Field Data Sheet" (Appendix A-3, Form 1), which duplicates the "header" information on the physical/chemical field sheet.

9. Record the percentage of each habitat type in the reach. Note the sampling gear used, and comment on conditions of the sampling, e.g., high flows, treacherous rocks, difficult access to stream, or anything that would indicate adverse sampling conditions.

Chapter 7: Benthic Macroinvertebrate Protocols

10. Document observations of aquatic flora and fauna. Make qualitative estimates of macroinvertebrate composition and relative abundance as a cursory estimate of ecosystem health and to check adequacy of sampling.

11. Perform habitat assessment (Appendix A-1, Form 2) after sampling has been completed; walking the reach helps ensure a more accurate assessment. Conduct the habitat assessment with another team member, if possible.

12. Return samples to laboratory and complete log-in form (Appendix A-3, Form 2).

QUALITY CONTROL (QC) IN THE FIELD

1. Sample labels must be properly completed, including the sample identification code, date, stream name, sampling location, and collector's name, and placed into the sample container. The outside of the container should be labeled with the same information. Chain-of-custody forms, if needed, must include the same information as the sample container labels.

2. After sampling has been completed at a given site, all nets, pans, etc. that have come in contact with the sample should be rinsed thoroughly, examined carefully, and picked free of organisms or debris. Any additional organisms found should be placed into the sample containers. The equipment should be examined again prior to use at the next sampling site.

3. Replicate (1 duplicate sample) 10% of the sites to evaluate precision or repeatability of the sampling technique or the collection team.

7.2   MULTIHABITAT APPROACH: D–FRAME DIP NET

Streams in many states vary from high gradient, cobble dominated to low gradient streams with sandy or silty sediments. Therefore, a method suitable to sampling a variety of habitat types is desired in these cases. The method that follows is based on Mid-Atlantic Coastal Streams Workgroup recommendations designed for use in streams with variable habitat structure (MACS 1996) and was used for statewide stream bioassessment programs by Florida DEP (1996) and Massachusetts DEP (1995). This method focuses on a multihabitat scheme designed to sample major habitats in proportional representation within a sampling reach. Benthic

FIELD EQUIPMENT/SUPPLIES NEEDED FOR BENTHIC MACROINVERTEBRATE SAMPLING —MULTI-HABITAT APPROACH

- standard D-frame dip net, 500 μ opening mesh, 0.3 m width (~ 1.0 ft frame width)
- sieve bucket, with 500 μ opening mesh
- 95% ethanol
- sample containers, sample container labels
- forceps
- pencils, clipboard
- Benthic Macroinvertebrate Field Data Sheet[*]
- first aid kit
- waders (chest-high or hip boots)
- rubber gloves (arm-length)
- camera
- Global Positioning System (GPS) Unit

[*] It is helpful to copy fieldsheets onto water-resistant paper for use in wet weather conditions

---

macroinvertebrates are collected systematically from all available instream habitats by kicking the substrate or jabbing with a D-frame dip net. A total of 20 jabs (or kicks) are taken from all major habitat types in the reach resulting in sampling of approximately 3.1 m² of habitat. For example, if the habitat in the sampling reach is 50% snags, then 50% or 10 jabs should be taken in that habitat. An organism-based subsample (usually 100, 200, 300, or 500 organisms) is sorted in the laboratory and identified to the lowest practical taxon, generally genus or species.

7.2.1 Habitat Types

The major stream habitat types listed here are in reference to those that are colonized by macroinvertebrates and generally support the diversity of the macroinvertebrate assemblage in stream ecosystems. Some combination of these habitats would be sampled in the multihabitat approach to benthic sampling.

Cobble (hard substrate) - Cobble will be prevalent in the riffles (and runs), which are a common feature throughout most mountain and piedmont streams. In many high-gradient streams, this habitat type will be dominant. However, riffles are not a common feature of most coastal or other low-gradient streams. Sample shallow areas with coarse (mixed gravel, cobble or larger) substrates by holding the bottom of the dip net against the substrate and dislodging organisms by kicking the substrate for 0.5 m upstream of the net.

Snags - Snags and other woody debris that have been submerged for a relatively long period (not recent deadfall) provide excellent colonization habitat. Sample submerged woody debris by jabbing in medium-sized snag material (sticks and branches). The snag habitat may be kicked first to help dislodge organisms, but only after placing the net downstream of the snag. Accumulated woody material in pool areas are considered snag habitat. Large logs should be avoided because they are generally difficult to sample adequately.

Vegetated banks - When lower banks are submerged and have roots and emergent plants associated with them, they are sampled in a fashion similar to snags. Submerged areas of undercut banks are good habitats to sample. Sample banks with protruding roots and plants by jabbing into the habitat. Bank habitat can be kicked first to help dislodge organisms, but only <u>after</u> placing the net downstream.

Submerged macrophytes - Submerged macrophytes are seasonal in their occurrence and may not be a common feature of many streams, particularly those that are high-gradient. Sample aquatic plants that are rooted on the bottom of the stream in deep water by drawing the net through the vegetation from the bottom to the surface of the water (maximum of 0.5 m each jab). In shallow water, sample by bumping or jabbing the net along the bottom in the rooted area, avoiding sediments where possible.

Sand (and other fine sediment) - Usually the least productive macroinvertebrate habitat in streams, this habitat may be the most prevalent in some streams. Sample banks of unvegetated or soft soil by bumping the net along the surface of the substrate rather than dragging the net through soft substrates; this reduces the amount of debris in the sample.

7.2.2 Field Sampling Procedures for Multihabitat

1. A 100 m reach that is representative of the characteristics of the stream should be selected. Whenever possible, the area should be at least 100 m upstream from any road or bridge crossing to minimize its effect on stream velocity, depth and overall habitat quality. There should be no major tributaries discharging to the stream in the study area.

2. Before sampling, complete the physical/chemical field sheet (see Chapter 5; Appendix A-1, Form 1) to document site description, weather conditions, and land use. After sampling, review this information for accuracy and completeness.

3. Draw a map of the sampling reach. This map should include in-stream attributes

---

ALTERNATIVES FOR STREAM REACH DESIGNATION

*   **Fixed-distance designation**—A standard length of stream, such as a reach, is commonly used to obtain an estimate of natural variability. Conceptually, this approach should provide a mixture of habitats in the reach and provide, at a minimum, duplicate physical and structural elements such as a riffle/pool sequence.

*   **Proportional-distance designation**— Alternatively, a standard number of stream "widths" is used to measure the stream distance, e.g., 40 times the stream width is defined by EMAP for sampling (Klemm and Lazorchak 1995). This approach allows variation in the length of the reach based on the size of the stream.

(e.g., riffles, falls, fallen trees, pools, bends, etc.) and important structures, plants, and attributes of the bank and near stream areas. Use an arrow to indicate the direction of flow. Indicate the areas that were sampled for macroinvertebrates on the map. Approximate "river mile" to sampling reach for probable use in data management of the water resource agency. If available, use hand-held GPS for latitude and longitude determination taken at the furthest downstream point of the sampling reach.

4. Different types of habitat are to be sampled in approximate proportion to their representation of surface area of the total macroinvertebrate habitat in the reach. For example, if snags comprise 50% of the habitat in a reach and riffles comprise 20%, then 10 jabs should be taken in snag material and 4 jabs should be take in riffle areas. The remainder of the jabs (6) would be taken in any remaining habitat type. Habitat types contributing less than 5% of the stable habitat in the stream reach should not be sampled. In this case, allocate the remaining jabs proportionately among the predominant substrates. The number of jabs taken in each habitat type should be recorded on the field data sheet.

5. Sampling begins at the downstream end of the reach and proceeds upstream. A total of 20 jabs or kicks will be taken over the length of the reach; a single *jab* consists of forcefully thrusting the net into a productive habitat for a linear distance of 0.5 m. A *kick* is a stationary sampling accomplished by positioning the net and disturbing the substrate for a distance of 0.5 m upstream of the net.

6. The jabs or kicks collected from the multiple habitats will be composited to obtain a single homogeneous sample. Every 3 jabs, more often if necessary, wash the collected material by running clean stream water through the net two to three times. If clogging does occur that may hinder obtaining an appropriate sample, discard the material in the net and redo that portion of

---

Rapid Bioassessment Protocols for Use in Streams and Wadeable Rivers: Periphyton, Benthic Macroinvertebrates, and Fish, Second Edition

7-7

the sample in the same habitat type but in a different location. Remove large debris after rinsing and inspecting it for organisms; place any organisms found into the sample container. Do not spend time inspecting small debris in the field.

7. Transfer the sample from the net to sample container(s) and preserve in enough 95% ethanol to cover the sample. Forceps may be needed to remove organisms from the dip net. Place a label indicating the sample identification code or lot number, date, stream name, sampling location, and collector name into the sample container. The outside of the container should include the same information and the words "preservative: 95% ethanol". If more that one container is needed for a sample, each container label should contain all the information for the sample and should be numbered (e.g., 1 of 2, 2 of 2, etc.). This information will be recorded in the "Sample Log" at the biological laboratory (Appendix A-3, Form 2).

8. Complete the top portion of the "Benthic Macroinvertebrate Field Data Sheet" (Appendix A-3, Form 1), which duplicates the "header" information on the physical/chemical field sheet.

9. Record the percentage of each habitat type in the reach. Note the sampling gear used, and comment on conditions of the sampling, e.g., high flows, treacherous rocks, difficult access to stream, or anything that would indicate adverse sampling conditions.

10. Document observations of aquatic flora and fauna. Make qualitative estimates of macroinvertebrate composition and relative abundance as a cursory estimate of ecosystem health and to check adequacy of sampling.

11. Perform habitat assessment (Appendix A-1, Form 3) after sampling has been completed. Having sampled the various microhabitats and walked the reach helps ensure a more accurate assessment. Conduct the habitat assessment with another team member, if possible.

12. Return samples to laboratory and complete log-in forms (Appendix A-3, Form 2).

QUALITY CONTROL (QC) IN THE FIELD

1. Sample labels must be properly completed, including the sample identification code, date, stream name, sampling location, and collector's name and placed into the sample container. The outside of the container should be labeled with the same information. Chain-of-custody forms, if needed, must include the same information as the sample container labels.

2. After sampling has been completed at a given site, all nets, pans, etc. that have come in contact with the sample should be rinsed thoroughly, examined carefully, and picked free of organisms or debris. Any additional organisms found should be placed into the sample containers. The equipment should be examined again prior to use at the next sampling site.

3. Replicate (1 duplicate sample) 10% of the sites to evaluate precision or repeatability of sampling technique or collection team.

7.3 LABORATORY PROCESSING FOR MACROINVERTEBRATE SAMPLES

Macroinvertebrate samples collected by either intensive method, i.e., single habitat or multihabitat, are best processed in the laboratory under controlled conditions. Aspects of laboratory processing include subsampling, sorting, and identification of organisms.

All samples should be dated and recorded in the "Sample Log" notebook or on sample log form (Appendix A-3, Form 2) upon receipt by laboratory personnel. All information from the sample container label should be included on the sample log sheet. If more than one container was used, the number of containers should be indicated as well. All samples should be sorted in a single laboratory to enhance quality control.

7.3.1 Subsampling and Sorting

Subsampling benthic samples is not a requirement, and in fact, is frowned upon by certain scientists. Courtemanch (1996) provides an argument against subsampling, or to use a volume-based procedure if samples are to be subsampled. Vinson and Hawkins (1996) and Barbour and Gerritsen (1996) provide arguments for a fixed-count method, which is the preferred subsampling technique for RBPs.

> **LABORATORY EQUIPMENT/SUPPLIES NEEDED FOR BENTHIC MACROINVERTEBRATE SAMPLE PROCESSING**
>
> * log-in sheet for samples
> * standardized gridded pan (30 cm x 36 cm) with approximately 30 grids (6 cm x 6 cm)
> * 500 micron sieve
> * forceps
> * white plastic or enamel pan (15 cm x 23 cm) for sorting
> * specimen vials with caps or stoppers
> * sample labels
> * standard laboratory bench sheets for sorting and identification
> * dissecting microscope for organism identification
> * fiber optics light source
> * compound microscope with phase contrast for identification of mounted organisms (e.g., midges)
> * 70% ethanol for storage of specimens
> * appropriate taxonomic keys

Subsampling reduces the effort required for the sorting and identification aspects of macroinvertebrate surveys and provides a more accurate estimate of time expenditure (Barbour and Gerritsen 1996). The RBPs use a fixed-count approach to subsampling and sorting the organisms from the sample matrix of detritus, sand, and mud. *The following protocol is based on a 200-organism subsample, but it could be used for any subsample size (100, 300, 500, etc.).* The subsample is sorted and preserved separately from the remaining sample for quality control checks.

1. Prior to processing any samples in a lot (i.e., samples within a collection date, specific watershed, or project), complete the sample log-in sheet to verify that all samples have arrived at the laboratory, and are in proper condition for processing.

2. Thoroughly rinse sample in a 500 μm-mesh sieve to remove preservative and fine sediment. Large organic material (whole leaves, twigs, algal or macrophyte mats, etc.) not removed in the field should be rinsed, visually inspected, and discarded. If the samples have been preserved in alcohol, it will be necessary to soak the sample contents in water for about 15 minutes to hydrate the benthic organisms, which will prevent them from floating on the water surface during sorting. If the sample was stored in more than one container, the contents of all

Rapid Bioassessment Protocols for Use in Streams and Wadeable Rivers: Periphyton, Benthic Macroinvertebrates, and Fish, Second Edition

7-9

containers for a given sample should be combined at this time. Gently mix the sample by hand while rinsing to make homogeneous.

SUBSAMPLE PROCEDURE MODIFICATIONS

Subsampling procedures developed by Hilsenhoff (1987) and modified by Plafkin et al. (1989) were used in the original RBP II and RBP III protocols. As an improvement to the mechanics of the technique, Caton (1991) designed a sorting tray consisting of two parts, a rectangular plastic or plexiglass pan (36 cm x 30 cm) with a rectangular sieve insert. The sample is placed on the sieve, in the pan and dispersed evenly.

When a random grid(s) is selected, the sieve is lifted to temporarily drain the water. A "cookie-cutter" like metal frame 6 cm x 6 cm is used to clearly define the selected grid; debris overhanging the grid may be cut with scissors. A 6 cm flat scoop is used to remove all debris and organisms from the grid. The contents are then transferred to a separate sorting pan with water for removal of macroinvertebrates.

These modifications have allowed for rapid isolation of organisms within the selected grids and easy removal of all organisms and debris within a grid while eliminating investigator bias.

---

3. After washing, spread the sample evenly across a pan marked with grids approximately 6 cm x 6 cm. On the laboratory bench sheet, note the presence of large or obviously abundant organisms; *do not remove them from the pan*. However, Vinson and Hawkins (1996) present an argument for including these large organisms in the count, because of the high probability that these organisms will be excluded from the targeted grids.

4. Use a random numbers table to select 4 numbers corresponding to squares (grids) within the gridded pan. Remove all material (organisms and debris) from the four grid squares, and place the material into a shallow white pan and add a small amount of water to facilitate sorting. If there appear (through a cursory count or observation) to be 200 organisms ± 20% (cumulative of 4 grids), then subsampling is complete.

Any organism that is lying over a line separating two grids is considered to be on the grid containing its head. In those instances where it may not be possible to determine the location of the head (worms for instance), the organism is considered to be in the grid containing most of its body.

If the density of organisms is high enough that many more than 200 organisms are contained in the 4 grids, transfer the contents of the 4 grids to a second gridded pan. Randomly select grids for this second level of sorting as was done for the first, sorting grids one at a time until 200 organisms ± 20% are found. If picking through the entire next grid is likely to result in a subsample of greater than 240 organisms, then that grid may be subsampled in the same manner as before to decrease the likelihood of exceeding 240 organisms. That is, spread the contents of the last grid into another gridded pan. Pick grids one at a time until the desired number is reached. The total number of grids for each subsorting level should be noted on the laboratory bench sheet.

TESTING OF SUBSAMPLING

Ferraro et al. (1989) describe a procedure for calculating the "power-cost efficiency" (PCE), which incorporates both the number of samples and the cost (i.e. time or money) for each alternative sampling scheme. With this analysis, the optimal subsampling size is that by which the costs of increased effort are offset by the lowest theoretical number of samples predicted from the power analysis to provide reliable resolution (Barbour and Gerritsen 1996).

There are 4 primary steps in assessing the PCE of a suite of alternative subsampling strategies:

Step 1: For each subsampling strategy (i.e., 100-, 200-, 300- organism level, or other) collect samples at several reference and impaired stations. The observed differences in each of the core metrics is defined to be the magnitude of the difference desired to be detected. The difference is the "effect size" and is equivalent to the inverse coefficient of variation (CV).

Step 2: Assess the "cost" (c_i), in time or money, of each subsampling scheme i at each site. The cost can include labor hours for subsampling, sorting, identification, and documentation. Total cost of each subsampling alternative is the product of cost per site and required sample size.

Step 3: Conduct statistical power analyses to determine the minimum number of replicate samples (n_i) needed to detect the effect size with an acceptable probability of Type I ($\propto$; the probability that the null hypothesis [e.g., "sites are good"] is true and it is rejected. Commonly termed the significance level.) and Type II (β; the probability that the null hypothesis is false and it is accepted) error. Typically, $\propto$ and β are set at 0.05. This step may be deleted for those programs that already have an established number of replicate samples.

Step 4: Calculate the PCE for each sampling scheme by:

$$PCE_i = \frac{(n \, X \, c)_{min}}{(n_i \, X \, c_i)}$$

where $(n \, X \, c)_{min}$ = minimum value of $(n \, X \, c)$ among the i sampling schemes. The PCE formula is equivalent to the "power efficiency" ratio of the sample sizes attained by alternative tests under similar conditions (Ferraro et al. 1989) with the n's multiplied by the "cost" per replicate sample. Multiplying n by c puts efficiency on a total "cost" rather than on a sample size basis. The reciprocal of PCE_i is the factor by which the optimal subsampling scheme is more efficient than alternative scheme i. When PCE is determined for multiple metrics, the overall optimal subsampling scheme may be defined as that which ranks highest in PCE for most metrics of interest.

5.   Save the sorted debris residue in a separate container. Add a label that includes the words "sorted residue" in addition to all prior sample label information and preserve in 95% ethanol. Save the remaining unsorted sample debris residue in a separate container labeled "sample residue"; this container should include the original sample label. Length of storage and archival is determined by the laboratory or benthic section supervisor.

6.   Place the sorted 200-organism (± 20%) subsample into glass vials, and preserve in 70% ethanol. Label the vials inside with the sample identifier or lot number, date, stream name, sampling location and taxonomic group. If more than one vial is needed, each should be labeled separately and numbered (e.g., 1 of 2, 2 of 2). For convenience in reading the labels inside the

vials, insert the labels left-edge first. If identification is to occur immediately after sorting, a petri dish or watch glass can be used instead of vials.

7.   Midge (Chironomidae) larvae and pupae should be mounted on slides in an appropriate medium (e.g., Euperal, CMC-9); slides should be labeled with the site identifier, date collected, and the first initial and last name of the collector. As with midges, worms (Oligochaeta) must also be mounted on slides and should be appropriately labeled.

8.   Fill out header information on Laboratory Bench Sheet as in field sheets (see Chapter 5). Also check subsample target number. Complete back of sheet for subsampling/sorting information. Note number of grids picked, time expenditure, and number of organisms. If QC check was performed on a particular sample, person conducting QC should note findings on the back of the Laboratory Bench Sheet. Calculate sorting efficiency to determine whether sorting effort passes or fails.

9.   Record date of sorting and slide monitoring, if applicable, on Log-In Sheet as documentation of progress and status of completion of sample lot.

QUALITY CONTROL (QC) FOR SORTING

1.   Ten percent of the sorted samples in each lot should be examined by laboratory QC personnel or a qualified co-worker. (A lot is defined as a special study, basin study, entire index period, or individual sorter.) The QC worker will examine the grids chosen and tray used for sorting and will look for organisms missed by the sorter. Organisms found will be added to the sample vials. If the QC worker finds less than 10 organisms (or 10% in larger subsamples) remaining in the grids or sorting tray, the sample passes; if more than 10 (or 10%) are found, the sample fails. If the first 10% of the sample lot fails, a second 10% of the sample lot will be checked by the QC worker. Sorters in-training will have their samples 100% checked until the trainer decides that training is complete.

2.   After laboratory processing is complete for a given sample, all sieves, pans, trays, etc., that have come in contact with the sample will be rinsed thoroughly, examined carefully, and picked free of organisms or debris; organisms found will be added to the sample residue.

7.3.2 Identification of Macroinvertebrates

Taxonomy can be at any level, but should be done consistently among samples. In the original RBPs, two levels of identification were suggested — family (RBP II) and genus/species (RBP III) (Plafkin et al. 1989). Genus/species provides more accurate information on ecological/ environmental relationships and sensitivity to impairment. Family level provides a higher degree of precision among samples and taxonomists, requires less expertise to perform, and accelerates assessment results. In either case, only those taxonomic keys that have been peer-reviewed and are available to other taxonomists should be used. Unnamed species (i.e., species A, B, 1, or 2) may be ecologically informative, but may be inconsistently handled among taxonomists and will, thus, contribute to variability when a statewide database is being developed.

1.   Most organisms are identified to the lowest practical level (generally genus or species) by a qualified taxonomist using a dissecting microscope. Midges (Diptera: Chironomidae) are

---

mounted on slides in an appropriate medium and identified using a compound microscope. Each taxon found in a sample is recorded and enumerated in a laboratory bench notebook and then transcribed to the laboratory bench sheet for subsequent reports. Any difficulties encountered during identification (e.g., missing gills) are noted on these sheets.

2.   Labels with specific taxa names (and the taxonomist's initials) are added to the vials of specimens by the taxonomist. (Note that individual specimens may be extracted from the sample to be included in a reference collection or to be verified by a second taxonomist.) Slides are initialed by the identifying taxonomist. A separate label may be added to slides to include the taxon (taxa) name(s) for use in a voucher or reference collection.

3.   Record the identity and number of organisms on the Laboratory Bench Sheet (Appendix A-3, Form 3). Either a tally counter or "slash" marks on the bench sheet can be used to keep track of the cumulative count. Also, record the life stage of the organisms, the taxonomist's initials and the Taxonomic Certainty Rating (TCR) as a measure of confidence.

4.   Use the back of the bench sheet to explain certain TCR ratings or condition of organisms. Other comments can be included to provide additional insights for data interpretation. If QC was performed, record on the back of the bench sheet.

5.   For archiving samples, specimen vials, (grouped by station and date), are placed in jars with a small amount of denatured 70% ethanol and tightly capped. The ethanol level in these jars must be examined periodically and replenished as needed, before ethanol loss from the specimen vials takes place. A stick-on label is placed on the outside of the jar indicating sample identifier, date, and preservative (denatured 70% ethanol).

QUALITY CONTROL (QC) FOR TAXONOMY

1.   A voucher collection of all samples and subsamples should be maintained. These specimens should be properly labeled, preserved, and stored in the laboratory for future reference. A taxonomist (the reviewer) not responsible for the original identifications should spot check samples corresponding to the identifications on the bench sheet.

2.   The reference collection of each identified taxon should also be maintained and verified by a second taxonomist. The word "val." and the 1st initial and last name of the person validating the identification should be added to the vial label. Specimens sent out for taxonomic validations should be recorded in a "Taxonomy Validation Notebook" showing the label information and the date sent out. Upon return of the specimens, the date received and the finding should also be recorded in the notebook along with the name of the person who performed the validation.

3.   Information on samples completed (through the identification process) will be recorded in the "sample log" notebook to track the progress of each sample within the sample lot. Tracking of each sample will be updated as each step is completed (i.e., subsampling and sorting, mounting of midges and worms, taxonomy).

4.   A library of basic taxonomic literature is essential in aiding identification of specimens and should be maintained (and updated as needed) in the taxonomic laboratory (see attached list). Taxonomists should participate in periodic training on specific taxonomic groups to ensure accurate identifications.

7.4   BENTHIC METRICS

Benthic metrics have undergone evolutionary developments and are documented in the Invertebrate Community Index (ICI) (DeShon 1995), RBPs (Shackleford 1988, Plafkin et al. 1989, Barbour et al. 1992, 1995, 1996b, Hayslip 1993, Smith and Voshell 1997), and the benthic IBI (Kerans and Karr 1994, Fore et al. 1996). Metrics used in these indices evaluate aspects of both elements and processes within the macroinvertebrate assemblage. Although these indices have been regionally developed, they are typically appropriate over wide geographic areas with minor modification (Barbour et al. 1995).

The process for testing the efficacy and calibrating the metrics is described in Chapter 9. While the candidate metrics described here are ecologically sound, they may require testing on a regional basis. Those metrics that are most effective are those that have a response across a range of human influence (Fore et al. 1996, Karr and Chu 1999). Resh and Jackson (1993) tested the ability of 20 benthic metrics used in 30 different assessment protocols to discriminate between impaired and minimally impaired sites in California. The most effective measures, from their study, were the richness measures, 2 community indices (Margalef's and Hilsenhoff's family biotic index), and a functional feeding group metric (percent scrapers). Resh and Jackson emphasized that both the measures (metrics) and protocols need to be calibrated for different regions of the country, and, perhaps, for different impact types (stressors). In a study of 28 invertebrate metrics, Kerans and Karr (1994) demonstrated significant patterns for 18 metrics and used 13 in their final B-IBI (Benthic Index of Biotic Integrity). Richness measures were useful as were selected trophic and dominance metrics. One of the unique features of the fish IBI presently lacking in benthic indices is the ability to incorporate metrics on individual condition, although measures evaluating chironomid larvae deformities have recently been advocated (Lenat 1993).

Four studies that were published from 1995 through 1997 serve as a basis for the most appropriate candidates for metrics, because the metrics were tested in detail in these studies (DeShon 1995, Barbour et al. 1996b, Fore et al. 1996, Smith and Voshell 1997). These metrics have been evaluated for the ability to distinguish impairment and are recommended as the most likely to be useful in other regions of the country (Table 7-1). Other metrics that are currently in use in various states are listed in Table 7-2 and may be applicable for testing as alternatives or additions to the list in Table 7-1.

Taxa richness, or the number of distinct taxa, represents the diversity within a sample. Use of taxa richness as a key metric in a multimetric index include the ICI (DeShon 1995), the fish IBI (Karr et al. 1986), the benthic IBI (Kerans et al. 1992, Kerans and Karr, 1994), and RBP's (Plafkin et al. 1989, Barbour et al. 1996b). Taxa richness usually consists of species level identifications but can also be evaluated as designated groupings of taxa, often as higher taxonomic groups (i.e., genera, families, orders, etc.) in assessment of invertebrate assemblages. Richness measures reflect the diversity of the aquatic assemblage (Resh et al. 1995). The expected response to increasing perturbation is summarized, as an example, in Table 7-2. Increasing diversity correlates with increasing health of the assemblage and suggests that niche space, habitat, and food source are adequate to support survival and propagation of many species. Number of taxa measures the overall variety of the macroinvertebrate assemblage. No identities of major taxonomic groups are derived from the total taxa metric, but the elimination of taxa from a naturally diverse system can be readily detected. Subsets of "total" taxa richness are also used to accentuate key indicator groupings of organisms. Diversity or variety of taxa within these groups are good indications of the ability of the ecosystem to support varied taxa. Certain indices that focus on a pair-wise site comparison are also included in this richness category.

Table 7-1. Definitions of best candidate benthic metrics and predicted direction of metric response to increasing perturbation (compiled from DeShon 1995, Barbour et al. 1996b, Fore et al. 1996, Smith and Voshell 1997).

Category	Metric	Definition	Predicted response to increasing perturbation
Richness measures	Total No. taxa	Measures the overall variety of the macroinvertebrate assemblage	Decrease
	No. EPT taxa	Number of taxa in the insect orders Ephemeroptera (mayflies), Plecoptera (stoneflies), and Trichoptera (caddisflies)	Decrease
	No. Ephemeroptera Taxa	Number of mayfly taxa (usually genus or species level)	Decrease
	No. Plecoptera Taxa	Number of stonefly taxa (usually genus of species level)	Decrease
	No. Trichoptera Taxa	Number of caddisfly taxa (usually genus or species level)	Decrease
Composition measures	% EPT	Percent of the composite of mayfly, stonefly, and caddisfly larvae	Decrease
	% Ephemeroptera	Percent of mayfly nymphs	Decrease
Tolerance/Intolerance measures	No. of Intolerant Taxa	Taxa richness of those organisms considered to be sensitive to perturbation	Decrease
	% Tolerant Organisms	Percent of macrobenthos considered to be tolerant of various types of perturbation	Increase
	% Dominant Taxon	Measures the dominance of the single most abundant taxon. Can be calculated as dominant 2, 3, 4, or 5 taxa.	Increase
Feeding measures	% Filterers	Percent of the macrobenthos that filter FPOM from either the water column or sediment	Variable
	% Grazers and Scrapers	Percent of the macrobenthos that scrape or graze upon periphyton	Decrease
Habit measures	Number of Clinger Taxa	Number of taxa of insects	Decrease
	% Clingers	Percent of insects having fixed retreats or adaptations for attachment to surfaces in flowing water.	Decrease

Composition measures can be characterized by several classes of information, i.e., the identity, key taxa, and relative abundance. Identity is the knowledge of individual taxa and associated ecological patterns and environmental requirements (Barbour et al. 1995). Key taxa (i.e., those that are of special interest or ecologically important) provide information that is important to the condition of the targeted assemblage. The presence of exotic or nuisance species may be an important aspect of biotic interactions that relate to both identity and sensitivity. Measures of composition (or relative abundance) provide information on the make-up of the assemblage and the relative contribution of the

Rapid Bioassessment Protocols for Use in Streams and Wadeable Rivers: Periphyton, Benthic Macroinvertebrates, and Fish, Second Edition

7-15

populations to the total fauna (Table 7-2). Relative, rather than absolute, abundance is used because the relative contribution of individuals to the total fauna (a reflection of interactive principles) is more informative than abundance data on populations without a knowledge of the interaction among taxa (Plafkin et al. 1989, Barbour et al. 1995). The premise is that a healthy and stable assemblage will be relatively consistent in its proportional representation, though individual abundances may vary in magnitude. Percentage of the dominant taxon is a simple measure of redundancy (Plafkin et al. 1989). A high level of redundancy is equated with the dominance of a pollution tolerant organism and a lowered diversity. Several diversity indices, which are measures of information content and incorporate both richness and evenness in their formulas, may function as viable metrics in some cases, but are usually redundant with taxa richness and % dominance (Barbour et al. 1996b).

Table 7-2. Definitions of additional potential benthic metrics and predicted direction of metric response to increasing perturbation.

Category	Metric	Definition	Predicted response to increasing perturbation	References
Richness measures	No. *Pteronarcys* species	The presence or absence of a long-lived stonefly genus (2-3 year life cycle)	Decrease	Fore et al. 1996
	No. Diptera taxa	Number of "true" fly taxa, which includes midges	Decrease	DeShon 1995
	No. Chironomidae taxa	Number of taxa of chironomid (midge) larvae	Decrease	Hayslip 1993, Barbour et al. 1996b
Composition measures	% Plecoptera	Percent of stonefly nymphs	Decrease	Barbour et al. 1994
	% Trichoptera	Percent of caddisfly larvae	Decrease	DeShon 1995
	% Diptera	Percent of all "true" fly larvae	Increase	Barbour et al. 1996b
	% Chironomidae	Percent of midge larvae	Increase	Barbour et al. 1994
	% Tribe Tanytarsini	Percent of Tanytarisinid midges to total fauna	Decrease	DeShon 1995
	% Other Diptera and noninsects	Composite of those organisms generally considered to be tolerant to a wide range of environmental conditions	Increase	DeShon 1995
	% *Corbicula*	Percent of asiatic clam in the benthic assemblage	Increase	Kerans and Karr 1994
	% Oligochaeta	Percent of aquatic worms	Variable	Kerans and Karr 1994
Tolerance/ Intolerance measures	No. Intol. Snail and Mussel species	Number of species of molluscs generally thought to be pollution intolerant	Decrease	Kerans and Karr 1994
	% Sediment Tolerant organisms	Percent of infaunal macrobenthos tolerant of perturbation	Increase	Fore et al. 1996

Table 7-2. Definitions of additional potential benthic metrics and predicted direction of metric response to increasing perturbation (continued).

Category	Metric	Definition	Predicted response to increasing perturbation	References
	Hilsenhoff Biotic Index	Uses tolerance values to weight abundance in an estimate of overall pollution. Originally designed to evaluate organic pollution	Increase	Barbour et al. 1992, Hayslip 1993, Kerans and Karr 1994
Tolerance/ Intolerance measures (continued)	Florida Index	Weighted sum of intolerant taxa, which are classed as 1 (least tolerant) or 2 (intolerant). Florida Index = 2 X Class 1 taxa + Class 2 taxa	Decrease	Barbour et al. 1996b
	% Hydropsychidae to Trichoptera	Relative abundance of pollution tolerant caddisflies (metric could also be regarded as a composition measure)	Increase	Barbour et al. 1992, Hayslip 1993
Feeding measures	% Omnivores and Scavengers	Percent of generalists in feeding strategies	Increase	Kerans and Karr 1994
	% Ind. Gatherers and Filterers	Percent of collector feeders of CPOM and FPOM	Variable	Kerans and Karr 1994
	% Gatherers	Percent of the macrobenthos that "gather"	Variable	Barbour et al. 1996b
	% Predators	Percent of the predator functional feeding group. Can be made restrictive to exclude omnivores	Variable	Kerans and Karr 1994
	% Shredders	Percent of the macrobenthos that "shreds" leaf litter	Decrease	Barbour et al. 1992, Hayslip 1993
Life cycle measures	% Multivoltine	Percent of organisms having short (several per year) life cycle	Increase	Barbour et al. 1994
	% Univoltine	Percent of organisms relatively long-lived (life cycles of 1 or more years)	Decrease	Barbour et al. 1994

Tolerance/Intolerance measures are intended to be representative of relative sensitivity to perturbation and may include numbers of pollution tolerant and intolerant taxa or percent composition (Barbour et al. 1995). Tolerance is generally non-specific to the type of stressor. However, some metrics such as the Hilsenhoff Biotic Index (HBI) (Hilsenhoff 1987, 1988) are oriented toward detection of organic pollution; the Biotic Condition Index (Winget and Mangum 1979) is useful for evaluating sedimentation. The Florida Index (Ross and Jones 1979) is a weighted sum of intolerant taxa (insects and crustaceans) found at a site (Beck 1965) and functions similarly to the HBI (Hilsenhoff 1987) used in other parts of the country. The tolerance/intolerance measures can be independent of taxonomy or can be specifically tailored to taxa that are associated with pollution tolerances. For example, both the percent of Hydropsychidae to total Trichoptera and percent Baetidae to total Ephemeroptera are estimates of evenness within these insect orders that generally are considered to be sensitive to pollution. As these families (i.e., Hydropsychidae and Baetidae) increase in relative abundance, effects of pollution (usually organic) also increase. Density (number of

Rapid Bioassessment Protocols for Use in Streams and Wadeable Rivers: Periphyton, Benthic Macroinvertebrates, and Fish, Second Edition

7-17

individuals per some unit of area) is a universal measure used in all kinds of biological studies. Density can be classified with the trophic measures because it is an element of production; however, it is difficult to interpret because it requires careful quantification and is not monotonic in its response (i.e., density can either decrease or increase in response to pollution) and is usually linked to tolerance measures.

Feeding measures or trophic dynamics encompass functional feeding groups and provide information on the balance of feeding strategies (food acquisition and morphology) in the benthic assemblage. Examples involve the feeding orientation of scrapers, shredders, gatherers, filterers, and predators. Trophic dynamics (food types) are also included here and include the relative abundance of herbivores, carnivores, omnivores, and detritivores. Without relatively stable food dynamics, an imbalance in functional feeding groups will result, reflecting stressed conditions. Trophic metrics are surrogates of complex processes such as trophic interaction, production, and food source availability (Karr et al. 1986, Cummins et al. 1989, Plafkin et al. 1989). Specialized feeders, such as scrapers, piercers, and shredders, are the more sensitive organisms and are thought to be well represented in healthy streams. Generalists, such as collectors and filterers, have a broader range of acceptable food materials than specialists (Cummins and Klug 1979), and thus are more tolerant to pollution that might alter availability of certain food. However, filter feeders are also thought to be sensitive in low-gradient streams (Wallace et al. 1977). The usefulness of functional feeding measures for benthic macroinvertebrates has not been well demonstrated. Difficulties with the proper assignment to functional feeding groups has contributed to the inability to consider these reliable metrics (Karr and Chu 1997).

Habit measures are those that denote the mode of existence among the benthic macroinvertebrates. Morphological adaptation among the macroinvertebrate distinguishes the various mechanisms for maintaining position and moving about in the aquatic environment (Merritt et al. 1996). Habit categories include movement and positioning mechanisms such as skaters, planktonic, divers, swimmers, clingers, sprawlers, climbers, burrowers. Merritt et al. (1996) provide an overview of the habit of aquatic insects, which are the primary organisms used in these measures. Habit measures have been found to be more robust than functional feeding groups in some instances (Fore et al. 1996).

7.5 BIOLOGICAL RECONNAISSANCE (BioRecon) OR PROBLEM IDENTIFICATION SURVEY

The use of biological survey techniques can serve as a screening tool for problem identification and/or prioritizing sites for further assessment, monitoring, or protection. The application of biological surveys in site reconnaissance is intended to be expedient, and, as such, requires an experienced and well-trained biologist. Expediency in

---

FIELD EQUIPMENT/SUPPLIES NEEDED FOR BENTHIC MACROINVERTEBRATE SAMPLING —BIORECON

- standard D-frame dip net, 500 μ opening mesh, 0.3 meter width (~ 1.0 ft frame width)
- sieve bucket, with 500 μ opening mesh
- 95% ethanol
- sample containers
- sample container labels
- forceps
- field data sheets[*], pencils, clipboard
- first aid kit
- waders (chest-high or hip boots), rubber gloves (arm-length)
- camera
- Global Positioning System (GPS) Unit

[*] It is helpful to copy fieldsheets onto water-resistant paper for use in wet weather conditions

this technique is to minimize time spent in the laboratory and with analysis. The "turn-around" time from the biosurvey to an interpretation of findings is intended to be relatively short. The BioRecon is useful in discriminating obviously impaired and non-impaired areas from potentially affected areas requiring further investigation. Use of the BioRecon allows rapid screening of a large number of sites. Areas identified for further study can then either be evaluated using more rigorous bioassessment methods for benthic macroinvertebrates and/or other assemblages, or ambient toxicity methods.

Because the BioRecon involves limited data generation, its effectiveness depends largely on the experience of the professional biologist performing the assessment. The professional biologist should have assessment experience, a knowledge of aquatic ecology, and basic expertise in benthic macroinvertebrate taxonomy.

The BioRecon presented here is refined and standardized from the original RBP I (Plafkin et al. 1989), and is based on the technique developed by Florida DEP (1996), from which the approach derives its name. This biosurvey approach is based on a multihabitat approach similar to the more rigorous technique discussed in Section 7.2. The most productive habitats, i.e., those that contain the greatest diversity and abundance of macroinvertebrates, are sampled in the BioRecon. As a general rule, impairment is judged by richness measures, thereby emphasizing the presence or absence of indicator taxa. Biological attributes such as the relative abundance of certain taxa may be less useful than richness measures in the BioRecon approach, because samples are processed more quickly and in a less standardized manner.

7.5.1 Sampling, Processing, and Analysis Procedures

1. A 100 m reach representative of the characteristics of the stream should be selected. For the BioRecon, it is unlikely that the alternative reach designation approach (i.e., x times the stream width), will improve the resolution beyond a standard 100 m reach. Whenever possible, the area should be at least 100 meters upstream from any road or bridge crossing to minimize its effect on stream velocity, depth and overall habitat quality. There should be no major tributaries discharging to the stream in the study area.

2. Before sampling, complete the "Physical Characterization/Water Quality Field Data Sheet" (Appendix A-1, Form 1) to document site description, weather conditions, and land use. After sampling, review this information for accuracy and completeness.

3. The major habitat types (see 7.2.1 for habitat descriptions) represented in the reach are to be sampled for macroinvertebrates. A total of 4 jabs or kicks will be taken over the length of the reach. A minimum of 1 jab (or kick) is to be taken in each habitat. More than 1 jab may be desired in those habitats that are predominant. Habitat types contributing less than five percent of the stable habitat in the stream reach should not be sampled. Thus, allocate the remaining jabs proportionately among the predominant substrates. The number of jabs taken in each habitat type should be recorded on the field data sheet.

4. Sampling begins at the downstream end of the reach and proceeds upstream. A total of four jabs or kicks will be taken over the length of the reach; a single *jab* consists of forcefully thrusting the net into a productive habitat for a linear distance of 0.5 m. A *kick* is a stationary sampling accomplished by positioning the net and disturbing the substrate for a distance of 0.5 m upstream of the net.

5. The jabs or kicks collected from the multiple habitats will be composited into a sieve bucket to obtain a single homogeneous sample. If clogging occurs, discard the material in the net and redo that portion of the sample in the same habitat type but in a different location. Remove large debris after rinsing and inspecting it for organisms; place any organisms found into the sieve bucket.

6. Return to the bank with the sampled material for sorting and organism identifications. Alternatively, the material can be preserved in alcohol and returned to the laboratory for processing (see Step 7 in Section 7.1.1 for instructions).

7. Transfer the sample from the sieve bucket (or sample jar, if in laboratory) to a white enamel or plastic pan. A second, smaller, white pan may be used for the actual sorting. Place small aliquots of the detritus plus organisms in the smaller pan diluted with a minimal amount of site water (or tap water). Scan the detritus and water for organisms. When an organism is found, examine it with a hard lens, determine its identity to the lowest possible level (usually family or genus), and record it on the Preliminary Assessment Score Sheet (PASS) (Appendix A-3, Form 4) in the column labeled "tally." Place representatives of each taxon in a vial, properly labeled and containing alcohol.

QUALITY CONTROL (QC)

1. Sample labels must be properly completed, including the sample identification code date, stream name, sampling location, and collector's name and placed into the sample container. The outside of the container should be labeled with the same information. Chain-of-custody forms, if needed, must include the same information as the sample container labels.

2. After sampling has been completed at a given site, all nets, pans, etc. that have come in contact with the sample will be rinsed thoroughly, examined carefully, and picked free of organisms or debris. Any additional organisms found should be placed into the sample containers. The equipment should be examined again prior to use at the next sampling site.

3. A second biologist familiar with the recognition and taxonomy of the organisms should check the sample to ensure all taxa are encountered and documented.

---

8. If field identifications are conducted, verify in the lab and make appropriate changes for misidentifications.

9. Analysis is done by determining the value of each metric and comparing to a predetermined value for the associated stream class. These value thresholds should be sufficiently conservative so that "good" conditions or non-impairment is verified. Sites with metric values below the threshold(s) are considered "suspect" of impairment and may warrant further investigation. These simple calculations can be done directly on the PASS sheet.

7.6   TAXONOMIC REFERENCES FOR MACROINVERTEBRATES

The following references are provided as a list of taxonomic references currently being used around the United States for identification of benthic macroinvertebrates. Any of these references cited in the text of this document will also be found in Chapter 11 (Literature Cited).

Allen, R.K. 1978. The nymphs of North and Central American Leptohyphes. *Entomological Society of America* 71(4):537-558.

Allen, R.K. and G.F. Edmunds. 1965. A revision of the genus *Ephemerella* (Ephemeroptera: Ephemerellidae). VIII. The subgenus *Ephemerella* in North America. *Miscellaneous Publications of the Entomological Society of America* 4:243-282.

Allen, R.K. and G.F. Edmunds. 1963. A revision of the genus *Ephemerella* (Ephemeroptera: Ephemerellidae). VI. The subgenus *Seratella* in North America. *Annals of the Entomological Society of America* 56:583-600.

Allen, R.K. and G.F. Edmunds. 1963. A revision of the genus *Ephemerella* (Ephemeroptera: Ephemerellidae). VII. The subgenus *Eurylophella*. *Canadian Entomologist* 95:597-623.

Allen, R.K. and G.F. Edmunds. 1962. A revision of the genus *Ephemerella* (Ephemeroptera: Ephemerellidae). V. The subgenus *Drunella* in North America. *Miscellaneous Publications of the Entomological Society of America* 3:583-600.

Allen, R.K. and G.F. Edmunds. 1961. A revision of the genus *Ephemerella* (Ephemeroptera: Ephemerellidae). III. The subgenus *Attenuatella*. *Journal of the Kansas Entomological Society* 34:161-173.

Allen, R.K. and G.F. Edmunds. 1961. A revision of the genus *Ephemerella* (Ephemeroptera: Ephemerellidae). II. The subgenus *Caudatella*. *Annals of the Entomological Society of America* 54:603-612.

Allen, R.K. and G.F. Edmunds. 1959. A revision of the genus *Ephemerella* (Ephemeroptera: Ephemerellidae). I. The subgenus *Timpanoga*. *The Canadian Entomologist* 91:51-58.

Anderson, N.H. 1976. The distribution and biology of the Oregon Trichoptera. *Oregon Agricultural Experimental Station Technical Bulletin* 134:1-152.

Barr, C.B. and J.B. Chapin. 1988. The Aquatic Dryopoidea of Louisiana (Coleoptera:Psepheniae, Dryopidae, Elmidae). *Tulane Studies in Zoology and Botany* 26:89-164.

Baumann, R.W. 1975. Revision of the Stonefly Family Nemouridae (Plecoptera): A Study of the World Fauna at the Generic Level. *Smithsonian Contributions to Zoology* 211. 74 pp.

Baumann, R.W., A.R. Gaufin, and R.F. Surdick. 1977. The stoneflies (Plecoptera) of the Rocky Mountains. *Memoirs of the American Entomological Society* 31:1-208.

Beck, E.C. 1962. Five new Chironomidae (Diptera) from Florida. *Florida Entomologist* 45:89-92.

Beck, W.M., Jr. and E.C. Beck. 1970. The immature stages of some Chironomini (Chironomidae). *Quarterly Journal of the Academy of Biological Science* 33:29-42.

Beck, E.C. and W.M. Beck, Jr. 1969. Chironomidae (Diptera) of Florida. III. The *Harnischia* complex (Chironomidae). Bulletin of the Florida State Museum of Biological Sciences 13:277-313. Beck, W.M. and E.C. Beck. 1964. New Chironomidae from Florida. *Florida Entomologist.* 47:201-207.

Beck, W.M., Jr. and E.C. Beck. 1966. Chironomidae (Diptera) of Florida. I. Pentaneurini (Tanypodinae). *Bulletin of the Florida State Museum* 10: 305-379.

Bednarik, A.F. and W.P. McCafferty. 1979. Biosystematic revision of the genus *Stenonema* (Ephemeroptera:Heptageniidae). *Canadian Bulletin of Fisheries and Aquatic Sciences* 201:1-73.

Bergman, E.A. and W.L. Hilsenhoff. 1978. *Baetis* (Ephemeroptera:Baetidae) of Wisconsin. *The Great Lakes Entomologist* 11:125-35.

Berner, L. 1977. Distributional patterns of southeastern mayflies (Ephemeroptera). *Bulletin of the Florida State Museum of Biological Sciences* 22:1-55.

Berner, L. 1975. The Mayfly Family Leptophlebiidae in the Southeastern United States. *The Florida Entomologist* 58:137-156.

Berner, L. 1956. The genus *neoephemera* in North America (Ephemeroptera:Neoephemeridae). *Entomological Society of America* 49:33-42.

Berner, L. and M.L. Pescador. 1988. *The Mayflies of Florida*. University Presses of Florida. Pp. 415.

Boesel, M.W. 1985. A brief review of the genus *Polypedilum* in Ohio, with keys to the known stages of species occuring in Northeastern United States (Diptera:Chironomidae). *Ohio Journal of Science* 85:245-262.

Boesel, M.W. 1983. A review of the genus *Cricotopus* in Ohio, with a key to adults of species in the northeastern United States (Diptera:Chironomidae). *Ohio Journal of Science* 83:74-90.

Boesel, M.W. 1974. Observations on the Coelotanypodini of the northeastern states, with keys to the known stages (Diptera: Chironomidae: Tanypodinae). *Journal of the Kansas Entomology Society* 47:417-432.

Boesel, M.W. 1972. The early stages of *Ablabesmyia annulata* (Say) (Diptera:Chironomidae). *Ohio Journal of Science* 72:170-173.

Boesel, M.W. and R.W. Winner. 1980. Corynoneurinae of Northeastern United States, with a key to adults and observations on their occurrence in Ohio (Diptera:Chironomidae). *Journal of the Kansas Entomology Society* 53:501-508.

Brigham, A.R., W.U. Brigham, and A. Gnilka (eds.). 1982. *Aquatic insects and Oligochaetes of North and South Carolina*. Midwest Aquatic Enterprises, Mahomet, IL.

Brinkhurst, R.O. 1986. Guide to the freshwater microdrile Oligochaetes of North America. *Canada Special Publications Fisheries Aquatic Science* 84:1-259.

Brinkhurst, R.O. and B.G.M. Jamieson. 1971. *Aquatic Oligochaeta of the World.* Univ. Toronto Press, 860 pp.

Brittain, J.E. 1982. Biology of Mayflies. *Annual Review of Entomology* 27:119-147.

Brown, H.P. 1987. Biology of riffle beetles. *Annual Review of Entomology* 32:253-273.

Brown, H.P. 1976. *Aquatic dryopoid beetles (Coleoptera) of the United States.* USEPA. Water Pollution Control Research Series 18050 ELD04/72.

Brown, H.P. 1972. *Aquatic dryopoid beetles (Coleoptera) of the United States.* Biota of freshwater ecosystems identification manual no. 6. Water Pollution Control Research Series, EPA, Washington, D.C.

Brown, H.P. and D.S. White. 1978. Notes on Separation and Identification of North American Riffle Beetles (Coleoptera:Dryopoidea:Elmidae). *Entomological News* 89:1-13.

Burch, J.B. 1982. *Freshwater snails (Mollusca: Gastropoda) of North America.* EPA-600/3-82-026. USEPA, Office of Research and Development, Cincinnati, Ohio.

Burch, J.B. 1972. *Freshwater sphaeriacean clams (Mollusca: Pelecypoda) of North America.* EPA Biota of freshwater ecosystems identification manual No. 3. Water Pollution Control Research Series, EPA, Washington, DC.

Caldwell, B.A. 1986. Description of the immature stages and adult female of *Unniella multivirga* Saether (Diptera: Chironomidae) with comments on phylogeny. *Aquatic Insects* 8:217-222.

Caldwell, B.A. 1985. *Paracricotopus millrockensis,* a new species of Orthocladiinae (Diptera: Chironomidae) from the southeastern United States. *Brimleyana* 11:161-168.

Caldwell, B.A. 1984. Two new species and records of other chironomids from Georgia (Diptera: Chironomidae) with some observations on ecology. *Georgia Journal of Science* 42:81-96.

Caldwell, B.A. and A.R. Soponis. 1982. *Hudsonimyia parrishi,* a new species of Tanypodinae (Diptera: Chironomidae) from Georgia. *Florida Entomologist* 65:506-513.

Carle, F.L. 1978. A New Species of *Ameletus* (Ephemeroptera:Siphlonuriae) from Western Virginia. *Entomological Society of America* 71:581-584.

Carle, F.L. and P.A. Lewis. 1978. A new species of *Stenonema* (Ephemeroptera:Heptageniidae) from Eastern North America. *Annals of the Entomological Society of America* 71:285-288.

Clark, W. 1996. *Literature pertaining to the identification and distribution of aquatic macroinvertebrates of the Western U.S. with emphasis on Idaho.* Idaho Department of Health and Welfare, Division of Environmental Quality, Boise, Idaho.

Cranston, P.S. 1982. *A key to the larvae of the British Orthocladiinae (Chironomidae).* Freshwater Biological Association Scientific Publication No. 45:1-152.

Cranston, P.S. and D.D. Judd. 1987. *Metriocnemus* (Diptera: Chironomidae)-an ecological survey and description of a new species. *Journal New York Entomology Society* 95:534-546.

Cummins, K.W. and M.A. Wilzbach. 1985. *Field Procedures for Analysis of Functional Feeding Groups of Stream Macroinvertebrates*. Contribution 1611. Appalachian Environmental Laboratory, University of Maryland, Frostburg, Maryland.

Davis, J.R. 1982. New records of aquatic Oligochaeta from Texas, with observations on their ecological characteristics. *Hydrobiologia* 96:15-29.

Edmunds, G.F. and R.K. Allen. 1964. The Rocky Mountain species of *Epeorus* (Iron) Eaton (Ephemeroptera: Heptageniidae). *Journal of the Kansas Entomological Society*. 37:275-288.

Edmunds, G.F., Jr., S.L. Jensen, and L. Berner. 1976. *The mayflies of North and Central America*. University of Minnesota Press, Minneapolis.

Epler, J.H. 1988. Biosystematics of the genus *Dicrotendipes* Kieffer, 1913 (Diptera: Chironomidae: Chironominae) of the world. *Memoirs of the American Entomology Society* 36:1-214.

Epler, J.H. 1987. Revision of the nearctic *Dicrotendipes* Kieffer, 1913 (Diptera: Chironomidae). *Evolutionary Monographs*: 1-101.

Etnier, D.A. and G.A. Schuster. 1979. An annotated list of Trichoptera (Caddisflies) of Tennessee. *Journal of the Tennessee Academy of Science* 54:15-22.

Faulkner, G.M. and D.C. Tarter. 1977. Mayflies, or Ephemeroptera, of West Virginia with emphasis on the nymphal stage. *Entomological News* 88:202-206.

Ferrington, L.C. 1987. *Collection and identification of floating exuviae of Chironomidae for use in studies of surface water quality*. SOP No. FW 130A. U.S. Environmental Protection Agency, Region VII, Kansas City, Kansas.

Flint, O.S. 1984. *The genus Brachycentrus in North America, with a proposed phylogeny of the genera of Brachycentridae (Trichoptera)*. Smithsonian Contributions to Zoology.

Flint, O.S. Jr. 1964. Notes on some nearctic Psychomyiidae with special reference to their larvae (Trichoptera). *Proceeding of the United States National Museum* 115:467-481.

Flint, O.S. Jr. 1962. The immature stages of *Paleagapetus celsus* Ross (Trichoptera: Hydroptilidae). *Bulletin of the Brooklyn Entomological Society* LVII:40-44.

Flint, O.S. Jr. 1962, Larvae of the caddis fly genus *Rhyacophila* in Eastern North America (Trichoptera: Rhyacophilidae). *Proceedings of the United States National Museum* 113:465-493.

Flint, O.S. Jr. 1960. Taxonomy and biology of nearctic limnephilid larvae (Trichoptera), with special reference to species in Eastern United States. *Entomologicia Americana* XL:1-117.

Flowers, R.W. 1980. Two new genera of nearctic Heptageniidae (Ephemeroptera). *The Florida Entomologist*. 63:296-307.

Flowers, R.W. and W.L. Hilsenhoff. 1975. Heptageniidae (Ephemeroptera) of Wisconsin. *The Great Lakes Entomologist* 8:201-218.

Floyd, M.A. 1995. *Larvae of the caddisfly genus Oecetis (Trichoptera: Leptocerida) in North America.* Bulletin of the Ohio Biological Survey.

Fullington, K.E. and K.W. Stewart. 1980. Nymphs of the stonefly genus *Taeniopteryx* (Plecoptera: Taeniopterygidae) of North America. *Journal of the Kansas Entomological Society* 53(2):237-259.

Givens, D.R. and S.D. Smith. 1980. A synopsis of the western Arctopsychinae (Trichoptera: Hydropsychidae). *Melanderia* 35:1-24.

Grodhaus, G. 1987. Endochironmus Kieffer, Tribelos Townes, Synendotendipes, n. ge., and Endotribelos, n. gen. (Diptera: Chironomidae) of the nearctic region. *Journal of the Kansas Entomological Society* 60:167-247.

Hamilton, A.L. and O.A. Saether. 1969. A classification of the nearctic Chironomidae. *Journal of the Fisheries Research Board of Canada* Technical Report 124:1-42.

Hatch, M.H. 1965. *The beetles of the Pacific Northwest, Part IV, Macrodactyles, Palpicornes, and Heteromera.* University of Washington Publications in Biology, Volume 16.

Hatch, M.H. 1953. *The beetles of the Pacific Northwest, Part I, Introduction and Adephaga.* University of Washington Publications in Biology, Volume 16.

Hilsenhoff, W.L. 1973. Notes on *Dubiraphia* (Coleoptera: Elmidae) with descriptions of five new species. *Annals of the Entomolgical Society of America* 66:55-61.

Hitchcock, S.W. 1974. Guide to the insects of Connecticut: Part VII. The Plecoptera or stoneflies of Connecticut. *State Geological and Natural History Survey of Connecticut Bulletin* 107:191-211.

Hobbs, H.H., Jr. 1981. The crayfishes of Georgia. *Smithsonian Contribution in Zoology* 318:1-549.

Hobbs, H.H., Jr. 1972. *Crayfishes (Astacidae) of North and Middle America.* Biota of freshwater ecosystems identification manual no. 9. Water Pollution Control Research Series, E.P.A., Washington, D.C.

Holsinger, J.R. 1972. *The freshwater amphipod crustaceans (Gammaridae) of North America.* Biota of freshwater ecosystems identification manual no. 5. Water Pollution Control Research Series, E.P.A., Washington, D.C.

Hudson, P.A. 1971. The Chironomidae (Diptera) of South Dakota. *Proceedings of the South Dakota Academy of Sciences* 50:155-174.

Hudson, P.A., D.R. Lenat, B.A. Caldwell, and D. Smith. 1990. Chironomidae of the southeastern United States: a checklist of species and notes on biology, distribution, and habitat. *Fish and Wildlife Research.* 7:1-46.

Rapid Bioassessment Protocols for Use in Streams and Wadeable Rivers: Periphyton, Benthic Macroinvertebrates, and Fish, Second Edition

7-25

Hudson, P.L., J.C.Morse, and J.R. Voshell. 1981. Larva and pupa of *Cernotina spicata. Annals of the Entomological Society of America* 74:516 -519

Jackson, G.A. 1977. Nearctic and palaearctic *Paracladopelma* Harnisch and *Saetheria* n.ge. (Diptera:Chironomidae). *Journal of the Fisheries Research Board of Canada* 34:1321-1359.

Jensen, S.L. 1966. *The mayflies of Idaho.* Unpublished Master's Thesis, University of Utah.

Kenk, R. 1972. *Freshwater planarians (Turbellaria) of North America.* Biota of freshwater ecosystems identification manual no. 1. Water Pollution Control Research Series, U.S. Environmental Protection Agency, Washington, D.C.

Kirchner, R.F. and B.C. Kondratieff. 1985. The nymph of *Hansonoperla appalachia* Nelson (Plecoptera: Perlidae). *Proceedings of the Entomological Society of Washington.* 87(3):593-596.

Kirchner, R.F. and P.P. Harper. 1983. The nymph of *Bolotoperla rossi* (Frison) (Plecoptera: Taeniopterygidae: Brachypterinae). *Journal of the Kansas Entomological Society* 56(3): 411-414.

Kirk, V.M. 1970. A list of beetles of South Carolina, Part 2-Mountain, Piedmont, and Southern Coastal Plain. *South Carolina Agricultural Experiment Station Technical Bulletin* 1038:1-117.

Kirk, V.M. 1969. A list of beetles of South Carolina, Part 1-Northern Coastal Plain. *South Carolina Agricultural Experiment Station Technical Bulletin* 1033:1-124.

Klemm, D.J. 1982. *Leeches (Annelida:Hirudinea) of North America.* EPA-600/3-82-025. Office of Research and Development, Cincinnati, Ohio.

Klemm, D.J. 1972. *Freshwater leeches (Annelida: Hirudinea) of North America.* Biota of freshwater ecosystems identification manual no. 8. Water Pollution Control Research Series, U.S. Environmental Protection Agency, Washington, D.C.

Kondratieff, B.C. 1981. Seasonal distributions of mayflies (Ephemeroptera) in two piedmont rivers in Virginia. *Entomological News* 92:189-195.

Kondratieff, B.C. and R.F. Kirchner. 1984. New species of *Taeniopteryx* (Plecoptera: Taeniopterygidae) from South Carolina. *Annals of the Entomological Society of America* 77(6):733-736.

Kondratieff, B.C. and R.F. Kirchner. 1982. *Taeniopteryx nelsoni,* a new species of winter stonefly from Virginia (Plecoptera: Taeniopterygidae). *Journal of the Kansas Entomological Society* 55(1):1-7.

Kondratieff, B.C. and J.R. Voshell, Jr. 1984. The north and Central American species of *Isonychia* (Epnemeroptera: Oligoneuriidae). *Transactions of the American Entomological Society.* 110:129-244.

Kondratieff, B.C. and J.R. Voshell, Jr. 1983. A checklist of mayflies (Ephemeroptera) of Virginia, with a review of pertinent taxonomic literature. *University of Georgia Entomology Society* 18:213-279.

Kondratieff, B.C, R.F. Kirchner and K.W. Stewart. 1988. A review of *Perlinella* Banks (Plecoptera: Perlidae). *Annals of the Entomological Society of America* 81(1):19-27.

Kondratieff, B.C., J.W.W. Foster, III, and J.R. Voshell, Jr. 1981. Description of the Adult of *Ephemerella berneri* Allen and Edmunds (Ephemeroptera: Ephemerellidae). *Biological Notes.* 83(2):300-303.

Kondratieff, B.C., R.F. Kirchner and J.R. Voshell Jr. 1981. Nymphs of *Diploperla*. *Annals of the Entomological Society of America* 74:428-430.

Lago, P.K. & S.C. Harris. 1987. The *Chimarra* (Trichoptera: Philopotamidae) of eastern North America with descriptions of three new species. *Journal of the New York Entomological Society* 95:225-251.

Larson, D.J. 1989. Revision of North American *Agabus* (Coleoptera: Dytiscidae): introduction, key to species groups, and classification of the *ambiguus-, tristis-,* and *arcticus*-groups. *The Canadian Entomologist* 121:861-919.

Lenat, D.R. and D.L. Penrose. 1987. New distribution records for North Carolina macroinvertebrates. *Entomological News* 98:67-73.

LeSage, L. and A.D. Harrison. 1980. Taxonomy of *Cricotopus* species (diptera: Chironomidae) from Salem Creek, Ontario. *Proceedings of the Entomological Society of Ontario* 111:57-114.

Lewis, P.A. 1974. Three new Stenonema species from Eastern North America (Heptageniidae: Ephemeroptera). *Proceedings of the Entomological Society of Washington* 76:347-355.

Loden, M.S. 1978. A revision of the genus *Psammoryctides* (Oligochaeta: Tubificidae) in North America. *Proceedings of the Biological Society of Washington* 91:74-84.

Loden, M.S. 1977. Two new species of *Limnodrilus* (Oligochaeta: Tubificidae) from the Southeastern United States. *Transactions of the American Microscopal Society* 96:321-326.

Mackay, R.J. 1978. Larval identification and instar association in some species of *Hydropsyche* and *Cheumatopsyche* (Trichoptera: Hydropsychidae). *Annals of the Entomological Society of America* 71:499-509.

Mason, P.G. 1985. The larvae and pupae of *Stictochironomus marmoreus* and *S. quagga* (Diptera: Chironomidae). *Canadian Entomologist* 117:43-48.

Mason, P.G. 1985. The larvae of *Tvetenia vitracies* (Saether) (Diptera: Chironomidae). *Proceedings of the Entomological Society of Washington* 87:418-420.

McAlpine, J.F., B.V. Peterson, G.E. Shewell, H.J. Teskey, J.R. Vockeroth, and D.M. Wood (coords.). 1989. *Manual of nearctic Diptera, Vol. 3.* Research Branch of Agriculture Canada, Monograph 28.

McAlpine, J.F., B.V. Peterson, G.E. Shewell, H.J. Teskey, J.R. Vockeroth, and D.M. Wood (coords.). 1987. *Manual of nearctic Diptera, Vol. 2.* Research Branch of Agriculture Canada, Monograph 28.

McAlpine, J.F., B.V. Peterson, G.E. Shewell, H.J. Teskey, J.R. Vockeroth, and D.M. Wood (coords.). 1981. *Manual of nearctic Diptera, Vol. 1.* Research Branch of Agriculture Canada, Monograph 27.

McCafferty, W.P. 1990. A new species of *Stenonema* (Ephemeroptera: Heptageniidae) from North Carolina. *Proceedings of the Entomological Society of Washington* 92:760-764.

McCafferty, W.P. 1984. The relationship between North and Middle American *Stenonema* (Ephemeroptera: Heptageniidae). *The Great Lakes Entomologist* 17:125-128.

McCafferty, W.P. 1977. Newly associated larvae of three species of *Heptagenia* (Ephemeroptera: Heptageniidae). *Journal of the Georgia Entomology Society.* 12(4):350-358.

McCafferty, W.P. 1977. Biosystematic of *Dannella* and Related Subgenera of *Ephemerella* (Ephemeroptera: Ephemerellidae). *Annals of the Entomological Society of America* 70:881-889.

McCafferty, W.P. 1975. The burrowing mayflies (Epnemeroptera: Ephemeroidea) of the United States. *Transactions of the American Entomological Society* 101:447-504.

McCafferty, W.P. and Y.J. Bae. 1990. *Anthopotamus,* a new genus for North American species previously known as *Potamanthus* (Ephemeroptera: Potamanthidae). *Entomological News* 101(4):200-202.

McCafferty, W.P., M.J. Wigle, and R.D. Waltz. 1994. Contributions to the taxonomy and biology of *Acentrella turbida* (McDunnough) (Ephemeroptera: Baetidae). *Pan-Pacific Insects* 70:301-308.

Merritt, R.W. and K.W. Cummins (editors). 1996. *An introduction to the aquatic insects of North America, 3rd ed.* Kendall/Hunt Publishing Company, Dubuque, Iowa.

Merritt, R.W., D.H. Ross, and B.V. Perterson. 1978. Larval ecology of some lower Michigan blackflies (Diptera: Simuliidae) with keys to the immature stages. *Great Lakes Entomologist* 11:177-208.

Milligan, M.R. 1986. Separation of *Haber speciosus* (Hrabe) (Oligochaeta: Tubificidae) from its congeners, with a description of a new form from North America. *Proceedings of the Biological Society of Washington* 99:406-416.

Moore, J.W. and I.A. Moore. 1978. Descriptions of the larvae of four species of *Procladius* from Great Slave Lake (Chironomidae: Diptera). *Canadian Journal of Zoology* 56:2055-2057.

Morihara, D.K. and .W.P. McCafferty. 1979. The *Baetis* larvae of North America (Ephemeroptera: Baetidae). *Transactions of the American Entomological Society* 105:139-221.

Murray, D.A. and P. Ashe. 1981. A description of the larvae and pupa of *Eurycnemus crassipes* (panzer) (Diptera: Chironomidae) *Entomologica Scandinavica* 12:357-361.

Nelson, H.G. 1981. Notes on Nearctic Helichus (Coleoptera: Drypodidae). *Pan-Pacific Entomologist Vol 57:*226-227.

Oliver, D.R. 1982. *Xylotopus,* a new genus of Orthocladiinae (Diptera: Chironomidae). *Canadian Entomologist* 114:163-164.

Oliver, D.R. 1981. Description of Euryhapsis new genus including three new species (Diptera: Chironomidae). *Canadian Entomologist* 113:711-722.

Oliver, D.R. 1977. Bicinctus-group of the genus *Cricotopus* Van der Wulp (Diptera: Chironomidae) in the nearctic with a description of a new species. *Journal of the Fisheries Research Board of Canada* 34:98-104

Oliver, D.R. 1971. Description of *Einfeldia synchrona* n.sp. (Diptera: Chironomidae) *Canadian Entomologist* 103:1591-1595.

Oliver, D.R. and R.W. Bode. 1985. Description of the larvae and pupa of *Cardiocladius albiplumus* Saether (Diptera: Chironomidae). *Canadian Entomologist* 117:803-809.

Oliver, D.R. and M.E. Roussel. 1982. The larvae of *Pagastia* Oliver (Diptera: Chironomidae) with descriptions of the three nearctic species. *Canadian Entomologist* 114:849-854.

Parker, C.R. and G.B. Wiggins. 1987. *Revision of the caddisfly genus Psilotreta (Trichoptera: Odontoceridae)* Royal Ontario Museum Life Sciences Contributions 144. 55pp.

Pennak, R.W. 1989. *Freshwater invertebrates of the United States, 3rd ed.* J. Wiley & Sons, New York.

Pescador, M.L. 1985. Systematics of the nearctic genus *Pseudiron* (Ephemeroptera: Heptageniidae: Pseudironinae). *The Florida Entomologist* 68:432-444.

Pescador, M. L. and L. Berner. 1980. The mayfly family Baetiscidae (Ephemeroptera). Part II Biosystematics of the Genus *Baetisca. Transactions American Entomological Society* 107:163-228.

Pescador, M.L. and W.L. Peters. 1980. A Revisions of the Genus *Homoeoneuria* (Ephemeroptera: Oligoneuriidae). *Transactions of the American Entomological Society* 106:357-393.

Plotnikoff, R.W. 1994. *Instream biological assessment monitoring protocols: benthic macroinvertebrates.* Washington State Department of Ecology, Environmental Investigations and Laboratory Services, Olympia, Washington, Ecology Publication No. 94-113.

Provonsha, A.V. 1991. A revision of the genus *Caenis* in North America (Ephemeroptera: Caenidae). *Transactions of the American Entomological Society* 116:801-884.

Provonsha, A.V. 1990. A revision of the genus *Caenis* in North America (Ephemeroptera: Caenidae). *Transactions of the American Entomological Society* 116(4):801-884.

Resh, V.H. 1976. The biology and immature stages of the caddisfly genus *Ceraclea* in eastern North America (Trichoptera: Leptoceridae). *Annals of the Entomological Society of America* 69:1039-1061.

Ricker, W.E. and H.H. Ross. 1968. North American species of *Taeniopteryx* (Plecoptera, Insecta). *Journal Fisheries Research Board of Canada.* 25(7):1423-1439.

Roback, S.S. 1987. The immature chironomids of the eastern United States IX. Pentaneurini - Genus *Labrundinia* with the description of a new species from Kansas. *Proceedings of the Academy of Natural Sciences in Philadelphia* 138:443-465.

Roback, S.S. 1986. The immature chironomids of the eastern United States VII. Pentaneurini - Genus *Nilotanypus,* with description of some neotropical material. *Proceedings of the Academy of Natural Sciences in Philadelphia* 139:159-209.

Roback, S.S. 1986. The immature chironomids of the Eastern United States VII. Pentaneurini - Genus *Monopelopia,* with redescriptions of the male adults and description of some neotropical material. *Proceedings of the Academy of Natural Sciences in Philadelphia* 138:350-365.

Roback, S.S. 1985. The immature chironomids of the eastern United States VI. Pentaneurini - Genus *Ablabesmyia. Proceedings of the Academy of Natural Sciences in Philadelphia* 137:153-212.

Roback, S.S. 1983. *Krenopelopia hudsoni:* a new species from the Eastern United States (Diptera: Chironomidae: Tanypodinae). *Proceedings of the Academy of Natural Sciences in Philadelphia* 135:254-260.

Roback, S.S. 1981. The immature chironomids of the Eastern United States V. Pentaneurini-Thienemannimyia group. *Proceedings of the Academy of Natural Sciences in Philadelphia* 133:73-128.

Roback, S.S. 1980. The immature chironomids of the Eastern United States IV. Tanypodinae-Procladiini. *Proceedings of the Academy of Natural Sciences in Philadelphia* 132:1-63.

Roback, S.S. 1978. The immature chironomids of the eastern United States III. Tanypodinae-Anatopyniini, Macropelopiini and Natarsiini. *Proceedings of the Academy of Natural Sciences in Philadelphia* 129:151-202.

Roback, S.S. 1977. The immature chironomids of the eastern United States II. Tanypodinae - Tanypodini. *Proceedings of the Academy of Natural Sciences in Philadelphia* 128:55-87.

Roback, S.S. 1976. The immature chironomids of the eastern United States I. Introduction and Tanypodinae-Coelotanypodini. *Proceedings of the Academy of Natural Sciences in Philadelphia* 127:147-201.

Roback, S.S. 1975. New Rhyacophilidae records with some water quality data. *Proceedings of the Academy of Natural Sciences of Philadelphia* 127:45-50.

Roback, S.S. and W.P. Coffman. 1983. Results of the Catherwood Bolivian-Peruvian Altiplano expedition Part II. Aquatic Diptera including montane Diamesinae and Orthocladiinae (Chironomidae) from Venezuela. *Proceedings of the Academy of Natural Sciences in Philadelphia 135:9-79.*

Roback, S.S. and L.C. Ferrington, Jr. 1983. The immature stages of *Thienemannimyia barberi* (Coquillett) (Diptera: Chironomidae: Tanypodinae). *Freshwater Invertebrate Biology* 5:107-111.

Ruiter, D.E. 1995. The adult *Limnephilus* Leach (Trichoptera: Limnephiliae) of the New World. *Bulletin of the Ohio Biological Survey,* New Series 11 no. 1.

Saether, O.A. 1983. The larvae of *Prodiamesinae* (Diptera: Chironomidae) of the holarctic region - keys and diagnoses. *Entomologica Scandinavica Supplement* 19:141-147.

Saether, O.A. 1982. Orthocladiinae (Diptera: Chironomidae) from the SE U.S.A., with descriptions of *Plhudsonia, Unniella* and *Platysmittia* n. genera and *Atelopodella* n. subgen. *Entomologica Scandinanvia Supplement* 13:465-510.

Saether, O.A. 1980. Glossary of chironomid morphology terminology (Diptera: Chironomidae) *Entomologica Scandinanvica Supplement* 14:1-51.

Saether, O.A. 1977. Taxonomic studies on Chironomidae: *Nanocladius, Pseudochironomus,* and the *Harnischia* complex. *Bulletin of the Fisheries Research Board of Canada* 196:1-143.

Saether, O.A. 1976. Revision of *Hydrobaenus, Trissocladius, Zalutschia, Paratrissocladius,* and some related genera (Diptera: Chironomidae). *Bulletin of the Fisheries Research Board of Canada* 195:1-287.

Saether, O.A. 1975. Nearctic and Palaearctic *Heterotrissocladius* (Diptera:Chironomidae). *Bulletin of the Fisheries Research Board of Canada* 193:1-67.

Saether, O.A. 1975. Twelve new species of *Limnophyes* Eaton, with keys to nearctic males of the genus (Diptera: Chironomidae). *Canadian Entomologist* 107:1029-1056.

Saether, O.A. 1975. Two new species of *Protanypus* Kieffer, with keys to nearctic and palaearctic species of the genus (Diptera: Chironomidae). *Journal of the Fisheries Research Board of Canada* 32:367-388.

Saether, O.A. 1973. Four species of *Bryophaenocladius* Thien., with notes on other Orthocladiinae (Diptera: Chironomidae). *Canadian Entomologist* 105:51-60.

Saether, O.A. 1971. Four new and unusual Chironomidae (Diptera). *Canadian Entomologist* 103:1799-1827.

Saether, O.A. 1971. Nomenclature and phylogeny of the genus *Harnischia* (Diptera: Chironomidae). *Canadian Entomologist* 103:347-362.

Saether, O.A. 1971. Notes on general morphology and terminology of the Chironomidae (Diptera). *Canadian Entomologist* 103:1237-1260.

Saether, O.A. 1969. Some nearctic Podonominae, Diamesinae, and Orthocladiinae (Diptera: Chironomidae) *Bulletin of the Fisheries Research Board of Canada* 170:1-154.

Sawyer, R.T. and R.M. Shelley. 1976. New records and species of leeches (Annelida: Hirudinea) from North and South Carolina. *Journal of Natural History* 10:65-97.

Schefter, P.W. and G.B. Wiggins. 1986. *A systematic study of a the nearctic larvae of the Hydropsyche morosa group (Trichoptera: Hydropsychidae).* Miscellaneous Publications of the Royal Ontario Museum, Toronto, Canada.

Schmid, F. 1970. Le genre *Rhyacophila* et le famille des Rhyacophilidae (Trichoptera). *Memoirs of the Entomological Society of Canada* 66:1-230.

Schuster, G.A. and D.A. Etnier. 1978. *A manual for the identification of the larvae of the caddisfly genera Hydropsyche Pictet and Symphitopsyche Ulmer in eastern and central North America* (Trichoptera: Hydropsychidae). EPA-600/4-78-060.

Sherberger, F.F. and J.B. Wallace. 1971. Larvae of the southeastern species of *Molanna. Journal of the Kansas Entomological Society* 44:217-224.

Simpson, K.W. 1982. A guide to the basic taxonomic literature for the genera of North American Chironomidae (Diptera) - Adults, pupae, and larvae. *New York State Museum Bulletin* No.447: 1-43.

Simpson, K. W. and R.W. Bode. 1980. Common larvae of Chironomidae (Diptera) from New York State streams and rivers. *New York State Museum Bulletin* 439:1-105.

Smith, S.D., unpublished 1995. *Revision of the genus Rhyacophilia* (Trichoptera: Rhyacophilidae). Central Washington University, Ellensburg, Washington.

Smith, S.D. 1985. Studies of Nearctic *Rhyacophila* (Trichoptera: Rhyacophilidae): Synopsis of *Rhyacophila Nevadensis* Group. *Pan-Pacific Entomologist* 61:210-217.

Smith, S.D. 1968. The *Rhyacophila* of the Salmon River drainage of Idaho with special reference to larvae. *Annals of the Entomological Society of America* 61:655-674.

Soponis, A.R. and C.L. Russell. 1982. Identification of instars and species in some larval *Polypedilum* (Diptera: Chironomidae). *Hydrobiologia* 94:25-32.

Stark, B.P. 1986. The nearctic species of *Agnetina* (Plecoptera: Perlidae). *Journal of the Kansas Entomological Society.* 59(3):437-445.

Stark, B.P. 1983. A review of the genus *Soliperla* (Plecoptera: Peltoperlidae). *Great Basin Naturalist* 43:30-44.

Stark, B.P. and C.H. Nelson. 1994. Systematics, phylogeny, and zoogeography of the genus *Yoraperla* (Plecoptera: Peltoperliae). *Entomologica Scandinavica* 25:241-273.

Stark, B.P. and D.H. Ray. 1983. A Revision of the Genus *Helopicus* (Plecoptera: Perlodidae). *Freshwater Invertebrate Biology* 2(1):16-27.

Stark, B.P. and K.W. Stewart. 1982. *Oconoperla,* a new genus of North American Perlodinae (Plecoptera: Perlodidae). *Proceedings of the Entomological Society of Washington.* 84(4):747-752.

Stark, B.P. and K.W. Stewart. 1981. The nearctic genera of Peltoperlidae (Plecoptera). *Journal of the Kansas Entomological Society* 54:285-311.

Stark, B.P. and S.W. Szczytko. 1981. Contributions to the Systematics of *Paragnetina* (Plecoptera: Perlidae). *Journal of the Kansas Entomological Society* 54(3):625-648.

Stewart, K.W. and B.P. Stark. 1988. Nymphs of North American stonefly genera (Plecoptera). Thomas Say Foundation Series, *Entomological Society of America* 12:1-460.

Stewart, K.W. and B.P. Stark. 1984. Nymphs of North American Perlodinae genera (Plecoptera: erlodidae). *The Great Basin Naturalist* 44(3):373-415.

Stimpson, K.S. , D.J. Klemm and J.K. Hiltunen. 1982. *A guide to the freshwater Tubificidae (Annelida: Clitellata: Oligochaeta) of North America.* EPA-600/3-82-033, 61 pp.

Sublette, J.E. 1964. Chironomidae (Diptera) of Louisiana I. Systematics and immature stages of some lentic chironomids of West-central Louisiana. *Tulane Studies in Zoology* 11:109-150.

Szczytko, S.W. and K.W. Stewart. 1979. The genus *Isoperla* of western North America; holomorphology and systematics, and a new stonefly genus Cascadoperla. *Memoirs of the American Entomological Society* 32:1-120.

Thompson, F. G. 1983. *An identification manual of the freshwater snails of Florida.* Florida State Museum, Gainesville, Florida.

Thorp, J.H. and A.P. Covich (editors). 1991. *Ecology and Classification of North American Freshwater Invertebrates.* Academic Press, New York, New York.

Torre-Bueno, J.R. de la. 1989. *The Torre-Bueno Glossary of Entomology, Revised Edition.* The New York Entomological Society, New York.

Traver, J.R. 1937. Notes on mayflies of the Southeastern states (Ephemeroptera). *Journal of the Elisha Mitchell Scientific Society* 53:27-86.

Traver, J.R. 1933. Mayflies of North Carolina Part III. The Heptageniinae. *Journal of the Elisha Mitchell Scientific Society* 48:141-206.

Usinger, R.L. (editor). 1956. *Aquatic insects of California.* University of California Press, Berkeley, California.

Vineyard, R.N. and G.B. Wiggins. 1987. Seven new species from North America in the caddisfly genus *Neophylax* (Trichoptera: Limnephilidae). *Annals of the Entomological Society* 80:62-73.

Waltz, R.D. & W.P. McCafferty. 1987. Systematics of *Pseudocloeon, Acentrella, Baetiella,* and *Liebebiella,* new genus (Ephemeroptera: Baetidae). *Journal of New York Entomology Society.* 95(4):553-568.

Waltz, R.D., W.P. McCafferty, and J.H. Kennedy. 1985. *Barbaetis:* a new genus of eastern nearctic Mayflies (Ephemeroptera: Baetidae). *The Great Lakes Entomologist:*161-165.

Weaver, J.S., III. 1988. *A synopsis of the North American Lepidostomatidae (Trichoptera).* Contributions to the American Entomological Institute 24.

Weaver, J.S. III, and T.R. White. 1981. Larval description of *Rhyacophila appalachia* Morse and Ross (Trichoptera: Rhyacophilidae). *Journal of the Georgia Entomological Society* 16:269-271.

Wetzel, M.J. 1987. *Limnodrilus tortilipenis,* a new North American species of freshwater Tubificidae (Annelida:Clitellata:Oligochaeta). *Proceedings of the Biological Society of Washington* 100:182-185.

White, D S. 1978. A revision of the nearctic *Optioservus* (Coleoptera: Elmidae), with descriptions of new species. *Systematic Entomology* 3:59-74.

Wiederholm, T. (editor). 1986. Chironomidae of the Holartic region. Keys and diagnoses. Part 2. Pupae. *Entomologica Scandinavica Supplement* 28: 1-482.

Wiederholm, T. (editor). 1983. Chironomidae of the holarctic region. Keys and diagnoses, Part 1, Larvae. *Entomologica Scandinavica Supplement no. 19,* 1-457.

Wiggins, G.B. 1995. *Larvae of the North American caddisfly genera (Trichoptera), 2nd ed.* University of Toronto Press, Toronto, Canada.

Wiggins, G.B. 1977. *Larvae of the North American caddisfly genera (Trichoptera).* University of Toronto Press, Toronto, Canada.

Wiggins, G.B. 1965. Additions and revisions to the genera of North American caddisflies of the family Brachycentridae with special reference to the larval stages (Trichoptera). *Canadian Entomologist* 97:1089-1106.

Wiggins, G.B. and J.S. Richardson. 1989. Biosystematics of *Eocosmoecus,* a new Nearctic caddisfly genus (Trichoptera: Limnephilidae: dicosmoecinae). *Journal of the North American Benthological Society* 8:355-369.

Wiggins, G.B. and J.S. Richardson. 1982. Revision and synopsis of the caddisfly genus *Dicosmoecus* (Trichoptera: Limnephilidae: Dicosmoecinae). *Aquatic Insects* 4:181- 217.

Wold, J.L. 1974. *Systematics of the genus Rhyacophila (Trichoptera: Rhyacophilidae).* Unpublished Master's Thesis, Oregon State University, Corvallis, Oregon.

Wolf, W.G. and J.F. Matta. 1981. Notes on nomenclature and classification of *Hydroporus* subgenera with the description of a new genus of Hydroporinia (Coleoptera: Dytiscidae). *Pan-Pacific Entomologist* 57:149-175.

Yamamoto, T. and G.B. Wiggins. 1964. A comparative study of the North American species in the caddisfly genus *Mystacides* (Trichoptera: Leptoceridae). *Canadian Journal of Zoology* 42:1105-1210.

Young, F. N. 1954. *The water beetles of Florida.* University of Florida Press, Gainesville, Florida.

8 FISH PROTOCOLS

Monitoring of the fish assemblage is an integral component of many water quality management programs, and its importance is reflected in the aquatic life use-support designations of many states. Narrative expressions such as "maintaining coldwater fisheries", "fishable" or "fish propagation" are prevalent in state standards. Assessments of the fish assemblage must measure the overall structure and function of the ichthyofaunal community to adequately evaluate biological integrity and protect surface water resource quality. Fish bioassessment data quality and comparability are assured through the utilization of qualified fisheries professionals and consistent methods.

The Rapid Bioassessment Protocol (RBP) for fish presented in this document, is directly comparable to RBP V in Plafkin et al. (1989). The principal evaluation mechanism utilizes the technical framework of the Index of Biotic Integrity (IBI) — a fish assemblage assessment approach developed by Karr (1981). The IBI incorporates the zoogeographic, ecosystem, community and population aspects of the fish assemblage into a single ecologically-based index. Calculation and interpretation of the IBI involves a sequence of activities including: fish sample collection; data tabulation; and regional modification and calibration of metrics and expectation values. This concept has provided the overall multimetric index framework for rapid bioassessment in this document. A more detailed description of this approach for fish is presented in Karr et al. (1986) and Ohio EPA (1987). Regional modification and applications are described in Leonard and Orth (1986), Moyle et al. (1986), Hughes and Gammon (1987), Wade and Stalcup (1987), Miller et al. (1988), Steedman (1988), Simon (1991), Lyons (1992a), Simon and Lyons (1995), Lyons et al. (1996), and Simon (1999).

The RBP for fish involves careful, standardized field collection, species identification and enumeration, and analyses using aggregated biological attributes or quantification of the numbers (and in some cases biomass, see Section 8.3.3, Metric 13) of key species. The role of experienced fisheries scientists in the adaptation and application of the RBP and the taxonomic identification of fishes cannot be overemphasized. The fish RBP survey yields an objective discrete measure of the condition of the fish assemblage. Although the fish survey can usually be completed in the field by qualified fish biologists, difficult species identifications will require laboratory confirmation. Data provided by the fish RBP can serve to assess use attainment, develop biological criteria, prioritize sites for further evaluation, provide a reproducible impact assessment, and evaluate status and trends of the fish assemblage.

Fish collection procedures must focus on a multihabitat approach — sampling habitats in relative proportion to their local representation (as determined during site reconnaissance). Each sample reach should contain riffle, run and pool habitat, when available. Whenever possible, the reach should be sampled sufficiently upstream of any bridge or road crossing to minimize the hydrological effects on overall habitat quality. Wadeability and accessability may ultimately govern the exact placement of the sample reach. A habitat assessment is performed and physical/chemical parameters measured concurrently with fish sampling to document and characterize available habitat specifics within the sample reach (see Chapter 5: Habitat Assessment and Physicochemical Characterization).

8.1   FISH COLLECTION PROCEDURES: ELECTROFISHING

All fish sampling gear types are generally considered selective to some degree; however, electrofishing has proven to be the most comprehensive and effective *single* method for collecting stream fishes. Pulsed DC (direct current) electrofishing is the method of choice to obtain a representative sample of the fish assemblage at each sampling station. However, electrofishing in any form has been banned from certain salmonid spawning streams in the northwest. As with any fish sampling method, the proper scientific collection permit(s) must be obtained before commencement of any electrofishing activities. The accurate identification of each fish collected is essential, and species-level identification is required (including hybrids in some cases, see Section 8.3.3, Metric 11). Field identifications are acceptable; however, voucher specimens must be retained for laboratory verification, particularly if there is any doubt about the correct identity of the specimen (see Section 8.2). Because the collection methods used are not consistently effective for young-of-the-year fish and because their inclusion may seasonally skew bioassessment results, fish less than 20 millimeters total length will not be identified or included in standard samples.

ELECTROFISHING CONFIGURATION AND FIELD TEAM ORGANIZATION

All field team members must be trained in electrofishing safety precautions and unit operation procedures identified by the electrofishing unit manufacturer. Each team member must be insulated from the water and the electrodes; therefore, chest waders and rubber gloves are required. Electrode and dip net handles must be constructed of insulating materials (e.g., woods, fiberglass). Electrofishers/electrodes must be equipped with functional safety switches (as installed by virtually all electrofisher manufacturers). Field team members must not reach into the water unless the electrodes have been removed from the water or the electrofisher has been disengaged.

It is recommended that at least 2 fish collection team members be certified in CPR (cardiopulmonary resuscitation). *Many* options exist for electrofisher configuration and field team organization; however, procedures will always involve pulsed DC electrofishing and a minimum 2-person team for sampling streams and wadeable rivers. Examples include:

- Backpack electrofisher with 2 hand-held electrodes mounted on fiberglass poles, one positive (anode) and one negative (cathode). One crew member, identified as the electrofisher unit operator, carries the backpack unit and manipulates both the anode and cathode poles. The anode may be fitted with a net ring (and shallow net) to allow the unit operator to net specimens. The remaining 1 or 2 team members net fish with dip nets and are responsible for specimen transport and care in buckets or livewells.

- Backpack electrofisher with 1 hand-held anode pole and a trailing or floating cathode. The electrofisher unit operator manipulates the anode with one hand, and has a second hand free for use of a dip net. The remaining 1 or 2 team members also aid in the netting of specimens, and in addition are responsible for specimen transport in buckets or livewells.

- Tote barge (pramunit) electrofisher with 2 hand-held anode poles and a trailing/floating cathode (recommended for large streams and wadeable rivers). Two team members are each equipped with an anode pole and a dip net. Each is responsible for electrofishing and the netting of specimens. The remaining team member will follow, pushing or pulling the barge through the sample reach. A livewell is maintained within the barge and/or within the sampling reach but outside the area of electric current.

The safety of all personnel and the quality of the data is assured through the adequate education, training, and experience of all members of the fish collection team. At least 1 biologist with training and experience in electrofishing techniques and fish taxonomy *must* be involved in each sampling event. Laboratory analyses are conducted and/or supervised by a fisheries professional trained in fish taxonomy. Quality assurance and quality control must be a continuous process in fisheries monitoring and assessment, and must include all program aspects (i.e., field sampling, habitat measurement, laboratory processing, and data recording).

Tote barge (pram unit) Electrofishing

Backpack Electrofishing

8.1.1 Field Sampling Procedures

1. A representative stream reach (see Alternatives for Stream Reach Designation, next page) is selected and measured such that primary physical habitat characteristics of the stream are included within the reach (e.g., riffle, run and pool habitats, when available). The sample reach should be located away from the influences of major tributaries and

FIELD EQUIPMENT/SUPPLIES NEEDED FOR FISH SAMPLING—ELECTROFISHING

- appropriate scientific collection permit(s)
- backpack or tote barge-mounted electrofisher
- dip nets
- block nets (i.e., seines)
- elbow-length insulated waterproof gloves
- chest waders (equipped with wading cleats, when necessary)
- polarized sunglasses
- buckets/livewells
- jars for voucher/reference specimens
- waterproof jar labels
- 10% buffered formalin (formaldehyde solution)
- measuring board (500 mm minimum, with 1 mm increments)[a]
- balance (gram scale)[b]
- tape measure (100 m minimum)
- fish Sampling Field Data Sheet[c]
- applicable topographic maps
- copies of field protocols
- pencils, clipboard
- first aid kit
- Global Positioning System (GPS) Unit

[a] Needed only if program/study requires length frequency information
[b] Needed only if total biomass and/or the Index of Well-Being are included in the assessment process (see Section 8.3.3, Metric 13).
[c] It is helpful to copy fieldsheets onto water-resistant paper for use in wet weather conditions.

Rapid Bioassessment Protocols for Use in Streams and Wadeable Rivers: Periphyton, Benthic Macroinvertebrates, and Fish, Second Edition

8-3

bridge/road crossings (e.g., sufficiently upstream to decrease influences on overall habitat quality). The exact location (i.e., latitude and longitude) of the downstream limit of the reach must be recorded on each field data sheet. (If a Global Positioning System unit is used to provide location information, the accuracy or design confidence of the unit should be noted.) A habitat assessment and physical/ chemical characterization of water quality should be performed within the same sampling reach (see Chapter 5: Habitat Assessment and Physicochemical Characterization).

2. Collection via electrofishing begins at a shallow riffle, or other physical barrier at the downstream limit of the sample reach, and terminates at a similar barrier at the upstream end of the reach. In the absence of physical barriers, block nets should be set at the upstream and downstream ends of the reach prior to the initiation of any sampling activities.

ALTERNATIVES FOR STREAM REACH DESIGNATION

The collection of a representative sample of the fish assemblage is essential, and the appropriate sampling station length for obtaining that sample is best determined by conducting pilot studies (Lyons 1992b, Simonson et al. 1994, Simonson and Lyons 1995). Alternatives for the designation of stream sampling reaches include:

- **Fixed-distance designation**—A standard length of stream, e.g., a 150-200-meter reach (Ohio EPA 1987), 100-meter reach (Massachusetts DEP 1995) may be used to obtain a representative sample. Conceptually, this approach should provide a mixture of habitats in the reach and provide, at a minimum, duplicate physical and structural elements such as riffle/pool sequences.

- **Proportional-distance designation**— A standard number of stream channel "widths" may be used to measure the stream study reach, e.g., 40 times the stream width is defined by Environmental Monitoring & Assessment Program (EMAP) for sampling (Klemm and Lazorchak 1995). This approach allows variation in the length of the reach based on the size of the stream. Application of the proportional-distance approach in large streams or wadeable rivers may require the establishment of sampling program time and/or distance maxima (e.g., no more than 3 hours of electrofishing or 500-meter reach per sampling site, [Klemm et al. 1993]).

---

3. Fish collection procedures commence at the downstream barrier. A minimum 2-person fisheries crew proceeds to electrofish in an upstream direction using a side-to-side or bank-to-bank sweeping technique to maximize area coverage. All wadeable habitats within the reach are sampled via a single pass, which terminates at the upstream barrier. Fish are held in livewells (or buckets) for subsequent identification and enumeration.

4. Sampling efficiency is dependent, at least in part, on water clarity and the field team's ability to see and net the stunned fish. Therefore, each team member should wear polarized sunglasses, and sampling is conducted only during periods of optimal water clarity and flow.

5. All fish (greater than 20 millimeters total length) collected within the sample reach must be identified to species (or subspecies). Specimens that cannot be identified with certainty in the field are preserved in a 10% formalin solution and stored in labeled jars for subsequent laboratory identification (see Section 8.2). A representative voucher collection must be retained for unidentified specimens, very small specimens, new locality records, and/or a particular region. In addition to the unidentified specimen jar, a voucher collection of a

---

subsample of each species identified in the field should be preserved and labeled for subsequent laboratory verification, if necessary. Obviously, species of special concern (e.g., threatened, endangered) should be noted and released *immediately* on site. Labels should contain (at a minimum) location data (verbal description and coordinates), date, collectors' names, and sample identification code and/or station numbers for the particular sampling site. Young-of-the-year fish less than 20 millimeters (total length) are not identified or included in the sample, and are released on site. Specimens that can be identified in the field are counted, examined for external anomalies (i.e., deformities, eroded fins, lesions, and tumors), and recorded on field data sheets. An example of a "Fish Sampling Field Data Sheet" is provided in Appendix A-4, Form 1. Space is available for optional fish length and weight measurements, should a particular program/study require length frequency or biomass data. However, these data *are not required* for the standard multimetric assessment. Space is allotted on the field data sheets for the *optional* inclusion of measurements (nearest millimeter total length) and weights (nearest gram) for a subsample (to a maximum 25 specimens) of each species. Although fish length and weight measurements are optional, recording a range of lengths for species encountered may be a useful routine measure. Following the data recording phase of the procedure, specimens that have been identified and processed in the field are released on site to minimize mortality.

6. The data collection phase includes the completion of the top portion of the "Fish Sampling Field Data Sheet" (Appendix A-4, Form 1),

---

QUALITY CONTROL (QC) IN THE FIELD

1. Quality control must be a continuous process in fish bioassessment and should include all program aspects, from field collection and preservation to habitat assessment, sample processing, and data recording. Field validation should be conduced at selected sites and will involve the collection of a duplicate sample taken from an adjacent reach upstream of the initial sampling site. The adjacent reach should be similar to the initial site with respect to habitat and stressors. Sampling QC data should be evaluated following the first year of sampling in order to determine a level of acceptable variability and the appropriate duplication frequency.

2. Field identifications of fish *must* be conducted by qualified/trained fish taxonomists, familiar with local and regional ichthyofauna. Questionable records are prevented by: (a) requiring the presence of at least one experienced/trained fish taxonomist on every field effort, and (b) preserving selected specimens (e.g., Klemm and Lazorchak 1995 recommend a subsample of a maximum 25 voucher specimens of each species) and those that cannot by readily identified in the field for laboratory verification and/or examination by a second qualified fish taxonomist (see Section 8.2). Specimens must be properly preserved and labeled (refer to Section 8.1.1, number 5). When needed, chain-of-custody forms must be initiated following sample preservation, and must include the same information as the sample container labels.

3. All field equipment must be in good operating condition, and a plan for routine inspection, maintenance, and/or calibration must be developed to ensure consistency and quality of field data. Field data must be complete and legible, and should be entered on standardized field data forms and/or digital recorders. While in the field, the field team should possess sufficient copies of standardized field data forms and chains-of-custody for all anticipated sampling sites, as well as copies of all applicable Standard Operating Procedures (SOPs).

Rapid Bioassessment Protocols for Use in Streams and Wadeable Rivers: Periphyton, Benthic Macroinvertebrates, and Fish, Second Edition

8-5

which duplicates selected information from the physical/chemical field sheet. Information regarding the sample collection procedures must also be recorded. This includes method of fish capture, start time, ending time, duration of sampling, maximum and mean stream widths. The percentage of each habitat type in the reach is estimated and documented on the data sheet. Comments should include sampling conditions, e.g., visibility, flow, difficult access to stream, or anything that may prove to be valuable information to consider for future sampling events or by personnel unfamiliar with the site.

8.2   LABORATORY IDENTIFICATION AND VERIFICATION

Fish records of questionable quality are prevented by preserving specimens (that cannot be readily identified in the field) for laboratory examination and/or a voucher collection for laboratory verification. Specimens must be properly preserved (e.g., 10% formalin for tissue fixing and 70% ethanol for long-term storage) and labeled (using museum-grade archival labels/paper, and formalin/alcohol-proof pen or pencil). Labels should contain (at a minimum) site location data (i.e., verbal description and site coordinates), collection date, collector's names, species identification (for fishes identified in the field), species totals, and sample identification code and/or station number. All samples received in the laboratory should be tracked using a sample log-in procedure (Appendix A-4, Form 2). Laboratory fisheries professionals *must* be capable of identifying fish to the lowest possible taxonomic level (i.e., species or subspecies) and should have access to suitable regional taxonomic references (see Section 8.4) to aid in the identification process. Laboratories that do not typically identify fish, or trained fisheries professionals that have difficulty identifying a particular specimen or group of fish, should contact a taxonomic specialist (i.e., a recognized authority for that particular taxonomic group). Taxonomic nomenclature *must* be kept consistent and current. Common and scientific names of fishes from the United States and Canada are listed in Robins et al. (1991).

8.3   DESCRIPTION OF FISH METRICS

QUALITY CONTROL (QC) FOR TAXONOMY

1. A representative voucher collection must be retained for unidentified specimens, small specimens, and new locality records. In addition, a second voucher jar should be retained for a subsample of each species identified in the field (e.g., Klemm and Lazorchak 1995 recommend a subsample of 25 voucher specimens of each species). The vouchers must be properly preserved, labeled, and stored in the laboratory for future reference (see Section 8.2).

2. Voucher collections should be verified by a second qualified fish taxonomist, i.e., a professional other than the taxonomist responsible for the original field identifications. The word "validated" and the name of the taxonomist that validated the identification should be added to each voucher label. Specimens sent from the laboratory to taxonomic specialists should be recorded in a "Taxonomy Validation Notebook" (see Chapter 7), noting the label information and date sent. Upon return of the specimens, the date received and findings should also be recorded in the notebook (and the voucher label), along with the name of the person who performed the validation.

3. Information on samples completed (through the identification/validation process) will be tracked in a "Sample Log" notebook, to track the progress of each sample (Appendix A-4, Form 2). Sample log entries will be updated as each step is completed (e.g., receipt, identification, validation, archive).

4. A library of taxonomic literature is essential for the aid and support of identification/verification activities, and must be maintained (and updated as needed) in the laboratory. A list of selected taxonomic references is provided in Section 8.4.

Through the IBI, Karr et al. (1986) provided a consistent theoretical framework for analyzing fish assemblage data. The IBI is an aggregation of 12 biological metrics that are based on the fish assemblage's taxonomic and trophic composition and the abundance and condition of fish. Such multiple-parameter indices are necessary for making objective evaluations of complex systems. The IBI was designed to evaluate the quality of small Midwestern warmwater streams but has been modified for use in many regions (e.g., eastern and western United States, Canada, France) and in different ecosystems (e.g., rivers, impoundments, lakes, and estuaries).

The metrics attempt to quantify a biologist's best professional judgment (BPJ) of the quality of the fish assemblage. The IBI utilizes professional judgment, but in a prescribed manner, and it includes quantitative standards for discriminating the condition of the fish assemblage (Figure 8-1). BPJ is involved in choosing both the most appropriate population or assemblage element that is representative of each metric and in setting the scoring criteria. This process can be easily and clearly modified, as opposed to judgments that occur after results are calculated. Each metric is scored against criteria based on expectations developed from appropriate regional reference sites. Metric values

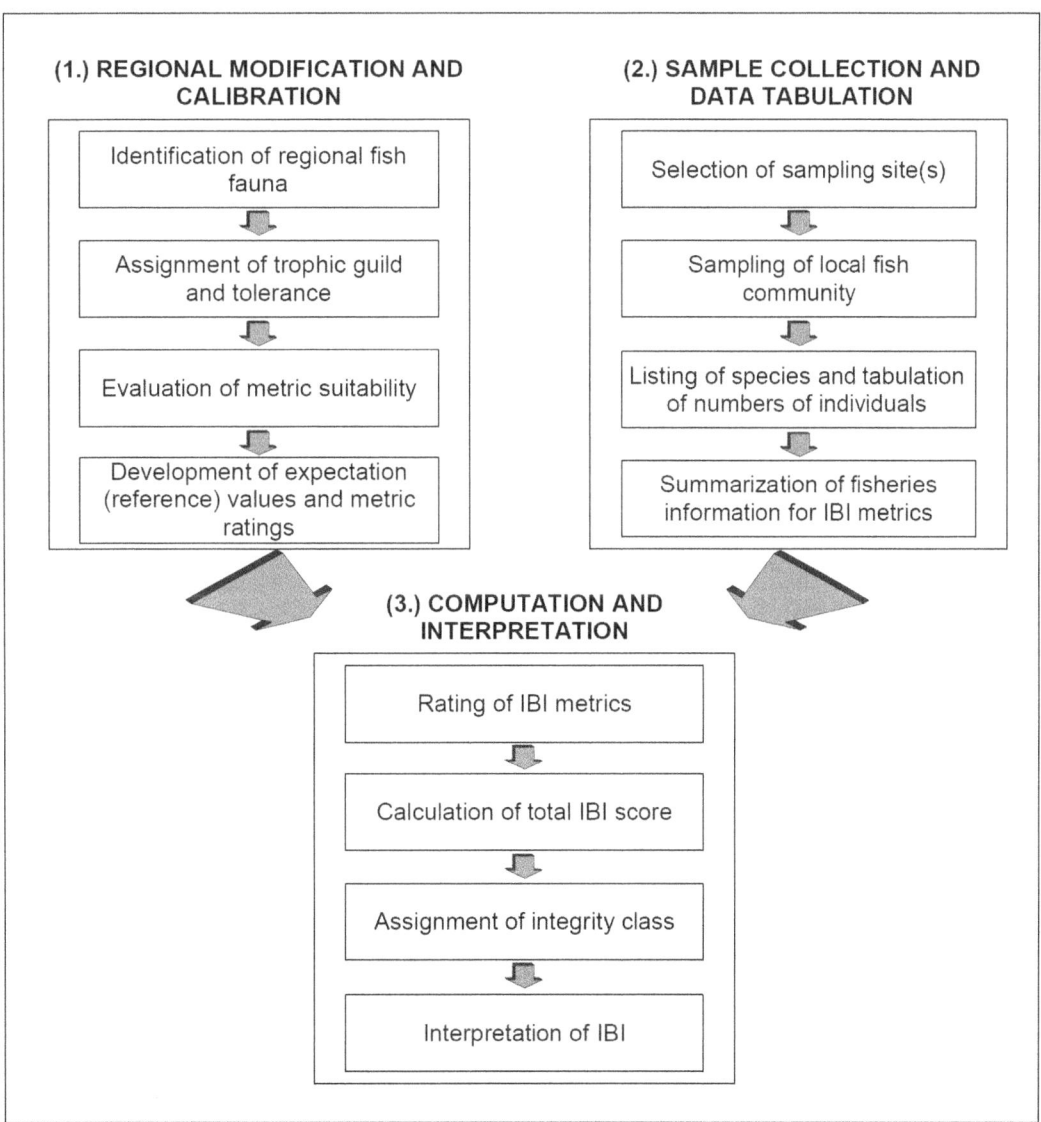

Figure 8-1. Sequence of activities involved in calculating and interpreting the Index of Biotic Integrity (adapted from Karr et al. 1986).

approximating, deviating slightly from, or deviating greatly from values occurring at the reference sites are scored as 5, 3, or 1, respectively. The scores of the 12 metrics are added for each station to give an IBI ranging from a maximum of 60 (excellent) to a minimum of 12 (very poor). Trophic and tolerance classifications of selected fish species are listed in Appendix C. Additional classifications can be derived from information in State and regional fish texts, by objectively assessing a large statewide database, or by contacting authors/originators of regional IBI programs or pilot studies. Use of the IBI by water resource agencies may result in further modifications. Many modifications have occurred (Miller et al. 1988) without changing the IBI's basic theoretical foundations.

The IBI serves as an integrated analysis because individual metrics may differ in their relative sensitivity to various levels of biological condition. A description and brief rationale for each of the 12 IBI metrics is outlined below. The original metrics described by Karr (1981) for Illinois streams are followed by substitutes used in or proposed for different geographic regions and stream sizes. Because of zoogeographic differences, different families or species are evaluated in different regions, with regional substitutes occupying the same general habitat or niche. The source for each substitute is footnoted below. Table 8-1 presents an overview of the IBI metric alternatives and their sources for various areas of the United States and Canada.

> **EXAMPLES OF SOURCES FOR METRIC ALTERNATIVES**
>
> Karr et al. (1986)
> Leonard and Orth (1986)
> Moyle et al. (1986)
> Fausch and Schrader (1987)
> Hughes and Gammon (1987)
> Ohio EPA (1987)
> Miller et al. (1988)
> Steedman (1988)
> Simon (1991)
> Lyons (1992a)
> Barbour et al. (1995)
> Simon and Lyons (1995)
> Hall et al. (1996)
> Lyons et al. (1996)
> Roth et al. (1997)
> Simon (1999)

8.3.1 Species Richness and Composition Metrics

These metrics assess the species richness component of diversity and the health of resident taxonomic groupings and habitat guilds of fishes. Two of the metrics assess assemblage composition in terms of tolerant or intolerant species.

Metric 1. Total number of fish species Substitutes (Table 8-1): Total number of resident native fish species and salmonid age classes.

This number decreases with increased degradation; hybrids and introduced species are not included. In coldwater streams supporting few fish species, the age classes of the species found represent the suitability of the system for spawning and rearing. The number of species is strongly affected by stream size at most small warmwater stream sites, but not at large river sites (Karr et al. 1986, Ohio EPA 1987).

Metric 2. Number and identity of darter species Substitutes (Table 8-1): Number and identity of sculpin species, benthic insectivore species, salmonid juveniles (individuals); number of sculpins (individuals); percent round-bodied suckers, sculpin and darter species.

These species are sensitive to degradation resulting from siltation and benthic oxygen depletion because they feed and reproduce in benthic habitats (Kuehne and Barbour 1983, Ohio EPA 1987). Many smaller species live within the rubble interstices, are weak swimmers, and spend their entire lives in an area of 100-400 m^2 (Matthews 1986, Hill and Grossman 1987). Darters are appropriate in most

Mississippi Basin streams; sculpins and yearling trout occupy the same niche in western streams. Benthic insectivores and sculpins or darters are used in small Atlantic slope streams that have few sculpins or darters, and round-bodied suckers are suitable in large midwestern rivers.

Metric 3. Number and identity of sunfish species. Substitutes (Table 8-1): Number and identity of cyprinid species, water column species, salmonid species, headwater species, and sunfish and trout species.

Table 8-1. Fish IBI metrics used in various regions of North America.[a]

Alternative IBI Metrics	Midwestern United States	Central Appalachians	Sacramento-San Joaquin	Colorado Front Range	Western Oregon	Ohio	Ohio Headwater Sites	Northeastern United States	Ontario	Central Corn Belt Plain	Wisconsin-Warmwater	Wisconsin-Coldwater	Maryland Coastal Plain	Maryland Non-Tidal
1. Total Number of Species	X	X	X	X				X		X			X	X
#native fish species					X	X	X		X		X			
# salmonid age classes[b]				X	X									
2. Number of Darter Species	X	X		X				X		X	X			
# sculpin species					X									
# benthic insectivore species								X						
# darter and sculpin species							X							
# darter, sculpin, and madtom species											X			
# salmonid juveniles (individuals)[b]			X		X			X						
% round-bodied suckers						X[c]								
# sculpins (individuals)			X											
# benthic species													X	X
3. Number of Sunfish Species	X			X				X		X	X			
# cyprinid species					X									
# water column species								X						
# sunfish and trout species									X					
# salmonid species			X								X			
# headwater species								X						
% headwater species								X		X				
4. Number of Sucker Species	X			X	X			X		X	X			
# adult trout species[b]			X	X										
# minnow species			X					X		X				
# sucker and catfish species									X					
5. Number of Intolerant Species	X			X	X	X		X			X	X	X	X
# sensitive species								X		X				
# amphibian species			X											
presence of brook trout									X					
% stenothermal cool and cold water species												X		
% of salmonid ind. as brook trout												X		
6. % Green Sunfish	X													
% common carp				X										
% white sucker			X					X						
% tolerant species							X	X		X	X	X	X	X
% creek chub		X												
% dace species									X					
% eastern mudminnow													X	

Table 8-1. Fish IBI metrics used in various regions of North America.[a]

Alternative IBI Metrics	Midwestern United States	Central Appalachians	Sacramento-San Joaquin	Colorado Front Range	Western Oregon Ohio	Ohio Headwater Sites	Northeastern United States	Ontario	Central Corn Belt Plain	Wisconsin-Warmwater	Wisconsin-Coldwater	Maryland Coastal Plain	Maryland Non-Tidal
7. % Omnivores	X			X		X	X	X	X	X			
% generalist feeders		X											
% generalists, omnivores, and invertivores													X
8. % Insectivorous Cyprinids	X											X	
% insectivores					X		X			X	X	X	X[e]
% specialized insectivores		X	X										
# juvenile trout			X										
% insectivorous species					X	X							
9. % Top Carnivores	X					X		X	X	X	X		
% catchable salmonids					X								
% catchable trout				X									
% pioneering species							X			X		X	
Density catchable wild trout				X									
10. Number of Individuals (or catch per effort)	X	X	X	X	X	X[d]	X[d]		X	X	X[d]	X	
Density of individuals								X					X
% abundance of dominant species												X	X
Biomass (per m²)													X[f]
11. % Hybrids	X						X						
% introduced species			X	X									
% simple lithophills						X				X	X		X
# simple lithophills species							X						
% native species		X											
% native wild individuals		X											
% silt-intolerant spawners												X	
12. % Diseased Individuals (deformities, eroded fins, lesions, and tumors)	X	X		X	X	X	X	X	X	X		X	X

Note: X = metric used in region. Many of these variations are applicable elsewhere.

a   Taken from Karr et al. (1986), Leonard and Orth (1986), Moyle et al. (1986), Fausch and Schrader (1987), Hughes and Gammon (1987), Ohio EPA (1987), Miller et al. (1988), Steedman (1988), Simon (1991), Lyons (1992a), Barbour et al. (1995), Simon and Lyons (1995), Hall et al. (1996), Lyons et al. (1996), Roth et al. (1997).

b   Metric suggested by Moyle et al. (1986) or Hughes and Gammon (1987) as a provisional replacement metric in small western salmonid streams.

c   Boat sampling methods only (i.e., larger streams/rivers).

d   Excluding individuals of tolerant species.

e   Non-coastal Plain streams only.

f   Coastal Plain streams only.

These pool species decrease with increased degradation of pools and instream cover (Gammon et al. 1981, Angermeier 1987, Platts et al. 1983). Most of these fishes feed on drifting and surface invertebrates and are active swimmers. The sunfishes and salmonids are important sport species. The sunfish metric works for most Mississippi Basin streams, but where sunfish are absent or rare, other groups are used. Cyprinid species are used in coolwater western streams; water column species occupy the same niche in northeastern streams; salmonids are suitable in coldwater streams; headwater species serve for midwestern headwater streams; and trout and sunfish species are used in southern Ontario streams. Karr et al. (1986) and Ohio EPA (1987) found the number of sunfish species to be dependent on stream size in small streams, but Ohio EPA (1987) found no relationship between stream size and sunfish species in medium to large streams, nor between stream size and headwater species in small streams.

Metric 4. Number and identity of sucker species. Substitutes (Table 8-1): Number of adult trout species, number of minnow species, and number of suckers and catfish.

These species are sensitive to physical and chemical habitat degradation and commonly comprise most of the fish biomass in streams. All but the minnows are longlived species and provide a multiyear integration of physicochemical conditions. Suckers are common in medium and large streams; minnows dominate small streams in the Mississippi Basin; and trout occupy the same niche in coldwater streams. The richness of these species is a function of stream size in small and medium sized streams, but not in large (e.g., non-wadeable) rivers.

Metric 5. Number and identity of intolerant species. Substitutes (Table 8-1): Number and identity of sensitive species, amphibian species, and presence of brook trout.

This metric distinguishes high and moderate quality sites using species that are intolerant of various chemical and physical perturbations. Intolerant species are typically the first species to disappear following a disturbance. Species classified as intolerant or sensitive should only represent the 5-10 percent most susceptible species, otherwise this becomes a less discriminating metric. Candidate species are determined by examining regional ichthyological books for species that were once widespread but have become restricted to only the highest quality streams. Ohio EPA (1987) uses number of sensitive species (which includes highly intolerant and moderately intolerant species) for headwater sites because highly intolerant species are generally not expected in such habitats. Moyle (1976) suggested using amphibians in northern California streams because of their sensitivity to silvicultural impacts. This also may be a promising metric in Appalachian streams which may naturally support few fish species. Steedman (1988) found that the presence of brook trout had the greatest correlation with IBI score in Ontario streams. The number of sensitive and intolerant species increases with stream size in small and medium sized streams but is unaffected by size of large (e.g., non-wadeable) rivers.

Metric 6. Proportion of individuals as green sunfish. Substitutes (Table 8-1): Proportion of individuals as common carp, white sucker, tolerant species, creek chub, and dace.

This metric is the reverse of Metric 5. It distinguishes low from moderate quality waters. These species show increased distribution or abundance despite the historical degradation of surface waters, and they shift from incidental to dominant in disturbed sites. Green sunfish are appropriate in small midwestern streams; creek chubs were suggested for central Appalachian streams; common carp were suitable for a coolwater Oregon river; white suckers were selected in the northeast and Colorado where green sunfish are rare to absent; and dace (*Rhinichthys* species) were used in southern Ontario. To avoid weighting the metric on a single species, Karr et al. (1986) and Ohio EPA (1987) suggest using a small number of highly tolerant species (e.g., alternative Metric 6— percent abundance of tolerant species).

Chapter 8: Fish Protocols

8.3.2 Trophic Composition Metrics

These three metrics assess the quality of the energy base and trophic dynamics of the fish assemblage. Traditional process studies, such as community production and respiration, are time consuming to conduct and the results are equivocal; distinctly different situations can yield similar results. The trophic composition metrics offer a means to evaluate the shift toward more generalized foraging that typically occurs with increased degradation of the physicochemical habitat.

Metric 7. Proportion of individuals as omnivores. Substitutes (Table 8-1): Proportion of individuals as generalist feeders.

The percent of omnivores in the community increases as the physical and chemical habitat deteriorates. Omnivores are defined as species that consistently feed on substantial proportions of plant and animal material. Ohio EPA (1987) excludes sensitive filter feeding species such as paddlefish and lamprey ammocoetes and opportunistic feeders like channel catfish. In areas where few species fit the true definition of omnivore, the proportion of generalized feeders may be substituted (Leonard and Orth 1986).

Metric 8. Proportion of individuals as insectivorous cyprinids. Substitutes (Table 8-1): Proportion of individuals as insectivores, specialized insectivores, insectivorous species, and number of juvenile trout.

Invertivores, primarily insectivores, are the dominant trophic guild of most North American surface waters. As the invertebrate food source decreases in abundance and diversity due to habitat degradation (e.g., anthropogenic stressors), there is a shift from insectivorous to omnivorous fish species. Generalized insectivores and opportunistic species, such as blacknose dace and creek chub were excluded from this metric by Ohio EPA (1987). This metric evaluates the midrange of biological condition, i.e., low to moderate condition.

Metric 9. Proportion of individuals as top carnivores. Substitutes (Table 8-1): Proportion of individuals as catchable salmonids, catchable wild trout, and pioneering species.

The top carnivore metric discriminates between systems with high and moderate integrity. Top carnivores are species that feed, as adults, predominantly on fish, other vertebrates, or crayfish. Occasional piscivores, such as creek chub and channel catfish, are not included. In trout streams, where true piscivores are uncommon, the percent of large salmonids is substituted for percent piscivores. These species often represent popular sport fish such as bass, pike, walleye, and trout. Pioneering species are used by Ohio EPA (1987) in headwater streams typically lacking piscivores. Pioneering species predominate in unstable environments that have been affected by temporal desiccation or anthropogenic stressors, and are the first to reinvade sections of headwater streams following periods of desiccation.

8.3.3 Fish Abundance and Condition Metrics

The last 3 metrics indirectly evaluate population recruitment, mortality, condition, and abundance. Typically, these parameters vary continuously and are time consuming to estimate accurately. Instead of such detailed population attributes or estimates, general population parameters are evaluated. Indirect estimation is less variable and much more rapidly determined.

Metric 10. Number of individuals in sample. Substitutes (Table 8-1): Density of individuals.

This metric evaluates population abundance and varies with region and stream size for small streams. It is expressed as catch per unit effort, either by area, distance, or time sampled. Generally sites with lower integrity support fewer individuals, but in some nutrient poor regions, enrichment increases the number of individuals. Steedman (1988) addressed this situation by scoring catch per minute of sampling greater than 25 as a 3, and less than 4 as a 1. Unusually low numbers generally indicate toxicity, making this metric most useful at the low end of the biological integrity scale. Hughes and Gammon (1987) suggest that in larger streams, where sizes of fish may vary in orders of magnitude, total fish biomass may be an appropriate substitute or additional metric.

Metric 11. Proportion of individuals as hybrids. Substitutes (Table 8-1): Proportion of individuals as introduced species, simple lithophils, and number of simple lithophilic species.

This metric is an estimate of reproductive isolation or the suitability of the habitat for reproduction. Generally as environmental degradation increases the percent of hybrids and introduced species also increases, but the proportion of simple lithophils decreases. However, minnow hybrids are found in some high quality streams, hybrids are often absent from highly impacted sites, and hybridization is rare and difficult to detect. Thus, Ohio EPA (1987) substitutes simple lithophils for hybrids. Simple lithophils spawn where their eggs can develop in the interstices of sand, gravel, and cobble substrates without parental care. Hughes and Gammon (1987) and Miller et al. (1988) propose using percent introduced individuals. This metric is a direct measure of the loss of species segregation between midwestern and western fishes that existed before the introduction of midwestern species to western rivers.

THE INDEX OF WELL-BEING (IWB)

The Iwb (Gammon 1976, 1980, Hughes and Gammon 1987) incorporates two abundance and two diversity measures in an approximately equal fashion, thereby representing fish assemblage quality more realistically than a single diversity or abundance measure. The Iwb is calculated using the formula:

$$Iwb = 0.5 \ln N + 0.5 \ln B + \overline{H}_N + \overline{H}_B$$

where

N = number of individuals caught per unit distance sampled

B = biomass of individuals caught per unit distance

$\overline{H}$ = Shannon diversity index, calculated as:

$$\overline{H} = -\Sigma \frac{n_i}{N} \ln \left(\frac{n_i}{N}\right)$$

where

n_i = relative number or weight of the ith species

N = total number or weight of the sample

THE MODIFIED INDEX OF WELL-BEING (MIWB)

The MIwb (Ohio EPA 1987) retains the same formula as the Iwb; however, highly tolerant species, hybrids, and exotic species are eliminated from the abundance (i.e., number and biomass) components of the formula. This modification increases the sensitivity of the index to a wider array of environmental disturbances.

Metric 12. Proportion of individuals with disease, tumors, fin damage, and skeletal anomalies

This metric depicts the health and condition of individual fish. These conditions occur infrequently or are absent from minimally impacted reference sites but occur frequently below point sources and in areas where toxic chemicals are concentrated. They are excellent measures of the subacute effects of chemical pollution and the aesthetic value of game and nongame fish.

Metric 13. Total fish biomass (optional).

Hughes and Gammon (1987) suggest that in larger (e.g., non-wadeable) rivers where sizes of fish may vary in orders of magnitude this additional metric may be appropriate. Gammon (1976, 1980) and Ohio EPA (1987) developed an Index of Well-Being (Iwb) and Modified Index of Well-Being (MIwb), respectively, based upon both fish abundance and biomass measures. The combination of diversity and biomass measures is a useful tool for assessing fish assemblages in larger rivers (Yoder and Rankin 1995b). Ohio EPA (1987) found that the additional collection of biomass data (i.e., in addition to abundance information needed for the IBI) required to calculate the MIwb does not represent a significant expenditure of time, providing that subsampling techniques are applied (see Field Sampling Procedures 8.1.1).

Because the IBI is an adaptable index, the choice of metrics and scoring criteria is best developed on a regional basis through use of available publications (Karr et al. 1986, Ohio EPA 1987, Miller et al. 1988, Steedman 1988; Simon 1991, Lyons 1992a, Simon and Lyons 1995, Hall et al. 1996, Lyons et al. 1996, Roth et al. 1997, Simon 1999). Several steps are common to all regions. The fish species must be listed and assigned to trophic and tolerance guilds. Scoring criteria are developed through use of high quality historical data and data from minimally-impaired regional reference sites. This has been done for much of the country, but continued refinements are expected as more ecological data become available for the fish community.

8.4 TAXONOMIC REFERENCES FOR FISH

The following references are provided as a list of taxonomic references currently being used around the United States for identification of fish. Any of these references cited in the text of this document will also be found in Chapter 11 (Literature Cited).

Anderson, W.D. 1964. Fishes of some South Carolina coastal plain streams. *Quarterly Journal of the Florida Academy of Science* 27:31-54.

Bailey, R.M. 1956. *A revised list of the fishes of Iowa with keys for identification.* Iowa State Conservation Commission, Des Moines, Iowa.

Bailey, R.M. and M.O. Allum. 1962. *Fishes of South Dakota.* Miscellaneous Publications of the Museum of Zoology, University of Michigan, No. 119, 131pp.

Baxter, G.T. and J.R. Simon. 1970. *Wyoming fishes.* Wyoming Game and Fish Department. Bulletin No. 4, Cheyenne, Wyoming.

Baxter, G.T. and M.D. Stone. 1995. *Fishes of Wyoming.* Wyoming Game and Fish Department. Cheyenne, Wyoming.

Becker, G.C. 1983. *Fishes of Wisconsin.* University of Wisconsin Press, Madison, Wisconsin.

Behnke, R.J. 1992. *Native trout of western North America.* American Fisheries Society Monograph 6. American Fisheries Society. Bethesda, Maryland.

Rapid Bioassessment Protocols for Use in Streams and Wadeable Rivers: Periphyton, Benthic Macroinvertebrates, and Fish, Second Edition

8-15

Bond, C.E. 1973. *Keys to Oregon freshwater fishes*. Technical Bulletin 58:1-42. Oregon State University Agricultural Experimental Station, Corvallis, Oregon.

Bond, C.E. 1994. *Keys to Oregon freshwater fishes*. Oregon State University. Corvallis, Oregon.

Brown, C.J.D. 1971. *Fishes of Montana*. Montana State University, Bozeman, Montana.

Clay, W.M. 1975. *The fishes of Kentucky*. Kentucky Department of Fish and Wildlife Resources, Frankford, Kentucky.

Cook, F.A. 1959. *Freshwater fishes of Mississippi*. Mississippi Game and Fish Commission, Jackson, Mississippi.

Cooper, E.L. 1983. *Fishes of Pennsylvania and the northeastern United States*. Pennsylvania State Press, University Park, Pennsylvania.

Cross, F.B. and J.T. Collins. 1995. *Fishes of Kansas*. University of Kansas Press. Lawrence, Kansas.

Dahlberg, M.D. and D.C. Scott. 1971. The freshwater fishes of Georgia. *Bulletin of the Georgia Academy of Science* 19:1-64.

Douglas, N.H. 1974. *Freshwater fishes of Louisiana*. Claitors Publishing Division, Baton Rouge, Louisiana.

Eddy, S. and J.C. Underhill. 1974. *Northern fishes, with special reference to the Upper Mississippi Valley*. University of Minnesota Press, Minneapolis, Minnesota.

Etnier, D.A. and W.C. Starnes. 1993. *The fishes of Tennessee*. University of Tennessee Press, Knoxville, Tennessee.

Everhart, W.H. 1966. *Fishes of Maine*. Third edition. Maine Department of Inland Fisheries and Game, Augusta, Maine.

Everhart, W.H. and W.R. Seaman. 1971. *Fishes of Colorado*. Colorado Game, Fish, and Parks Division, Denver, Colorado.

Hankinson, T.L. 1929. Fishes of North Dakota. *Papers of the Michigan Academy of Science, Arts, and Letters* 10:439-460.

Hubbs, C. 1972. A checklist of Texas freshwater fishes. *Texas Parks and Wildlife Department Technical Service* 11:1-11.

Hubbs, C.L. and K.F. Lagler. 1964. *Fishes of the Great Lakes region*. University of Michigan Press, Ann Arbor, Michigan.

Jenkins, R.E. and N.M. Burkhead. 1994. *The freshwater fishes of Virginia*. American Fisheries Society. Bethesda, Maryland.

Kuehne, R.A. and R.W. Barbour. 1983. *The American darters*. University of Kentucky Press, Lexington, Kentucky.

La Rivers, I. 1994. *Fishes and fisheries of Nevada*. University of Nevada Press. Reno, Nevada.

Lee, D.S., C.R. Gilbert, C.H. Hocutt, R.E. Jenkins, D.E. McAllister, and J.R. Stauffer, Jr. 1980. *Atlas of North American freshwater fishes*. North Carolina Museum of Natural History, Raleigh, North Carolina.

Lee, D.S., S.P. Platania, C.R. Gilbert, R. Franz, and A. Norden. 1981. A revised list of the freshwater fishes of Maryland and Delaware. *Proceedings of the Southeastern Fishes Council* 3:1-10.

Loyacano, H.A. 1975. *A list of freshwater fishes of South Carolina*. Bulletin No. 580. South Carolina Agricultural Experiment Station.

Markle, D.F., D.L. Hill, and C.E. Bond. 1996. *Sculpin identification workshop and working guide to freshwater sculpins of Oregon and adjacent areas*. Oregon State University. Corvallis, Oregon.

McPhail, J.D. and C.C. Lindsey. 1970. *Freshwater fishes of northeastern Canada and Alaska*. Bulletin No. 173. Fisheries Research Board of Canada.

Menhinick, E.F. 1991. *The freshwater fishes of North Carolina*. University of North Carolina, Charlotte, North Carolina.

Miller, R.J. and H.W. Robinson. 1973. *The fishes of Oklahoma*. Oklahoma State University Press, Stillwater, Oklahoma.

Minckley, W.L. 1973. *Fishes of Arizona*. Arizona Game and Fish Department, Phoenix, Arizona.

Morris, J.L. and L. Witt. 1972. *The fishes of Nebraska*. Nebraska Game and Parks Commission, Lincoln, Nebraska.

Morrow, J.E. 1980. *The freshwater fishes of Alaska*. Alaska Northwest Publishing Company, Anchorage, Alaska.

Moyle, P.B. 1976. *Inland fishes of California*. University of California Press, Berkeley, California.

Mugford, P.S. 1969. *Illustrated manual of Massachusetts freshwater fish*. Massachusetts Division of Fish and Game, Boston, Massachusetts.

Page, L.M. 1983. *Handbook of darters*. TFH Publishing, Neptune, New Jersey.

Page, L.M. and B.M. Burr. 1991. *A field guide to freshwater fishes*. Houghton Mifflin Company, Boston, Massachusetts.

Pflieger, W.L. 1975. *The fishes of Missouri*. Missouri Department of Conservation, Columbia, Missouri.

Rapid Bioassessment Protocols for Use in Streams and Wadeable Rivers: Periphyton, Benthic Macroinvertebrates, and Fish, Second Edition

8-17

Robison, H.W. and T.M. Buchanan. 1988. *The fishes of Arkansas*. University of Arkansas Press, Fayetteville, Arkansas.

Rohde, F.C., R.G. Arndt, D.G. Lindquist, and J.F. Parnell. 1994. *Freshwater fishes of the Carolinas, Virginia, Maryland, and Delaware*. University of North Carolina Press. Chapel Hill, North Carolina.

Scarola, J.F. 1973. *Freshwater fishes of New Hampshire*. New Hampshire Fish and Game Department, Concord, New Hampshire.

Scott, W.B. and E.J. Crossman. 1973. *Freshwater fishes of Canada*. Bulletin No. 1984. Fisheries Research Board of Canada.

Sigler, W.F. and R.R. Miller. 1963. *Fishes of Utah*. Utah Game and Fish Department. Salt Lake City, Utah.

Sigler, W.F., and J.W. Sigler. 1996. *Fishes of Utah: A natural history*. University of Utah Press, Ogden, Utah..

Simon, T.P., J.O. Whitaker, J. Castrale, and S.A. Minton. 1992. Checklist of the vertebrates of Indiana. *Proceedings of the Indiana Academy of Science*.

Simpson, J.C. and R.L. Wallace. 1982. *Fishes of Idaho*. The University of Idaho Press, Moscow, Idaho.

Smith, C.L. 1985. *Inland fishes of New York*. New York State Department of Environmental Conservation, Albany, New York.

Smith, P.W. 1979. *The fishes of Illinois*. Illinois State Natural History Survey. University of Illinois Press, Urbana, Illinois.

Smith-Vaniz, W.F. 1987. *Freshwater fishes of Alabama*. Auburn University Agricultural Experiment Station, Auburn, Alabama.

Stauffer, J.R., J.M. Boltz, and L.R. White. 1995. *The fishes of West Virginia*. Academy of Natural Sciences of Philadelphia.

Stiles, E.W. 1978. *Vertebrates of New Jersey*. Edmund W. Stiles Publishers, Somerset, New Jersey.

Sublette, J.E., M.D. Hatch, and M. Sublette. 1990. *The fishes of New Mexico*. University of New Mexico Press, Albuquerque, New Mexico.

Tomelleri, J.R. and M.E. Eberle. 1990. *Fishes of the central United States*. University Press of Kansas, Lawrence, Kansas.

Trautman, M.B. 1981. *The fishes of Ohio*. Ohio State University Press, Columbus, Ohio.

Whitworth, W.R., P.L. Berrien, and W.T. Keller. 1968. *Freshwater fishes of Connecticut*. Bulletin No. 101. State Geological and Natural History Survey of Connecticut.

Wydoski, R.S. and R.R. Whitney. 1979. *Inland fishes of Washington.* University of Washington Press.

Rapid Bioassessment Protocols for Use in Streams and Wadeable Rivers: Periphyton, Benthic Macroinvertebrates, and Fish, Second Edition

8-19

This Page Intentionally Left Blank

9 BIOLOGICAL DATA ANALYSIS

States are faced with the challenge of not only developing tools that are both appropriate and cost-effective (Barbour 1997), but also the ability to translate scientific data for making sound management decisions regarding the water resource. The approach to analysis of biological (and other ecological) data should be straightforward to facilitate a translation for management application. This is not meant to reduce the rigor of data analysis but to ensure its place in making crucial decisions regarding the protection, mitigation, and management of the nation's aquatic resources. In fact, biological monitoring should combine biological insight with statistical power (Karr 1987). Karr and Chu (1999) state that a knowledge of regional biology and natural history (not a search for statistical relationships and significance) should drive both sampling design and analytical protocol.

A framework for bioassessment can be either an *a priori* or *a posteriori* approach to classifying sites and establishing reference condition. To provide a broad comparison of the 2 approaches, it is assumed that candidate reference sites are available from a wide distribution of streams. In the first stage, data collection is conducted at a range of reference sites (and non-reference or test sites) regardless of the approach. The differentiation of site classes into more homogeneous groups or classes may be based initially on *a priori* physicochemical or biogeographical attributes, or solely on *a posteriori* analysis of biology (Stage 2 as illustrated in Figure 9-1). Analysts who use multimetric indices tend to use *a priori* classification; and analysts who use one of the multivariate approaches tend to use *a posteriori*, multivariate classification. However, there is no reason *a priori* classification could not be used with multivariate assessments, and vice-versa.

Two data analysis strategies have been debated in scientific circles (Norris 1995, Gerritsen 1995) over the past few years — the multimetric approach as implemented by most water resource agencies in the United States (Davis et al. 1996), and a multivariate approach advocated by several water resource agencies in Europe and Australia (Wright et al. 1993, Norris and Georges 1993). The contrast and similarity of these 2 approaches are illustrated by Figure 9-1 in a 5-stage generic process of bioassessment development. While there are many forms of multivariate analyses, the 2 most common multivariate approaches are the Benthic Assessment of Sediment (BEAST) used in parts of Canada, the River Invertebrate Prediction and Classification System (RIVPACS) used in parts of England and its derivation, the Australian River Assessment System (AusRivAS) used in Australia.

The development of the reference condition from the range of reference sites (Figure 9-1, Stage 4), is formulated by a suite of biological metrics in the multimetric approach whereas the species composition data are the basis for models used in the multivariate approach. However, both multivariate techniques differ in their probability models. Once the reference condition is established, which serves as a benchmark for assessment, the final stage becomes the basis for the assessment and monitoring program. In this fifth and final stage (Figure 9-1), the multimetric approach uses established percentiles of the population distribution of the reference sites for the metrics to discriminate between impaired and minimally impaired conditions. Where a dose/response relationship can be established from sites having a gradient of conditions (reference sites unknown), an upper percentile of the metric is used to partition metric values into condition ranges. The BEAST multivariate technique uses a probability model based on taxa ordination space

Rapid Bioassessment Protocols for Use in Streams and Wadeable Rivers: Periphyton, Benthic Macroinvertebrates, and Fish, Second Edition

9-1

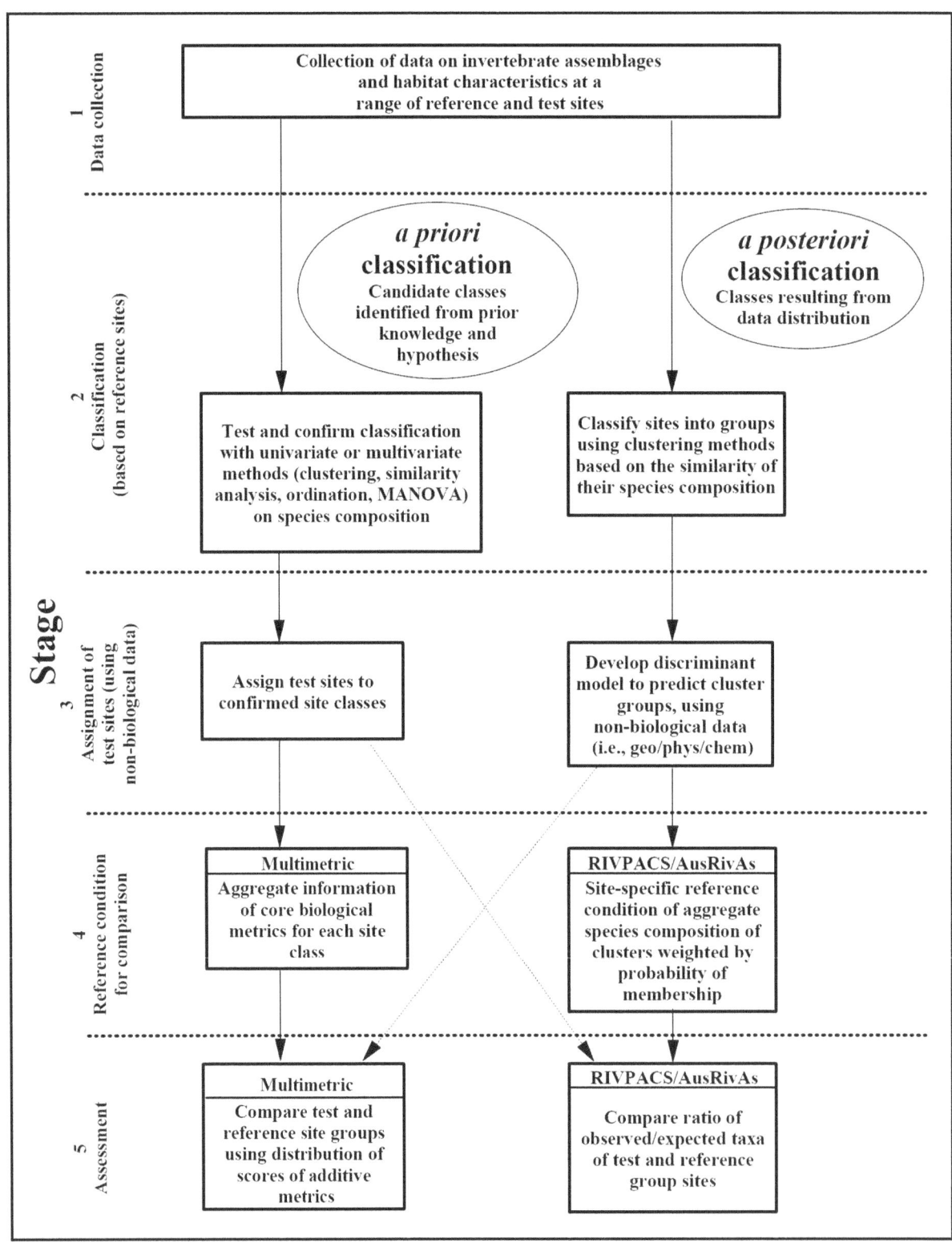

Figure 9-1. Comparison of the developmental process for the multimetric and multivariate approaches to biological data analysis (patterned after ideas based on Reynoldson, Rosenberg, and Resh, unpublished data).

and the "best fit" of the test site(s) to the probability ellipses constructed around the reference site classes (Reynoldson et al. 1995). The AusRivAS/RIVPACS model calculates the probability of expected taxa occurrence from the weighted reference site groups.

The bioassessment program in Maine is an example of a state that uses a multivariate analysis in the form of discriminant function models and applies these models to a variety of metrics. Decisions are made with regard to attainment (or non-attainment) of designated aquatic life uses. The approach used by Maine is based on characteristics of both the multivariate and multimetric approach. In this chapter, only the multimetric approach to biological data analysis is discussed in detail. Discussion of multivariate approaches is restricted to the overview of the discriminant function model used by Maine and the AusRivAS/RIVPACS technique.

9.1   THE MULTIMETRIC APPROACH

Performing data analysis for the Rapid Bioassessment Protocols (RBPs) or any other multimetric approach typically involves 2 phases: (1) Selection and calibration of the metrics and subsequent aggregation into an index according to homogenous site classes; and (2) assessment of biological condition at sites and judgment of impairment. The first phase is a developmental process and is only necessary as biological programs are being implemented. This process is essentially the characterizing of reference conditions that will form the basis for assessment. It is well-documented (Davis and Simon 1995, Gibson et al. 1996, Barbour et al. 1996b) and is summarized here. Developing the framework for reference conditions (i.e., background or natural conditions) is a process that is applicable to non-biological (i.e., physical and chemical) monitoring as well (Karr 1993, Barbour et al. 1996a).

The actual assessment of biological condition is ongoing and becomes cost-effective once Phase 1 has been completed, and the thresholds for determining attainment or non-attainment (impairment) have been established. The establishment of reference conditions (through actual sites or other means) is crucial to the determination of metric and index thresholds. These thresholds are essential elements in performing the assessment. It is possible that reference conditions (and resultant thresholds) will need to be established on a seasonal basis to accommodate year-round sampling and assessment. If data are available, a dose/response relationship between specific or cumulative stressors and biological condition will provide information on a gradient response, which can be a powerful means of determining impairment thresholds.

The 2 phases in data analysis for the multimetric approach are discussed separately in the following section. The reader is referred to supporting documentation cited throughout for more in-depth discussion of the concepts of multimetric assessment.

9.1.1 Metric Selection, Calibration, And Aggregation Into an Index

The development of biological indicators as part of a bioassessment program and as a framework for biocriteria is an iterative process where the site classification and metric selections are revisited at various stages of the analysis. However, once this process has been completed and the various technical issues have been addressed, continued monitoring becomes cost-effective. The conceptual process for proceeding from measurements to indicators to assessment of condition is illustrated in Figure 9-2 (Paulsen et al. 1991; Barbour et al., 1995; Gibson et al., 1996).

Index development outlined in this section requires a stream classification framework to partition natural variability and in which metrics are evaluated for scientific validity. The core metrics representing various attributes of the targeted aquatic assemblage can be either aggregated into an index or retained as individual measures.

Step 1. Classify the Stream Resource

> *Classification* is the partitioning of natural variability into groups or classes of stream sites that are relatively homogeneous with regard to physical, chemical, and biological attributes.

Site classification provides a framework for organizing and interpreting natural variability among streams; ecoregions are a principal example of a classification framework (Omernik 1995). However, classification variables can be at a coarser or finer scale than ecoregions or subecoregions, such as elevation and drainage area. Elevation was determined to be an important classification variable in montane regions of the country (Barbour et al. 1992, 1994, Spindler 1996). Spindler (1996) found that benthic data adhered more closely to elevation than to ecoregions. Ohio EPA (1987) found that stream size (or drainage area) was a covariate and not a determinant of stream classes. The number of fish species increased with stream size (Figure 9-3).

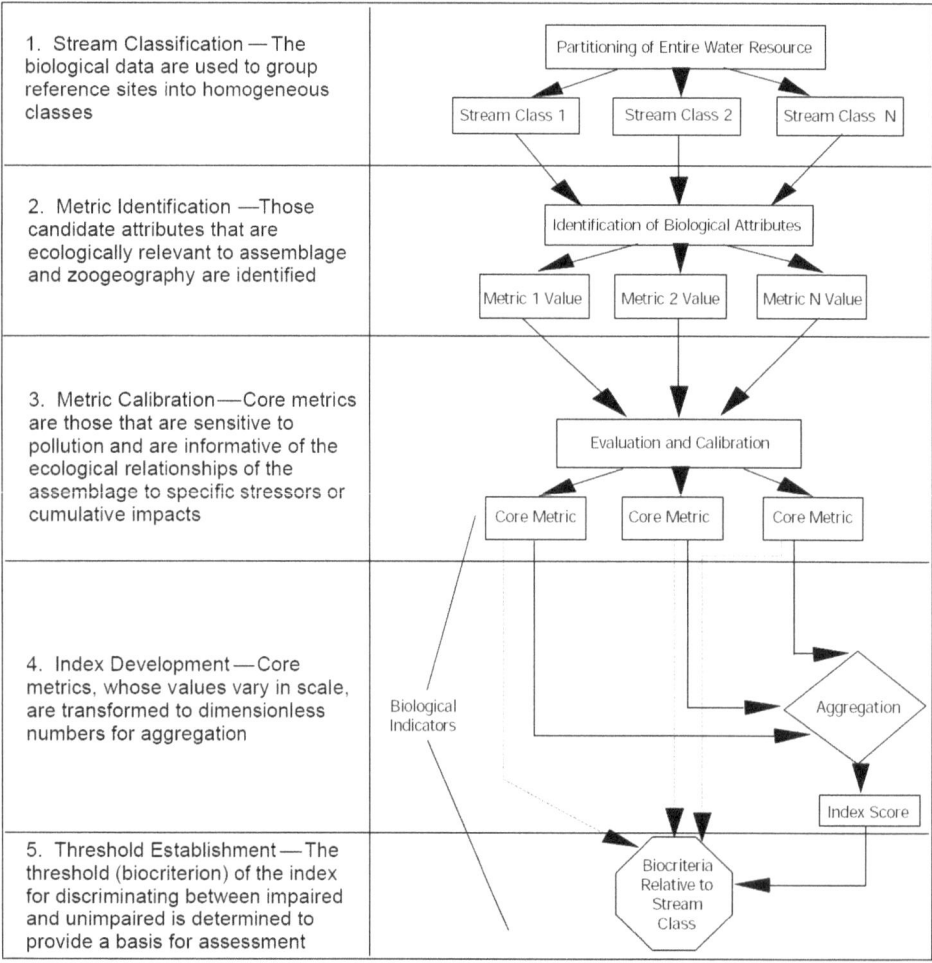

Figure 9-2. Process for developing assessment thresholds (modified from Paulsen et al. [1991] and Barbour et al. [1995]). Dotted lines indicate use of individual metric information to aid in the evaluation of biological condition and cause of impairment.

Chapter 9: Multimetric Data Analysis

Classification is best accomplished with reference sites that reflect the most natural and representative condition of the region. Candidate reference sites that are based on minimally degraded physical habitat and water chemistry are used as the basis for stream classification. Quantitative criteria for reference sites aid in a consistent framework for selection. An example of quantitative criteria for identifying reference sites in a statewide study for Maryland (Roth et al., 1997) is presented below (a reference site must meet all 12 criteria):

1. pH ≥ 6; if blackwater stream, then pH < 6 and DOC ≥ 8 mg/l

2. ANC ≥ 50 µeq/l

3. DO ≥ 4 ppm

4. nitrate ≤ 300 µeq/l

5. urban land use ≤ 20% of catchment area

6. forest land use ≥ 25% of catchment area

7. remoteness rating: optimal or suboptimal

8. aesthetics rating: optimal or suboptimal

9. instream habitat rating: optimal or suboptimal

10. riparian buffer width ≥ 15 m

11. no channelization

12. no point source discharges

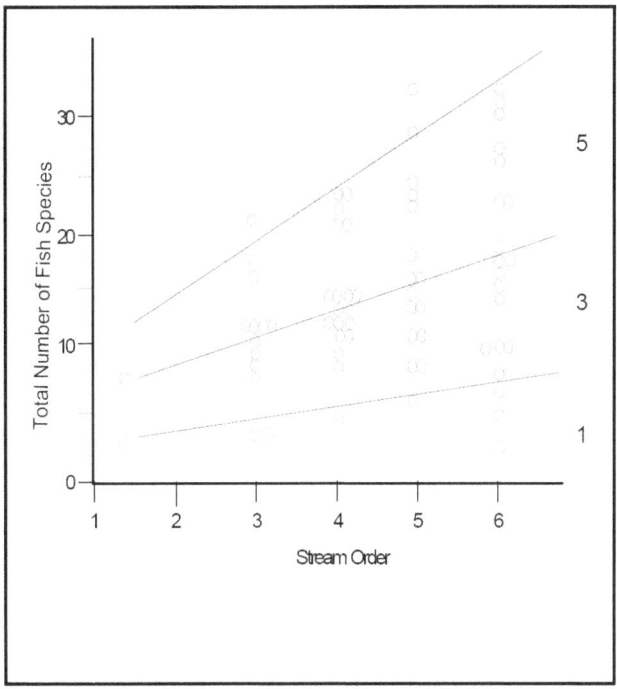

Figure 9-3. Species richness versus stream size (taken from Fausch et al. 1984).

Sites are initially classified according to distinctive geographic, physical, or chemical attributes. Refinement and confirmation of the site classes is accomplished using the biological data (Figure 9-4). Classification is used to determine whether the sampled sites should be placed into specific groups that will minimize variance *within* groups and maximize variance *among* groups. As an example, 3 ecoregionally based delineations (bioregions) were effective at partitioning the variability among reference sites in Florida (Figure 9-5).

Components of Step 1 include:

- ! Identify classification alternatives. Use physical and chemical parameters that are minimally influenced by human activity to identify classes for testing.

- ! Identify candidate reference sites that meet the criteria of most "natural" conditions of region.

- ! Test alternative classification schemes of subecoregion, stream type, elevation, etc., using multiple metric and non-

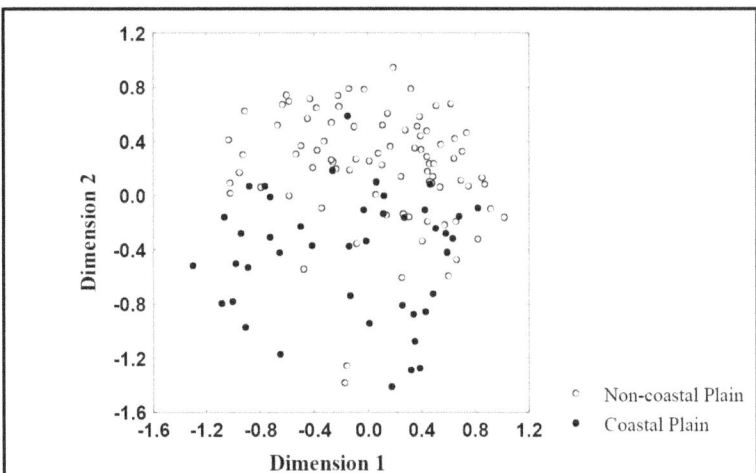

Figure 9-4. Results of mutivariate ordination on benthic macroinvertebrate data from "least impaired" streams from Maryland, using nonmetric multidimensional scaling (NMDS) of Bray-Curtis dissimilarity coefficients.

metric biological characteristics including measures such as species composition and EPT taxa (Figure 9-5). Several multivariate classification and ordination methods, and univariate descriptions and tests, can assist in this process (Reckhow and Warren-Hicks 1996, Gerritsen 1995, 1996, Barbour et al. 1996b).

- ! Evaluate classification alternatives and determine best distinction into groups or classes using biological data. By confirming resource classification based on biological data, site classes are identified that adequately partition variability.

Step 2. Identify Potential Measures For Each Assemblage

> A *metric* is a characteristic of the biota that changes in some predictable way with increased human influence.

Metrics allow the investigator to use meaningful indicator attributes in assessing the status of assemblages and communities in response to perturbation. The definition of a metric is a characteristic of the biota that changes in some predictable way with increased human influence (Barbour et al. 1995). For a metric to be useful, it must have the following technical attributes: (1) ecologically relevant to the biological assemblage or community under study and to the specified program objectives; (2) sensitive to stressors and provides a response that can be discriminated from natural variation. The purpose of using multiple metrics to assess biological condition is to aggregate and convey the information available regarding the elements and processes of aquatic communities.

All metrics that have ecological relevance to the assemblage under study and that respond to the targeted stressors are potential metrics for testing.

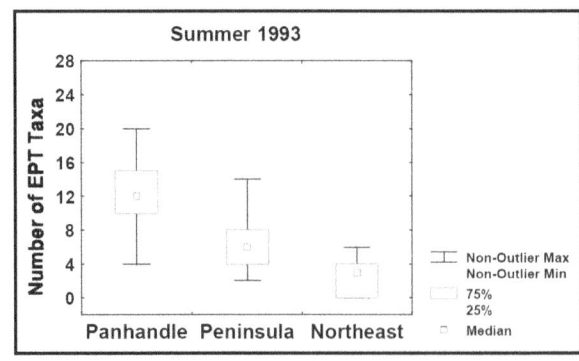

Figure 9-5. An example of a metric that illustrates classification of reference stream sites in Florida into bioregions.

From this "universe" of metrics, some will be eliminated because of insufficient data or because the range of values is not sufficient for discrimination between natural variability and anthropogenic effects. This step is to identify the candidate metrics that are most informative, and therefore, warrant further analysis.

The potential measures that are relevant to the ecology of streams within the region or state should be selected to ensure that various aspects of the elements and processes of the aquatic assemblage are addressed. Representative metrics should be selected from each of 4 primary categories: (1) richness measures for diversity or variety of the assemblage; (2) composition measures for identity and dominance; (3) tolerance measures that represent sensitivity to perturbation; and (4) trophic or habit measures for information on feeding strategies and guilds. Karr and Chu (1999) suggest that measures of individual health be used to supplement other metrics. Karr has expanded this concept to include metrics that are reflective of landscape level attributes, thus providing a more comprehensive multimetric approach to ecological assessment (Karr et al. 1987). See Table 9-1 for potential metrics that have been useful for periphyton, benthic macroinvertebrates, and fish are summarized in Chapters 6, 7, and 8, respectively.

Components of Step 2 include:

! Review value ranges of potential metrics, and eliminate those that have too many zero values in the population of reference sites to calculate the metric at a large enough proportion of sites.

! Use descriptive statistics (central tendency, range, distribution, outliers) to characterize metric performance within the population of reference sites of each site class.

! Eliminate metrics that have too high variability in the reference site population that they can not discriminate among sites of different condition. The potential for each measure is based on possessing enough information and a specific range of variability to discriminate among site classes and biological condition.

Step 3. Select Robust Measures

Core metrics are those that will discriminate between good and poor quality ecological conditions. It is important to understand the effects of various stressors on the behavior of specific metrics. Metrics that are responsive to specific pollutants or stressors, where the response is well-characterized, are most useful as a diagnostic tool. Core metrics are those that represent diverse aspects of structure, composition, individual health, or processes of the aquatic biota. Together they form the foundation for a sound, integrated analysis of the biotic condition to judge attainment of biological criteria.

The ability of a biological metric to *discriminate* between "known" reference conditions and "known" stressed conditions (defined by physical and chemical characteristics) is crucial in the selection of *core metrics* for future assessments.

Discriminatory ability of biological metrics can be evaluated by comparing the distribution of each metric at a set of reference sites with the distribution of metrics from a set of "known" stressed sites (defined by physical and chemical characteristics) within each site class. If there is minimal or no overlap between the distributions, then the metric can be considered to be a strong discriminator between reference and impaired conditions (Figure 9-6).

As was done with candidate reference sites (see Step 1), criteria are established to identify a population of "known" stressed sites based on physical and chemical measures of degradation. An example set of criteria established for Maryland streams for which failure indicated a stressed site for testing discriminatory power (Roth et al. 1997) is as follows:

! pH ≤ 5 and ANC ≤ 0 μeq/l (except for blackwater streams, DOC ≥ 8 mg/l)

! DO ≤ 2 ppm

! nitrate > 500 μM/l and DO < 3 ppm

! instream habitat rating poor and urban land use > 50% of catchment area
! instream habitat rating poor and bank stability rating poor

! instream habitat rating poor and channel alteration rating poor

Table 9-1. Some potential metrics for periphyton, benthic macroinvertebrates, and fish that could be considered for streams. Redundancy can be evaluated during the calibration phase to eliminate overlapping metrics.

	Richness Measures	Composition Measures	Tolerance Measures	Trophic/Habit Measures
Periphyton	• Total no. of taxa • No. of common nondiatom taxa • No. of diatom taxa	• % community similarity • % live diatoms • Diatom (Shannon) diversity index	• % tolerant diatoms • % sensitive taxa • % aberrant diatoms • % acidobiontic • % alkalibiontic • % halobiontic	• % motile taxa • Chlorophyll a • % saprobiontic • % eutrophic
Benthic Macroinvertebrate	• No. Total taxa • No. EPT taxa • No. Ephemeroptera taxa • No. Plecoptera taxa • No. Trichoptera taxa	• % EPT • % Ephemeroptera • % Chironomidae	• No. Intolerant Taxa • % Tolerant Organisms • Hilsenhoff Biotic Index (HBI) • % Dominant Taxon	• No. Clinger taxa • % Clingers • % Filterers • % Scrapers
Fish	• Total no. of native fish species • No. and identity of darter species • No. and identity of sunfish species • No. and identity of sucker species	• % pioneering species • Number of fish per unit of sampling effort related to drainage area	• No. and identity of intolerant species • % of individuals as tolerant species • % of individuals as hybrids • % of individuals with disease, tumors, fin damage, and skeletal anomalies	• % omnivores • % insectivores • % top carnivores

Step 3 can be separated into 2 elements that correspond to discrimination of core metrics (element 1) and determination of biological/physicochemical associations (element 2). Components of these elements include:

Element 1   Select core measures that are best for discriminating degraded condition

! Good (reference) designations of stream sites should be based on land use, physical and chemical quality, and habitat quality.

! Poor (stressed) designations of stream sites for testing impairment discriminations are also based on judgement criteria involving land use, physical and chemical and quality, and habitat quality.

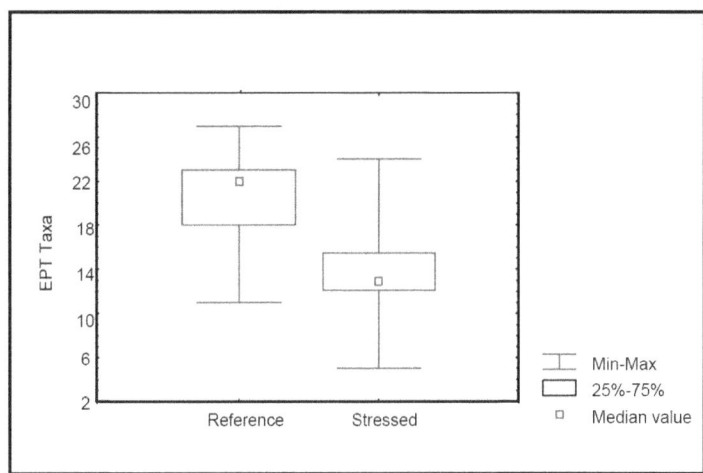

Figure 9-6. Example of discrimination, using the EPT index, between reference and stressed sites in Rocky Mountain streams, Wyoming.

! Determine which biological metrics best discriminate between the reference sites and sites with identified anthropogenic stressors.

! Those metrics having the strongest discriminatory power will provide the most confidence in assessing biological condition of unknown sites.

Element 2   Determine the associations/linkages between candidate biological and physicochemical measures

! Plot relationship of metric values against various stressor categories, e.g., chemical concentrations, habitat condition and other measured stressors.

! If desired, multivariate ordination models may be used to elucidate gradients of response of metrics to stressors.

! Monotonic relationships between metrics and stressors allow the use of extreme values (highest or lowest) as reference condition.

! Some metrics may not always be monotonic. For example, total biomass and taxa richness values may exceed the reference at intermediate levels of nutrient enrichment.

! Multiple metrics should be selected to provide a strong and predictable relationship with stream condition.

An *index* provides a means of integrating information from a composite of the various measures of biological attributes.

Step 4. Determine the best aggregation of core measures for indicating status and change in condition

The purpose of an index is to provide a means of integrating information from the various measures of biological attributes (or metrics). Metrics vary in their scale—they are integers, percentages, or dimensionless numbers. Prior to developing an integrated index for assessing biological condition, it is necessary to standardize core metrics via transformation to unitless scores. The standardization assumes that each metric has the same value and importance (i.e., they are weighted the same), and that a 50% change in one metric is of equal value to assessment as a 50% change in another.

Where possible, the scoring criterion for each metric is based on the distribution of values in the population of sites, which include reference streams; for example, the 95th percentile of the data distribution is commonly used (Figure 9-7) to eliminate extreme outliers. From this upper percentile, the range of the metric values can be standardized as a percentage of the 95th percentile value, or other (e.g., trisected or quadrisected), to provide a range of scores. Those values that are closest to the 95th percentile would receive higher scores, and those having a greater deviation from this percentile would have lower scores. For those metrics whose values *increase* in response to perturbation (see Table 7-2 for examples of "reverse" metrics for benthic macroinvertebrates) the 5th percentile is used to remove outliers and to form a basis for scoring.

Alternative methods for scoring metrics, as illustrated in Figure 9-7, are currently in use in various parts of the US for multimetric indexes. A "trisection" of the scoring range has been well-documented (Karr et al. 1986, Ohio EPA 1987, Fore et al. 1996, Barbour et al. 1996b). A "quadrisection" of the range has been found to be useful for benthic assemblages (DeShon 1995, Maxted et al. in press). More recent studies are finding that a standardization of all metrics as percentages of the 95th percentile value yields the most sensitive index, because information of the component metrics is retained (Hughes et al. 1998). Unpublished data from statewide databases for Idaho, Wyoming, Arizona, and West Virginia, are supportive of this third alternative for scoring metrics. Ideally, a composite of all sites representing a gradient of conditions is used. This situation is analogous to a determination of a dose/response relationship and depends on the ability of incorporating both reference and non-reference sites.

Aggregation of metric scores simplifies management and decision making so that a single index value is used to determine whether action is needed. Biological condition of waterbodies is judged based on the summed index value (Karr et al. 1986). If the index value is above a criterion, then the stream is judged as "optimal" or "excellent" in condition. The exact nature of the action needed (e.g., restoration, mitigation, pollution enforcement) is not determined by the index value, but by analyses of the component metrics, in addition to the raw data and integrated with other ecological information. Therefore, the index is not the sole determinant of impairment and diagnostics, but when used in concert with the component information, strengthens the assessment (Barbour et al. 1996a).

Components of Step 4 include:

! Determine scoring criteria for each metric (within each site class) from the appropriate percentile of the data distribution (Figure 9-7). If the metric is associated with a significant covariate such as watershed size, a

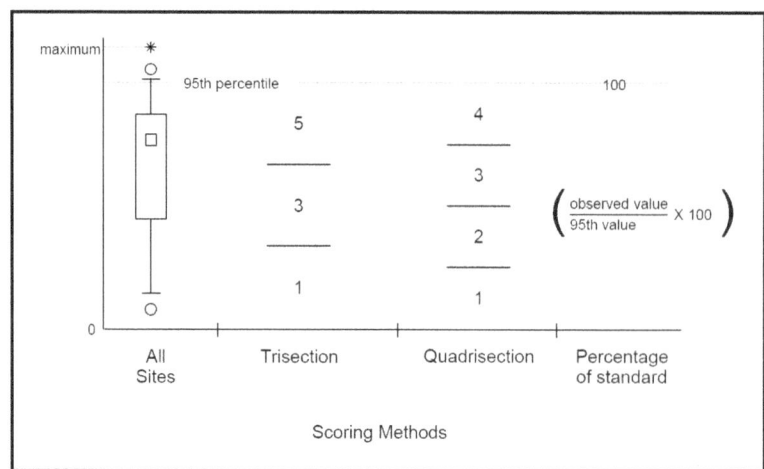

Figure 9-7. Basis of metric scores using the 95th percentile as a standard.

scatterplot of the metric and covariate (Figure 9-3) and a moving estimate of the appropriate percentile, are used to determine scoring criteria as a function of the covariate (e.g., Fausch et al. 1984, Plafkin et al. 1989).

! Test the ability of the final index to discriminate between populations of reference and anthropogenically affected (stressed) sites (Figure 9-8). Generally, indices (aggregate of metrics) discriminate better than individual metrics (e.g., total taxa is generally a weak metric because of inconsistency in taxonomic resolution). Those sites that are misclassified with regard to "reference" and "stressed" can be identified and evaluated for reassignment.

Step 5. Index thresholds for assessment and biocriteria

The multimetric index value for a site is a summation of the scores of the metrics and has a finite range within each stream class and index period depending on the maximum possible scores of the metrics (Barbour et al. 1996c). This range can be subdivided into any number of categories corresponding to various levels of impairment. Because the metrics are normalized to reference conditions and expectations for the stream classes, any decision on subdivision should reflect the distribution of the scores for the reference sites. For example, division of the Wyoming benthic IBI range (aggregation of metric scores) within each stream class provides 5 ordinal rating categories for assessment of impairment (Stribling et al. 1999, Figure 9-8).

Biocriteria are based on *thresholds* determined to differentiate impaired from non-impaired conditions. While these thresholds may be subjective, the performance of the *a priori* selected reference sites will ultimately verify the appropriateness of the threshold.

The 5 rating categories are used to assess the condition of both reference and non-reference sites. Most of the reference sites should be rated as *good* or *very good* in biological condition, which would be as expected. However, a few reference sites may be given the rating as *poor* sporadically among the collection dates. If a "reference" site consistently receives a fair or poor rating, then the site should be re-evaluated as to its proper assignment. Putative reference sites may be rated "poor" for several reasons:

! **Natural variability** — owing to seasonal, spatial, and random biological events, any reference site may score below the reference population 10th percentile. If due to natural variability, a low score should occur 10% of the time or less.

! **Impairment** — stressors that were not detected in previous sampling or surveys may occur at a

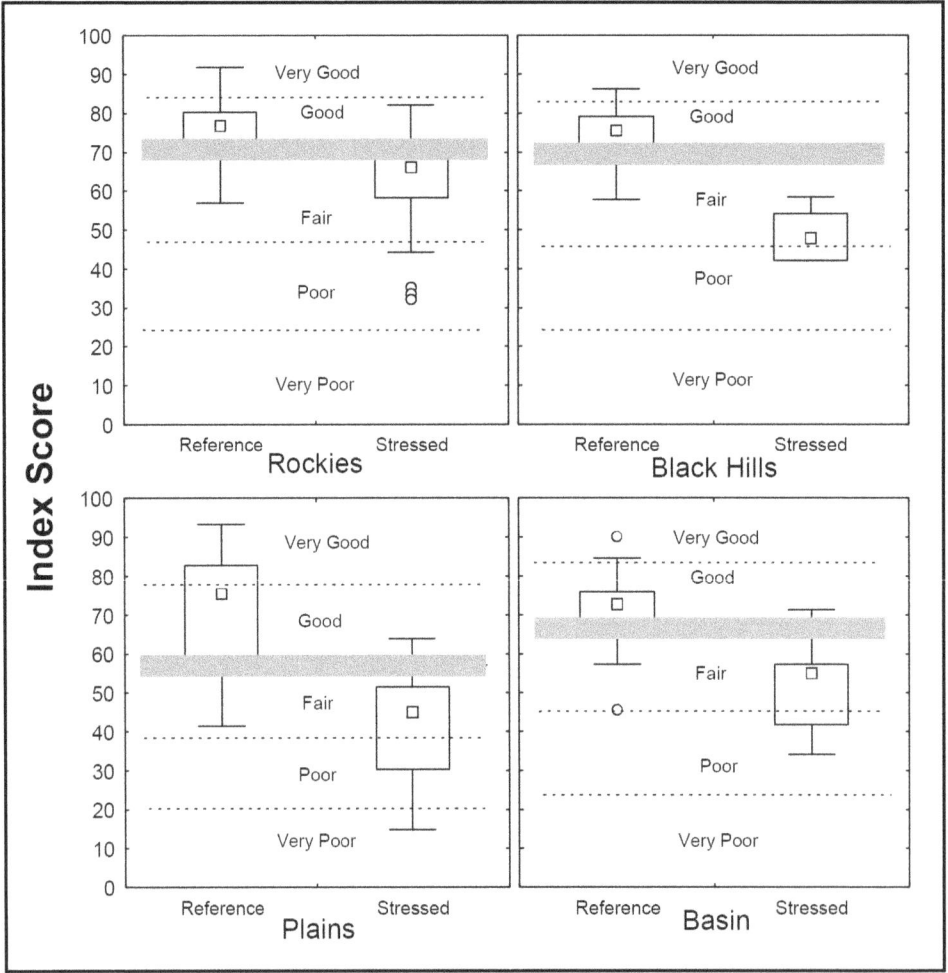

Figure 9-8. Discriminatory power analysis of the Wyoming Benthic Index of Biotic Integrity. The population of stressed sites was determined *a priori*. The 25th percentile of the reference distribution determined the threshold, or separation between "good" and "fair" condition ratings. All other condition ratings resulted from equidistant sectioning of the remaining index range. The shaded region represents the 90% confidence limits around a single observation (no replication) falling near the critical threshold.

Chapter 9: Multimetric Data Analysis

"reference" site; for example, episodic non-point-source pollution or historical contamination may be present at a site.

! **Non-representative site** — reference sites are intended to be representative of their class. If there are no anthropogenic stressors, yet a "reference" site consistently scores outside the range of the rest of the reference population the site may be a special or unique case, or it may have been misclassified and actually belong to another class of streams.

An understanding of variability is necessary to ensure that sites that are near the threshold are rated with known precision (discussed in more detail in Chapter 4). To account for variance associated with measurement error in an assessment, replication is required. The first step is to estimate the standard deviation of repeated measures of streams. The standard deviation is calculated as the root mean square error (RMSE) of an analysis of variance (ANOVA), where the sites are treatments in the ANOVA.

As an example, the question of precision was tested for the Wyoming Benthic IBI scores in the stream classes. This study showed that the 95% confidence interval (CI) around a single sample is ±8 points, on a scale of 100 (Table 9-2). What if a single site was sampled with no replication and found to be points below the biocriterion? The rightmost column (Table 9-2) shows that a triplicate sample is required for a 95% CI less than 5 points. These conclusions make 3 assumptions:

! measurement error is normally distributed,

! measurement error is not affected by subecoregion or impairment, and

! the sample standard deviation of repeated measures is an unbiased and precise estimate of population measurement error.

Components of Step 5 include:

! The range in possible scores for each stream class is the minimum number of metrics (if a score of 1 is assigned to greatest level of degradation) to the maximum aggregate of scores. Pentasect, quadrisect, or trisect this range, depending on how many biological condition categories are desired.

! Evaluate the validity of these biological condition categories by comparing the index scores of the reference and known stressed sites to those categories. If reference sites are not rated as good or very good, then some adjustment in either the biological condition designations or the listing of reference sites may be necessary.

! Test for confidence in multimetric analysis to determine biological condition for sites that fall within close proximity to threshold. Calculate precision and sensitivity values to determine repeatability and detectable differences that will be important in the confidence level of the assessment.

Rapid Bioassessment Protocols for Use in Streams and Wadeable Rivers: Periphyton, Benthic Macroinvertebrates, and Fish, Second Edition

9-13

Table 9-2. Statistics of repeated samples in Wyoming and the detectable difference (effect size) at 0.10 significance level. The index is on a 100 point scale (taken from Stribling et al. 1999).

Metric	Standard Deviation for Repeated Measures	Approx. Mean[a]	Approx. Coefficient of Variation (%)	Detectable Differences (p = 0.10)		
				Single Sample	Duplicate Samples	Triplicate Samples
Total Taxa	4.1	35.9	11.5	7 taxa	5 taxa	5 taxa
Ephemeroptera taxa	0.9	6.8	13.3	2 taxa	1 taxa	1 taxa
Plecoptera taxa	1.0	4.8	21.2	2 taxa	1 taxa	1 taxa
Trichoptera taxa	1.1	6.9	15.3	2 taxa	1 taxa	1 taxa
% non-insects	3.8	8.9	42.9	6.3 %	4.4 %	4.3 %
% diptera (non-chironomid)	1.3	5.1	25.0	2.1 %	1.5 %	1.4 %
HBI	0.27	3.43	7.85	0.44 units	0.31 units	0.26 units
% 5 dominant taxa	4.3	64.2	6.7	7.1 %	5.0 %	4.1 %
% scrapers	4.8	25.5	18.9	7.9 %	5.6 %	4.6 %
Index	2.0	70.0	2.9	3.3 units	2.3 units	1.9 units

a: Mean of 25 replicated sites; population means may differ.

9.1.2 Assessment of Biological Condition

Once the framework for bioassessment is in place, conducting bioassessments becomes relatively straightforward. Either a targeted design that focuses on site-specific problems or a probability-based design, which has a component of randomness and is appropriate for 305(b), area-wide, and watershed monitoring, can be done efficiently. Routine monitoring of reference sites should be based on a random selection procedure, which will allow cost efficiencies in sampling while monitoring the status of the reference condition of a state's streams. Potential reference sites of each stream class would be randomly selected for sampling, so that an unbiased estimate of reference condition can be developed. A randomized subset of reference sites can be resampled at some regular interval (e.g., a 4 year cycle) to provide information on trends in reference sites.

A reduced effort in monitoring reference sites allows more investment of time into assessing other stream reaches and problem sites. Through use of Geographical Information System (GIS) and station location codes, assessment sites throughout the state can be randomly selected for sampling as is being done for the reference sites. This procedure will provide a statistically valid means of estimating attainment of aquatic life use for the state's 305(b) reporting. In addition, the multimetric index will be helpful for targeted sampling at specific problem areas and judging biological condition with a procedure that has been calibrated regionally (Barbour et al. 1996c). To evaluate possible influences on the biological condition of sites, relationships among total bioassessment scores and physicochemical variables can be investigated. These relationships may indicate the influence of particular categories of stressors on the biological condition of individual sites. For example, a strong negative correlation between total bioassessment score and embeddedness would suggest that siltation from nonpoint sources could be affecting the biological condition at a site. Considerations relevant to assessment and diagnostics of biological condition are as follows:

> ! Evaluate the relationship of biological response signatures such as functional attributes (reproduction, feeding group responses, etc.) to specific stressors.

!   Hold physical habitat relationships constant and look for associations with other physical stressors (e.g., hydrologic modification, streambed stability), chemical stressors (e.g., point-source discharges or pesticide application to cropland), biological stressors (i.e., exotics), and landscape measures (e.g., impervious surface, Thematic mapper land use classes, human population census information, landscape ecology parameter of dominance, contagion, fractal dimension).

!   Explore the relationship between historical change in biota and change in landscape (e.g., use available historical data from the state or region).

9.2   DISCRIMINANT MODEL INDEX

Discriminant analysis may be used to develop a model that will divide, or discriminate, observations among two or more predetermined classes. Output of discriminant analysis is a function that is a linear combination of the input variables, and that obtains the maximum separation (discrimination) among the defined classes. The model may then be used to determine class membership of new observations. Thus, given a set of unaffected reference sites, and a set of degraded sites (due to toxicity, low DO, or habitat degradation), a discriminant function model can identify variables that will discriminate reference from degraded sites.

Developing biocriteria with a discriminant model requires a training data set to develop the discriminant model, and a confirmation data set to test the model. The training and confirmation data may be from the same biosurvey, randomly divided into two, or they may be two consecutive years of survey data, etc. All sites in each data set are identified by degradation class (e.g., reference vs stressed) or by designated aquatic life use class. To avoid circularity, identification of reference and stressed, or of designated use classes, should be made from non-biological information such as quality of the riparian zone and other habitat features; presence of known discharges and nonpoint sources, extent of impervious surface in the watershed, extent of land use practices, etc.

One or more discriminant function models are developed from the training set, to predict class membership from biological data. After development, the model is applied to the confirmation data set to determine its performance: The test determines how well the model can assign sites to classes, using independent data that were not used to develop the model. More information on discriminant analysis is in any textbook on multivariate statistics (e.g., Ludwig and Reynolds 1988, Jongman et al. 1987, Johnson and Wichern 1992).

An example of this approach is the hierarchical decision-making technique used by Maine DEP. It begins with statistical models (linear discriminant analysis) to make an initial prediction of the classification of an unknown sample by comparing it to characteristics of each class identified in the baseline database (Davies et al. 1993). The output from analysis by the primary statistical model is a list of probabilities of membership for each of four groups designated as classes A, B, C, and nonattainment (NA) of Class C (Table 9-3). Subsequent models are designed to distinguish between a given class and any higher classes as one group, and any lower classes as a second group.

One or more discriminant models to predict class membership are developed from the training set. The purpose of the discriminant analysis here is not to test the classification (the classification is administrative rather than scientific), but to assign test sites to one of the classes.

Stream biologists from Maine DEP assigned a training set of streams to four life use classes. In operational assessment, sites are evaluated with the two-step hierarchical models. The first stage linear discriminant model is applied to estimate the probability of membership of sites into one of the four classes (A, B, C, or NA). Second, the series of two-way models are applied to distinguish the membership between a given class and any higher classes, as one group. The model uses 31 quantitative measures of community structure, including the Hilsenhoff Biotic Index, Generic Species Richness, EPT, and EP values. Monitored test sites are then assigned to one of the four classes based on the probability of that result, and uncertainty is expressed for intermediate sites. The classification can be the basis for management action if a site has gone down in class, or for reclassification to a higher class if the site has improved.

Table 9-3. Maine's water quality classification system for rivers and streams, with associated biological standards (taken from Davies et al. 1993).

Aquatic Life Use Class	Management	Biological Standard	Discriminant Class
AA	High quality water for recreation and ecological interests. No discharges or impoundments permitted.	Habitat natural and free flowing. Aquatic life as naturally occurs.	A
A	High quality water with limited human interference. Discharges restricted to noncontact process water or highly treated wastewater equal to or better than the receiving water. Impoundments allowed.	Habitat natural. Aquatic life as naturally occurs.	A and AA are indistinguish-able because biota are "as naturally occurs."
B	Good quality water. Discharge of well treated effluent with ample dilution permitted.	Habitat minimally impaired. Ambient water quality sufficient to support life stages of all indigenous aquatic species. Only nondetrimental changes in community composition allowed.	B
C	Lowest water quality. Maintains the interim goals of the Federal Water Quality Act (fishable/swimmable). Discharge of well-treated effluent permitted.	Ambient water quality sufficient to support life stages of all indigenous fish species. Change in community composition may occur but structure and function of the community must be maintained.	C
NA			Not attaining Class C

Maine biocriteria thus establish a direct relationship between management objectives (the three aquatic life use classes and nonattainment) and biological measurements. The relationship is immediately viable for management and enforcement as long as the aquatic life use classes remain the same. If the classes are redefined, a complete reassignment of streams and a review of the calibration procedure would be necessary. This approach is detailed by Davies et al. (1993).

See Maine DEP's website for more information
http://www.state.me.us/dep/blwq/biohompg.htm

9.3 RIVER INVERTEBRATE PREDICTION AND CLASSIFICATION SCHEME (RIVPACS)

RIVPACS and its derivative, AusRivAS (Australian Rivers Assessment System) are empirical (statistical) models that predict the aquatic macroinvertebrate fauna that would be expected to occur at a site in the absence of environmental stress (Simpson et al. 1996). The AusRivAS models predict the invertebrate communities that would be expected to occur at test sites in the absence of impact. A comparison of the invertebrates predicted to occur at the test sites with those actually collected provides a measure of biological impairment at the tested sites. The predicted taxa list also provides a "target" invertebrate community to measure the success of any remediation measures taken to rectify identified impacts. The type of taxa predicted by the AusRivAS models may also provide clues as to the type of impact a test site is experiencing. This information can be used to facilitate further investigations e.g., the absence of predicted Leptophlebiidae may indicate an impact on a stream from trace metal input.

These models are the primary ecological assessment analysis techniques for Great Britain (Wright et al. 1993) and Australia (Norris 1995). The models are based on a stepwise progression of multivariate and univariate analyses and have been developed for several regions and various habitat types found in lotic systems. Regional applications of the AusRivAS model, in particular, have been developed for the Australian states and territories (Simpson et al. 1996), and for streams in the Sierra and Cascade mountain ranges in California (Hawkins and Norris 1997). Users of these models claim rapid turn around of results is possible and output can be tailored for a range of users including community groups, managers, and ecologists. These attributes make RIVPACS and AusRivAS likely candidate analysis techniques for rapid bioassessment programs.

Although the same procedures are used to build all AusRivAS models, each model is tailored to specific regions (or states) to provide the most accurate predictions for the season and habitat sampled. The stream habitats for which these models have been applied include the edge/backwater, main channel, riffle, pool, and macrophyte stands. The multihabitat sampling techniques used in many RBP programs have not yet been tested with a RIVPACS model. The models can be constructed for a single season, or data from several seasons may be combined to provide more robust predictions. To date the RIVPACS/AusRivAs models have only been developed for the benthic assemblage. Discussion of RIVPACS and AusRivAS is taken from the *Australian River Assessment System National River Health Program Predictive Model Manual* by Simpson et al. (1996). As is the case with the multimetric approach, a more thorough treatment of the RIVPACS/AusRivAS models can be obtained by referring to the citations of the supporting documentation provided in this discussion.

> The reader is directed to the AusRivAS website for more specific information and guidance regarding these multivariate techniques.
> **http://ausrivas.canberra.edu.au/ausrivas**

Rapid Bioassessment Protocols for Use in Streams and Wadeable Rivers: Periphyton, Benthic Macroinvertebrates, and Fish, Second Edition

9-17

This Page Intentionally Left Blank

10 DATA INTEGRATION AND REPORTING

Human impacts on the biological integrity of water resources are complex and cumulative (Karr 1998). Karr (1998) states that human actions jeopardize the biological integrity of water resources by altering one or more of five principal factors — physical habitat, seasonal flow of water, the food base of the system, interactions within the stream biota, and chemical quality of the water. These factors can be addressed in environmental management by shifting our focus from technology-based to water resource-based management strategies. This change in focus requires a commensurate shift from the measurement of pollutant loadings to a measurement of ecosystem health. Biological assessment addresses ecosystem health and cumulative impacts by concentrating on population and community level response rather than on discharger performance (Courtemanch 1995).

The translation of biological data into a report that adequately conveys the message of the assessment is a critical process. It is important to identify the intended audience(s) for the report and to bear in mind that users of the report will likely include groups (i. e. managers, elected officials, communities) who are not biologists. Reports must be coherent and easily understood in order for people to make informed decisions regarding the water resource. First, the data must be summarized and integrated, then clearly explained and presented. The use of a multimetric index provides a convenient, yet technically sound method for summarizing complex biological data for each assemblage (Karr et al. 1986, Plafkin et al. 1989). The procedures for developing the Multimetric Index for each assemblage is described in Chapter 9. The index itself is only an aggregation of contributory biological information and should not be used exclusive of its component metrics and data (Yoder 1991, Barbour et al. 1996a). However, the index and its component metrics serve as effective tools to communicate biological status of a water resource.

10.1 DATA INTEGRATION

Once indices and values are obtained for each assemblage, the question becomes how to interpret all of the results, particularly if the findings are varied and suggest a contradiction in assessment among the assemblages? Also, how are habitat data used to evaluate relationships with the biological data? These questions are among the most important that will be addressed in this chapter. The integration of chemical and toxicological data with biological data is not treated in depth here. It is briefly described in Chapter 3 and discussed in more detail elsewhere (Jackson 1992, USEPA 1997c).

10.1.1 Data Integration of Assemblages

USEPA advises incorporating more than 1 assemblage into biocriteria programs whenever practical. Surveying multiple assemblages provides a more complete assessment of biological condition since the various assemblages respond differently to certain stressors and restoration activities. For instance, Ohio EPA found, in a study of the Scioto River, that fish responded (recovered) more quickly than did benthos to restoration activities aimed at reducing the effects of cumulative impacts (i.e., impoundments, combined sewer overflows, wastewater treatment plants, urbanization) (Yoder and Rankin 1995a). Although significant improvement was observed in the condition of both assemblages in the river from 1980 to 1991, the benthic assemblage was still impaired in several reaches of the

Rapid Bioassessment Protocols for Use in Streams and Wadeable Rivers: Periphyton, Benthic Macroinvertebrates, and Fish, Second Edition

10-1

river; whereas, the fish assemblage met Ohio's warm water habitat criterion in 1991 for many of the same reaches. The use of both assemblages enhanced the agency's assessment of trend analysis for the Scioto River.

In addition, using more than 1 assemblage allows programs to more fully assess the occurrence of multiple stressors and seasonal variation in the intensity of the stressors (Gibson et al. 1996). Mount et al. (1984) found that benthic and fish assemblages responded differently to the same inputs in the Ottawa River in Ohio. Benthic diversity and abundance responded negatively to organic loading from a wastewater treatment plant and exhibited no observable response to chemical input from industrial effluent. Fish exhibited no response to the organic inputs and a negative response to metal concentrations in the water.

Integration of information from each assemblage should be done such that the results complement and supplement the assessment of the site. Trend analysis (monitoring changes over time) is useful to illustrate differences in response of the assemblages (Figure 10-1). In this example of the Scioto River (Figure 10-1), the improvement in the fish Index of Biotic Integrity (IBI) and the benthic macroinvertebrate Index of Community Integrity (ICI)

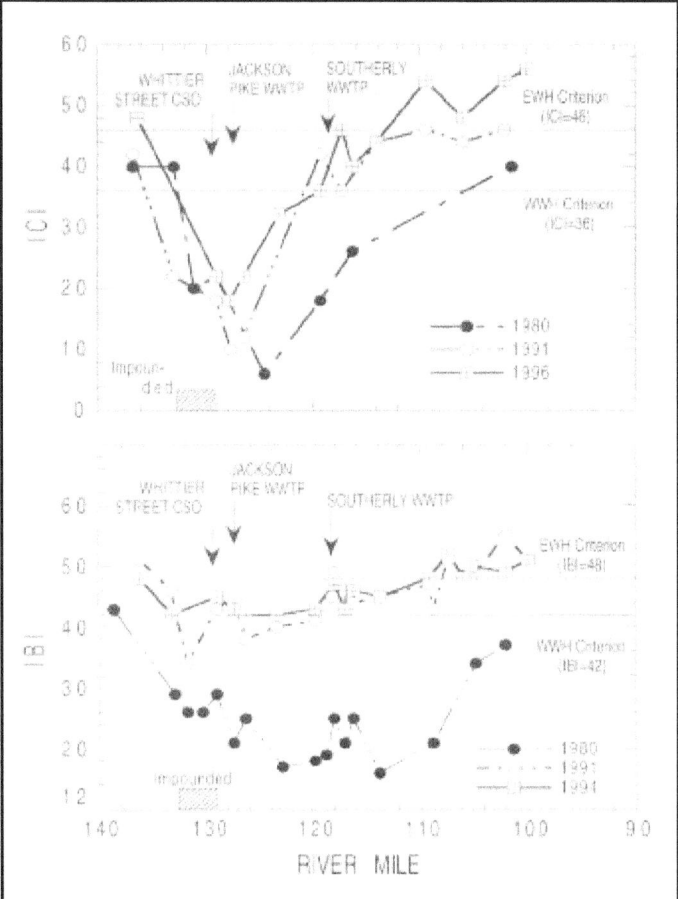

Figure 10-1. Cumulative frequency diagrams (CFD) for the IBI (upper) and the ICI (lower) comparing the pre-1988 and post-1988 status on a statewide basis from Ohio. In each case, estimated attainable level of future performance is indicated. The Warm Water Habitat (WWH) and Exceptional Warm Water Habitat (EWH) biological thresholds are given for each index.

assemblages can be seen over time (1980 and 1991) and over a length of the river (River Mile [RM] 140 to 90) (Yoder 1995a).

Biological attributes and indices can also be illustrated side-by-side to highlight differences and similarities in the results. Oftentimes, differences in the results are useful for diagnosing cause-and-effect.

10.1.2 Relationship Between Habitat and Biological Condition

Historically, non-chemical impacts to biotic systems have not been a major focus of the nation's water quality agencies. Yet there is clear evidence that habitat alteration is a primary cause of degraded aquatic resources (USEPA 1997c). Habitat degradation occurs as a result of hydrological flow modification, alteration of the system's energy base, or direct impact on the physical habitat structure. Preservation of an ecosystem's natural physical habitat is a fundamental requirement in maintaining diverse, functional aquatic communities in surface waters (Rankin 1995). Habitat quality is an

Chapter 10: Data Integration and Reporting

essential measurement in any biological survey because aquatic fauna often have very specific habitat requirements independent of water-quality composition (Barbour et al. 1996a). Diagnostic evaluations are enhanced when assessment of the habitat, flow regime, and energy base are incorporated into the interpretation of the biological condition (USEPA 1990b).

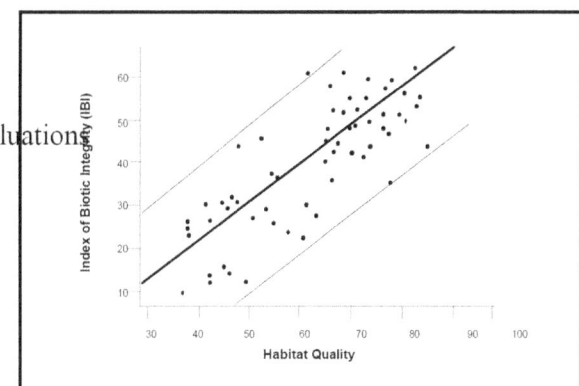

The relationship between habitat quality (as defined by site-specific factors, riparian quality, and upstream land use) and biological condition can be graphed, as illustrated in Figure 10-2 to enhance data interpretation. On the X-axis,

Figure 10-2. Relationship between the condition of the biological community and physical habitat.

habitat is shown to vary in quality from 30 points, which is poor (nonsupporting of an acceptable biological condition) to 85 points, which is good (comparable to the reference condition). Biological condition, represented by the fish IBI on the Y-axis, varies from 10 points (severely impaired) to 60 points (excellent). Interpretation of the relationship between habitat and biology as depicted by Figure 10-2 can be summarized by 4 points relating to specific areas of the graph.

1.   The upper right-hand corner of the curve is the ideal situation where optimal habitat quality and biological condition occur.

2.   The decrease in biological condition is proportional to a decrease in habitat quality.

3.   Perhaps the most important area of the graph is the lower right-hand corner where degraded biological condition can be attributed to something other than habitat quality (Barbour et al. 1996a).

4.   The upper left-hand corner is where optimal biological condition is not possible in a severely degraded habitat (Barbour et al. 1996a).

A relationship between biology and habitat should be substantiated with a large database sufficient to develop confidence intervals around a regression line. Rankin (1995) found that Ohio's visual-based habitat assessment approach, called the Qualitative Habitat Evaluation Index (QHEI), explained most of the variation in the IBI for the fish assemblage. However, Rankin also pointed out that covariate relationships between aggregate riparian quality and land use of certain subbasins could be used to partition natural variability. In one example, Rankin illustrated how high-quality patches of habitat structure in otherwise habitat-degraded stream reaches may harbor sensitive species, thus masking the effects of habitat alteration.

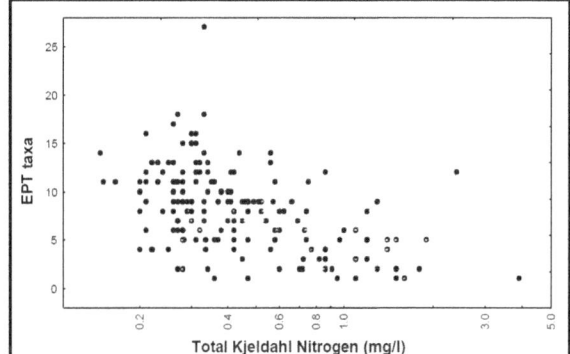

Figure 10-3. Data from a study of streams in Florida's Panhandle.

An informative approach to evaluating affects from specific or cumulative stressors is to

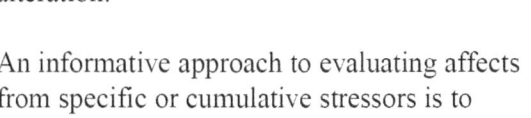

ascertain a gradient response of the aquatic community using a bivariate scatter plot. In one example provided by Florida DEP, a gradient response of the EPT taxa indicated a strong relationship to nitrogen in the stream (Figure 10-3).

When multiple data types (i.e., habitat, biological, chemical, etc.) are available, sun ray plots may be used to display the assessment results. As an example, the assessments of habitat, macroinvertebrates and fish are integrated for evaluating of the condition of individual stream sites in a Pennsylvania watershed (Snyder et al. 1998). The assessment scores for each of the triad data types are presented as a percentage of reference condition (Figure 10-4). The area enclosed by each sun ray plot can be measured to provide a comparison of the biological and habitat condition among the sites of interest (Snyder et al. 1998). This technique helps determine the extent of impairment and also which ecological components are most affected.

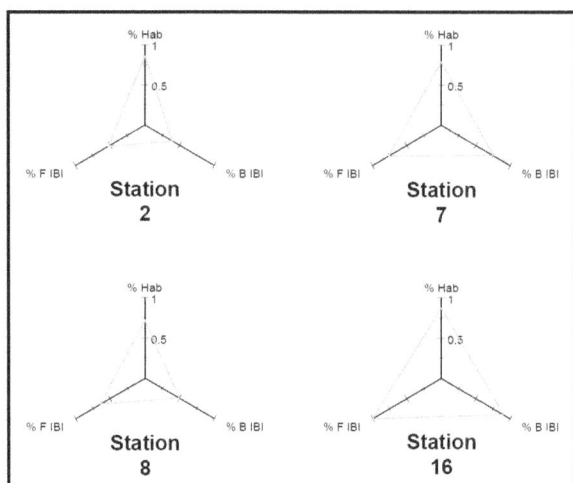

Figure 10-4. Comparison of integrated assessment (habitat, fish, and benthos) among stream sites in Pennsylvania. Station 16 is a reference site. (Taken from Snyder et al. 1998).

10.2   REPORTING

Historically, reports containing assessment results and recommendations for further action have been designed to address objectives and data uses relevant to the specific monitoring program. Increasingly, however, assessment reports are designed to reach a broader, non-scientific audience including water resource managers and the environmentally conscious public. Communicating the condition of biological systems, and the impact of human activities on those systems, is the ultimate purpose of biological monitoring (Karr and Chu 1999). Reporting style and format has become an important component in effectively communicating the findings of ecological assessments to diverse audiences. As pointed out by Karr and Chu (1999), effective communication can transform biological monitoring from a scientific exercise into a powerful tool for environmental decision making.

10.2.1 Graphical Display

Graphical displays are a fundamental tool for illustrating scientific information. Graphs reveal—more effectively than do strictly statistical tools—patterns of biological response. Patterns include "outliers," which may convey unique information that can help diagnose particular problems or reveal specific traits of a site (Karr and Chu 1999). Examples of some of the most useful graphical techniques are presented for specific biological program objectives:

1.   Stream classification — a graph should illustrate the distinction between and among site classes or groups. Two common graphical displays are bivariate scatter plots (used in non-metric multidimensional scaling) and cluster dendrograms.

Bivariate scatter plots—used for comparing the scatter or clustering of points given 2 dimensions. Can be used to develop regression lines or to incorporate 3 factors (3-dimensional) (Figure 10-5).

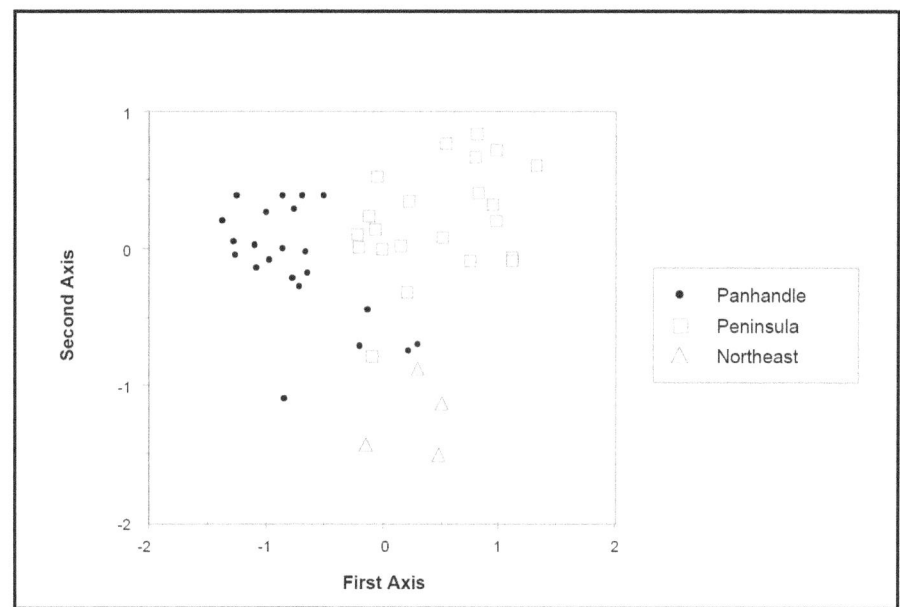

Figure 10-5. Use of multidimensional scaling on benthic data to ascertain stream classification. The first and second axes refer to the dimensions of combinations of data used to measure similarity (Taken from Barbour et al. 1996b).

Cluster dendrogram—used to illustrate the similarities and dissimilarities of sites in support of classes (Figure 10-6).

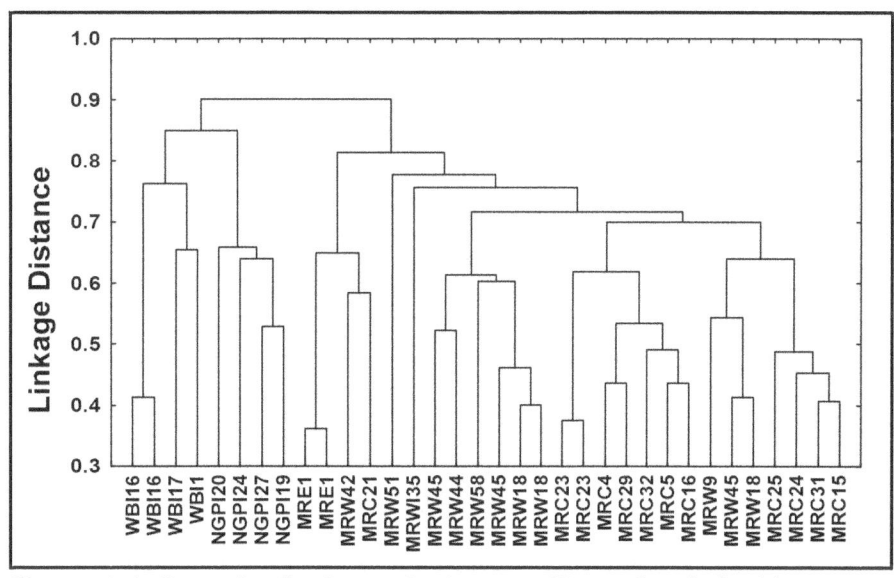

Figure 10-6. Example of a cluster dendrogram, illustrating similarities and clustering of sites (x-axis) using biological data.

2. Problem Identification and Status of Water Resource — The status of the condition of water resources requires consolidating information from many samples and can be illustrated in several ways.

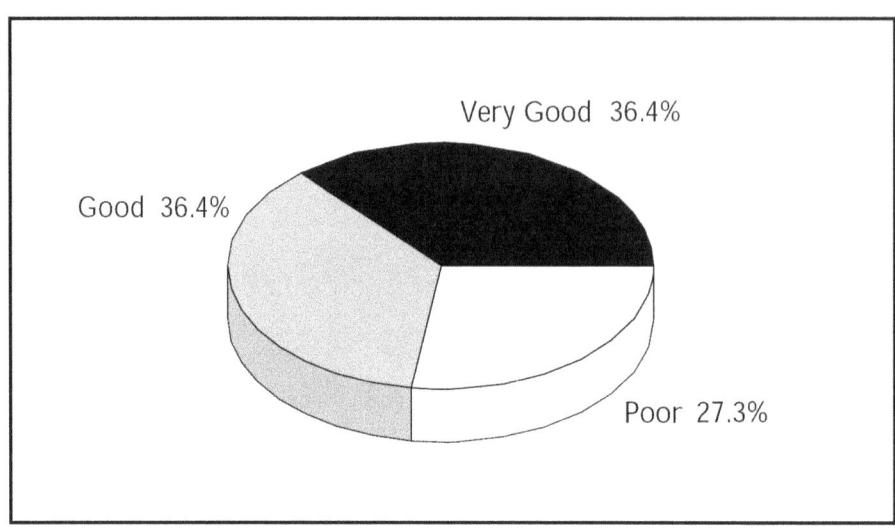

Figure 10-7. Results of the benthic assessment of streams in the Mattaponi Creek watershed of southern Prince George's County, Maryland. Percent of streams in each ecological condition category. (Taken from Stribling et al. 1996b).

Pie charts—used to illustrate proportional representation of the whole by its component parts. Can be sized according to magnitude or density (Figure 10-7)

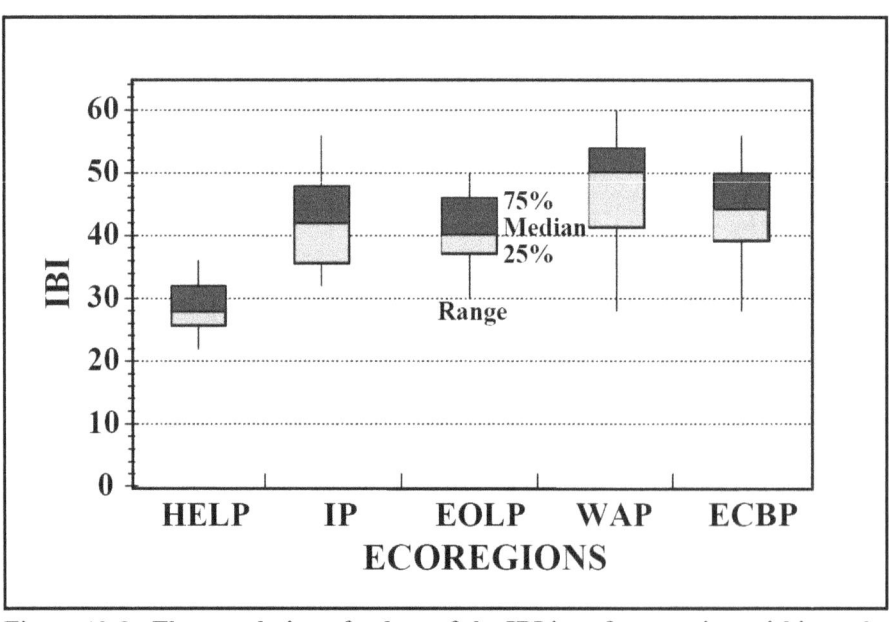

Box-and-whisker plots— used to illustrate population attributes (via percentile distribution) and provides some sense of variability (Figure 10-8).

Figure 10-8. The population of values of the IBI in reference sites within each of the ecoregions of Ohio. (Contributed by Ohio EPA).

Chapter 10: Data Integration and Reporting

Rapid Bioassessment Protocols for Use in Streams and Wadeable Rivers: Periphyton, Benthic Macroinvertebrates, and Fish, Second Edition

10-7

3.   Trend monitoring and assessment — Monitoring over a temporal or spatial scale requires a graphical display depicting trends, which may show improvement, degradation, or no change.

Line graphs—used to illustrate temporal or spatial trends that are contiguous. Assumes that linkage between points is linear (Figure 10-9).

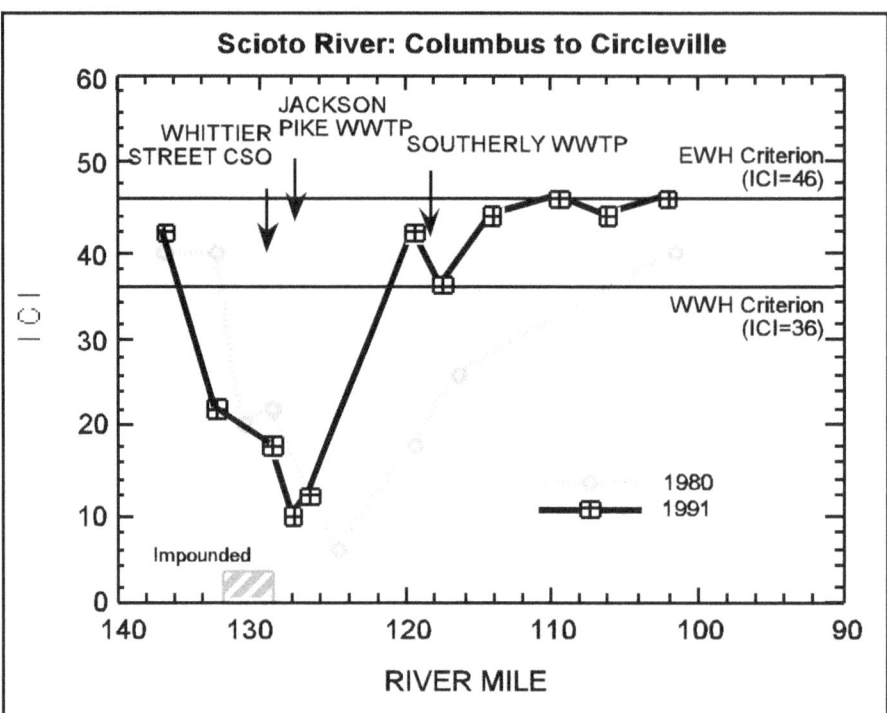

Figure 10-9. Spatial and temporal trend of Ohio's Invertebrate Community Index. The Scioto River - Columbus to Circleville. (Contributed by Ohio EPA).

Cumulative frequency diagram—illustrates an ordered accumulation of observations from lowest to highest value that allows one to determine status of resource at any given level (Figure 10-10).

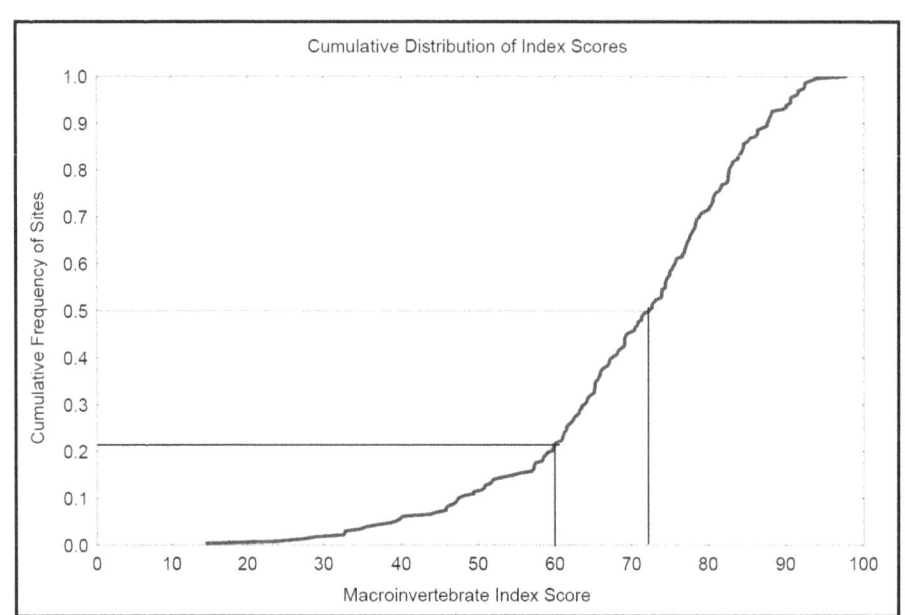

Figure 10-10. Cumulative distribution of macroinvertebrate index scores. 21% of sites scored at or below 60. The median index score is 75, where the cumulative frequency is 50%.

4. A determination of cause-and-effect — illustrating the source of impairment may not be a straightforward process. However, certain graphs lend themselves to showing comparative results in diagnosing problems.

Bar charts — used to display magnitude of values for discrete entities. Can be used to illustrate deviation from a value of central tendency (Figure 10-11).

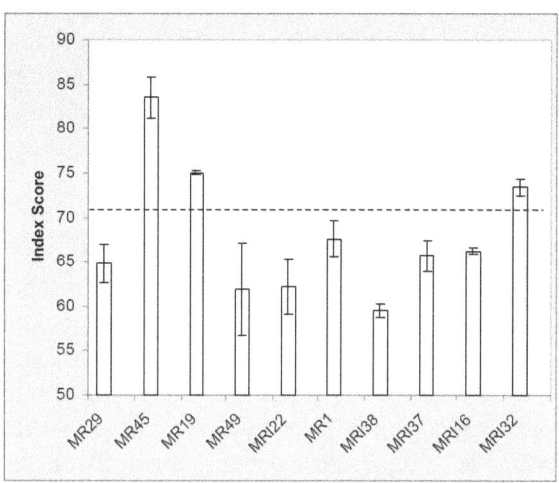

Figure 10-11. Biological assessment of sites in the Middle Rockies, showing mean and standard deviation of repeated measures and the assessment threshold (dashed line).

Sun Ray plots — used to compare more than 2 endpoints or data types. Most effective when reference condition is incorporated into axes or comparison (Figure 10-12).

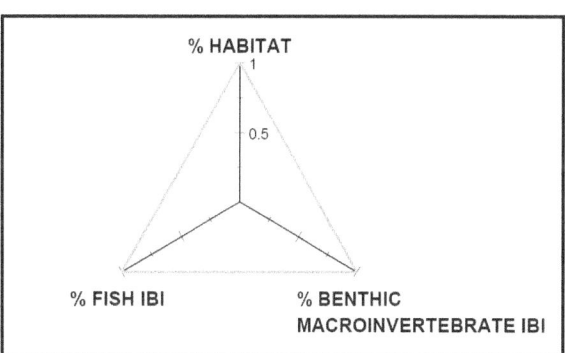

Figure 10-12. Integration of data from habitat, fish, and benthic assemblages.

Box-and-whisker plots— used to illustrate population attributes (via percentile distribution). Distinction among plots illustrates degree of similarity/differences (Figure 10-13).

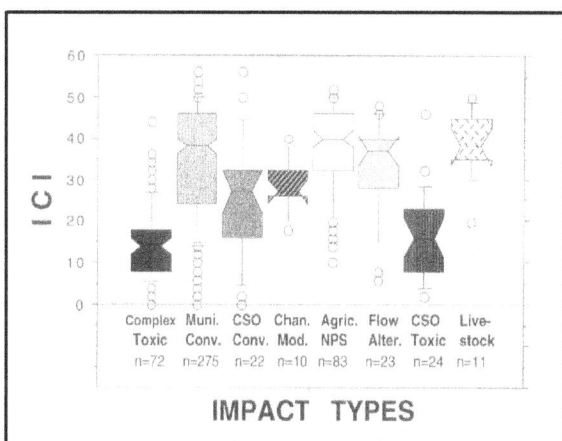

Figure 10-13. The response of the benthic macroinvertebrate assemblage (ICI) to various types of impacts (provided by Ohio EPA).

Rapid Bioassessment Protocols for Use in Streams and Wadeable Rivers: Periphyton, Benthic Macroinvertebrates, and Fish, Second Edition

10-9

10.2.2 Report Format

Two basic formats are recommended for reporting ecological assessments. Each of these formats is intended to highlight the scientific process, focus on study objectives, and judge the condition of the assessed sites. The first format is a summary report, targeted for use by managers in making decisions regarding the resource. This report format can also be an invaluable public information tool. The second report format is patterned after that of peer-reviewed journals and is primarily designed for informing a more technical audience.

The *Ecosummary* is an example of the first report format. It has an uncomplicated style and conveys various information including study results. The simplicity of this format quickly and effectively documents results and assists a non-technical audience in making informed decisions. An executive summary format is appropriate. An executive summary format is appropriate to present the "bottom line" assessment for the Ecosummary, which will be read by agency managers and decision-makers. Technical appendices or supplemental documentation should either accompany the report or be available to support the scientific integrity of the study.

These Ecosummaries are generally between 1-4 pages in length and lend themselves to quick and easy dissemination. Color graphics may be added to enhance the presentation or findings. An example of an Ecosummary format used by Florida Department of Environmental Protection (DEP) is illustrated in Figure 10-14. This 1-page report highlights the purpose of the study as well as the results and significance of the findings. A summary of the ecological data in the form of bar charts and tables may be provided on subsequent pages. Because this study follows prescribed methods and procedures, all of this documentation is not included in the report but is included in agency Standard Operating Procedures (SOPs).

The second format for reporting is a *scientific report*, which is structured similarly to a peer-reviewed journal. The report should be peer-reviewed by non-agency scientists to validate its scientific credibility. An abstract or executive summary should be prepared to highlight the essential findings. As in a peer-reviewed journal article, the methods and results are presented succinctly and clearly. The introductory text should outline the objectives and purpose of the study. A discussion of the results should include supporting literature to add credence to the findings, particularly if there is a discussion of suspected cause of impairment. Preparation of a report using this format will require more time than the Ecosummary. However, this report format is more inclusive of supportive information and will be more important in litigious situations.

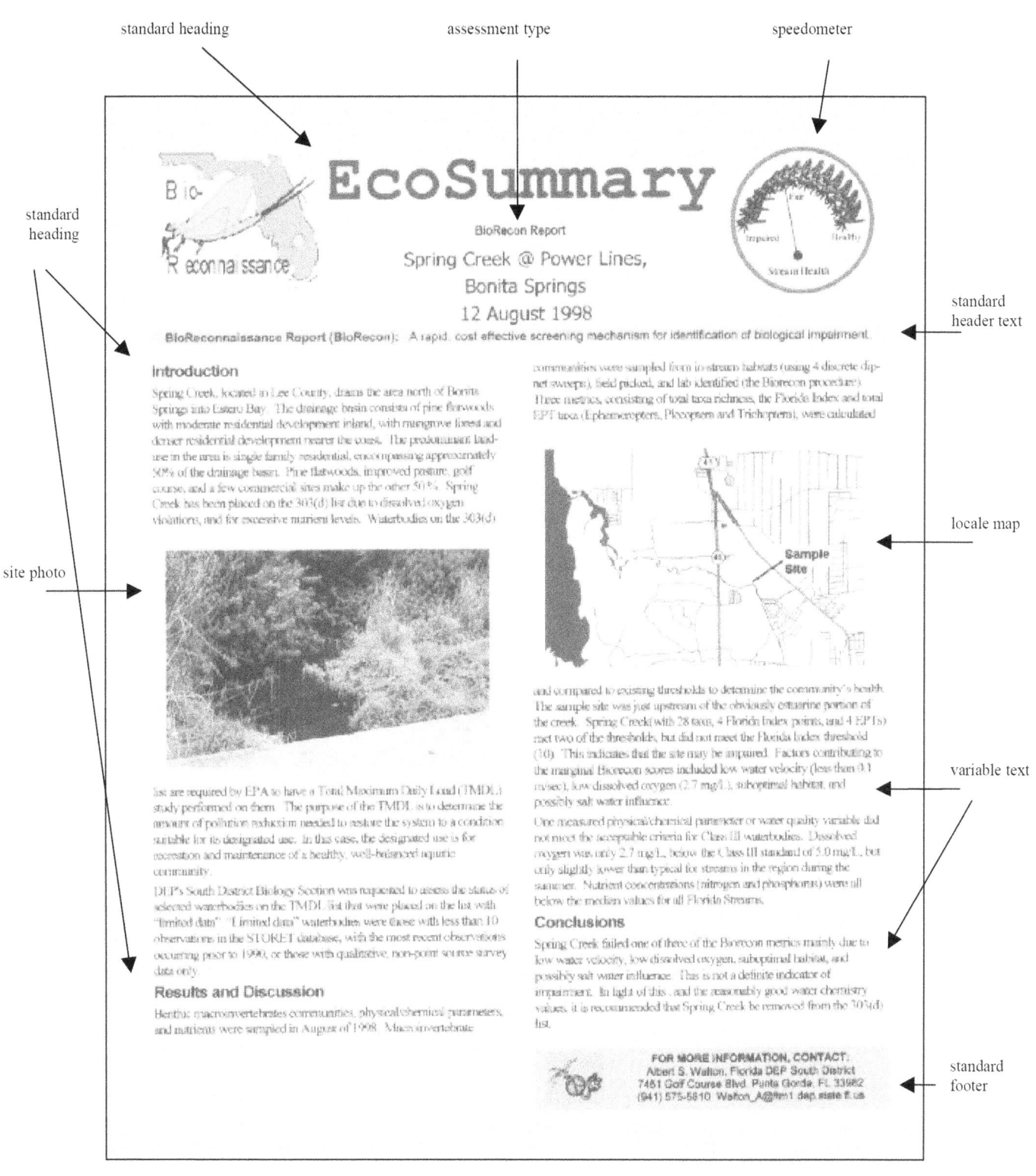

Figure 10-14. Guidance for Florida Ecosummary — A one-page bioassessment report. (Contributed by Florida DEP).

Rapid Bioassessment Protocols for Use in Streams and Wadeable Rivers: Periphyton, Benthic Macroinvertebrates, and Fish, Second Edition

10-11

11

LITERATURE CITED

Aloi, J.E. 1990. A critical review of recent freshwater periphyton field methods. *Canadian Journal of Fisheries and Aquatic Sciences* 47:656-670.

American Public Health Association (APHA), American Waterworks Association, and Water Pollution Control Federation. 1971. *Standard methods for the examination of water and wastewater*. American Public Health Association, Washington, D.C.

American Public Health Association (APHA). 1995. *Standard methods for the examination of water and wastewater*. American Public Health Association, American Water Works Association, and Water Pollution Control Federation. 19th edition, Washington, D.C.

American Society of Testing and Materials (ASTM). 1995. Biological effects and environmental fate. Volume 11.05. *Annual book of Standards: American Society of Testing and Materials*, Philadelphia, Pennsylvania.

Angermeier, P.L. and J.R. Karr. 1986. Applying an index of biotic integrity based on stream fish communities: Considerations in sampling and interpretation. *North American Journal of Fisheries Management* 6:418-429.

Angermeier, P.L. 1987. Spatiotemporal variation in habitat selection by fishes in small Illinois streams. Pages 52-60 *in* Matthews and Heins (eds.). *Community and Evolutionary Ecology of North American Stream Fishes*. University of Oklahoma Press, Norman, Oklahoma.

Armour, C.L., D.A. Duff, and W. Elmore. 1991. The effects of livestock grazing on riparian and stream ecosystems. *Fisheries* 16(1):7-11.

Bahls, L.L. 1993. *Periphyton bioassessment methods for Montana streams*. Montana Water Quality Bureau, Department of Health and Environmental Science, Helena, Montana.

Bahls, L.R., R. Burkantis, and S. Tralles. 1992. *Benchmark biology of Montana reference streams*. Department of Health and Environmental Science, Water Quality Bureau, Helena, Montana.

Bailey, R.G. 1976. *Ecoregions of the United States (Map scale 1:7,500,000)*. U.S. Department of Agriculture (USDA), Forest Service Ogden, Utah.

Bain, M.B. and J.M. Boltz. 1989. Regulated streamflow and warmwater stream fish: A general hypothesis and research agenda. *U.S. Fish and Wildlife Service Biological Report* 89(18):1-28.

Ball, J. 1982. *Stream Classification Guidelines for Wisconsin*. Wisconsin Department of Natural Resources Technical Bulletin. Wisconsin Department of Natural Resources, Madison, Wisconsin.

Rapid Bioassessment Protocols for Use in Streams and Wadeable Rivers: Periphyton, Benthic Macroinvertebrates, and Fish, Second Edition

11-1

Barbour, M.T. 1997. The re-invention of biological assessment in the U.S. *Human and Ecological Risk Assessment.* 3(6):933-940.

Barbour, M.T., and J.B. Stribling. 1991. Use of habitat assessment in evaluating the biological integrity of stream communities. *In* George Gibson, editor. *Biological criteria: Research and regulation, proceedings of a symposium, 12-13 December 1990, Arlington, Virginia.* Office of Water, U.S. Environmental Protection Agency, Washington, D.C. EPA-440-5-91-005.

Barbour, M.T., and J.B. Stribling. 1994. A technique for assessing stream habitat structure. Pages 156-178 in *Conference proceedings, Riparian ecosystems in the humid U.S.: Functions, values and management.* National Association of Conservation Districts, Washington, D.C. March 15-18, 1993, Atlanta, Georgia.

Barbour, M.T., and J. Gerritsen. 1996. Subsampling of benthic samples: A defense of the fixed-count method. *Journal of the North American Benthological Society* 15(3):386-391.

Barbour, M.T., J.L. Plafkin, B.P. Bradley, C.G. Graves, and R.W. Wisseman. 1992. Evaluation of EPA's rapid bioassessment benthic metrics: Metric redundancy and variability among reference stream sites. *Environmental Toxicology and Chemistry* 11(4):437-449.

Barbour, M.T., M.L. Bowman, and J.S. White. 1994. *Evaluation of the biological condition of streams in the Middle Rockies - Central ecoregion.* Prepared for Wyoming Department of Environmental Quality.

Barbour, M.T., J.B. Stribling, and J.R. Karr. 1995. Multimetric approach for establishing biocriteria and measuring biological condition. Pages 63-77 *in* W.S. Davis and T.P. Simon (editors). *Biological assessment and criteria. Tools for water resource planning and decision making.* Lewis Publishers, Boca Raton, Florida.

Barbour, M.T., J.M. Diamond, C.O. Yoder. 1996a. Biological assessment strategies: Applications and Limitations. Pages 245-270 *in* D.R. Grothe, K.L. Dickson, and D.K. Reed-Judkins (editors). *Whole effluent toxicity testing: An evaluation of methods and prediction of receiving system impacts*, SETAC Press, Pensacola, Florida.

Barbour, M.T., J. Gerritsen, G.E. Griffith, R. Frydenborg, E. McCarron, J.S. White, and M.L. Bastian. 1996b. A framework for biological criteria for Florida streams using benthic macroinvertebrates. *Journal of the North American Benthological Society* 15(2):185-211.

Barbour, M.T., J. Gerritsen, and J.S. White. 1996c. *Development of the stream condition index (SCI) for Florida.* Prepared for Florida Department of Environmental Protection, Tallahassee, Florida.

Barton, D.R., W.D. Taylor, and R.M. Biette. 1985. Dimensions of riparian buffer strips required to maintain trout habitat in southern Ontario streams. *North American Journal of Fisheries Management* 5:364-378.

Bauer, S.B., and T.A. Burton. 1993. *Monitoring protocols to evaluate water quality effects of grazing management on western rangeland streams.* U.S. Environmental Protection Agency, Region 10. Seattle, WA. EPA-910/R-93-017.

Beck, W.M., Jr. 1965. The Streams of Florida. *Bulletin of the Florida State Museum* 10(3):81-126.

Benke, A.C., T.C. Van Arsdall, Jr., and D.M. Gillespie. 1984. Invertebrate productivity in a subtropical blackwater river: The importance of habitat and life history. *Ecological Monographs* 54(1):25-63.

Rapid Bioassessment Protocols for Use in Streams and Wadeable Rivers: Periphyton, Benthic Macroinvertebrates, and Fish, Second Edition

11-3

Beschta, R.L. and W.S. Platts. 1986. Morphological features of small streams: Significance and function. *Water Resources Bulletin* 22(3):369-379.

Biggs, B. J. F. 1996. Patterns of benthic algae in streams. In *Algal Ecology: Freshwater Benthic Ecosystems*. R. J. Stevenson, M. Bothwell, and R. L. Lowe. pp. 31-55. Academic Press, San Diego, California, USA.

Bode, R.W. and M.A. Novak. 1995. Development and application of biological impairment criteria for rivers and streams in New York State. Pages 97-107 *in* W. S. Davis and T. P. Simon (editors). *Biological assessment and criteria: Tools for water resource planning and decision making.* Lewis Publishers, Ann Arbor, Michigan.

Brown, A.V. and P.P. Brussock. 1991. Comparisons of benthic invertebrates between riffles and pools. *Hydrobiologia* 220:99-108.

Brussock, P.P. and A.V. Brown. 1991. Riffle-pool geomorphology disrupts longitudinal patterns of stream benthos. *Hydrobiologia* 220:109-117.

Burton, T.A. and G.W. Harvey. 1990. *Estimating intergravel salmonid living space using the cobble embeddedness sampling procedure.* Water Quality Monitoring Protocols - Report No. 2. Idaho Department of Health and Welfare, Division of Environmental Quality, Water Quality Bureau, Boise, Idaho. September.

Cairns, J., Jr. 1982. *Artificial substrates.* Ann Arbor Science Publishers, Inc., Ann Arbor, Michigan.

Cairns, J., Jr. and R.L. Kaesler. 1971. Cluster analysis of fish in a portion of the Upper Potomac River. *Transactions of the American Fisheries Society* 100:750-756.

Cairns, J., Jr. and K.L. Dickson. 1971. A simple method for the biological assessment of the effects of waste discharges on aquatic bottom-dwelling organisms. *Journal of the Water Pollution Control Federation* 43:755-772.

Caton, L.W. 1991. Improving subsampling methods for the EPA "Rapid Bioassessment" benthic protocols. *Bulletin of the North American Benthological Society* 8(3):317-319.

Chessman, B.C. 1995. Rapid assessment of rivers using macroinvertebrates: A procedure based on habitat-specific sampling, family level identification and a biotic index. *Australian Journal of Ecology* (1995) 20:122-129.

Clements, W.H. 1987. The effect of rock surface area on distribution and abundance of stream insects. *Journal of Freshwater Ecology* 4(1):83-91.

Clifford, H.F. and R.J. Casey. 1992. Differences between operators in collecting quantitative samples of stream macroinvertebrates. *Journal of Freshwater Ecology* 7:271-276.

Cooper, C.M. and S. Testa III. 1999. Examination of revised rapid bioassessment protocols (RBP) in a watershed disturbed by channel incision. *Bulletin of the North American Benthological Society.* 16(1):198.

Cooper, J.M. and J.L. Wilhm. 1975. Spatial and temporal variability in productivity, species diversity, and pigment diversity of periphyton in a stream receiving domestic and oil refinery effluents. *Southwestern Naturalist* 19:413-428.

Corkum, L.D. 1989. Patterns of benthic invertebrate assemblages in rivers of northwestern North America. *Freshwater Biology* 21:191-205.

Courtemanch, D.L. 1995. Merging the science of biological monitoring with water resource management policy: Criteria development. Pages 315-325 *in* W.S. Davis and T.P. Simon (editors). *Biological assessment and criteria: Tools for water resource planning and decision making.* Lewis Publishers, Boca Raton, Florida.

Courtemanch, D.L. 1996. Commentary on the subsampling procedures used for rapid bioassessments. *Journal of the North American Benthological Society* 15:381-385.

Cox, E. J. 1996. *Identification of freshwater diatoms from live material.* Chapman & Hall, London.

Cuffney, T.G., M.E. Gurtz, and M.R. Meador, 1993a. *Guidelines for processing and quality assurance of benthic invertebrate samples collected as part of the National Water-Quality Assessment Program.* U.S. Geological Survey Open-File Report 93-407.

Cuffney, T.F., M.E. Gurtz, and M.R. Meador, 1993b. *Methods for collecting benthic invertebrate samples as part of the National Water-Quality Assessment Program.* U.S. Geological Survey Open-File Report 93-406.

Cummins, K.W. and M.J. Klug. 1979. Feeding ecology of stream invertebrates. *Annual Review of Ecology and Systematics* 10: 147-172.

Cummins, K.W., M.A. Wilzbach, D.M. Gates, J.B. Perry, and W.B. Taliaferro. 1989. Shredders and riparian vegetation. *Bioscience* 39(1):24-30.

Cushman, R.M. 1985. Review of ecological effects of rapidly varying flows downstream from hydroelectric facilities. *North American Journal of Fisheries Management* 5:330-339.

Davies, S.P., L. Tsomides, D.L. Courtemanch, and F. Drummond. 1993. *Maine Biological Monitoring and Biocriteria Development Program.* Maine Department of Environmental Protection, Bureau of Water Quality Control, Division of Environmental Evaluation and Lake Studies. Augusta, Maine.

Davis, W.S. and T.P. Simon (editors). 1995. *Biological assessment and criteria: Tools for water resource planning and decision making.* Lewis Publishers, Boca Raton, Florida.

Davis, W.S., B.D. Snyder, J.B. Stribling, and C. Stoughton. 1996. *Summary of State biological assessment programs for streams and rivers.* U.S. Environmental Protection Agency, Office of Planning, Policy, and Evaluation, Washington, D.C. EPA 230-R-96-007.

Descy, J.P. 1979. A new approach to water quality estimation using diatoms. *Nova Hedwigia* 64:305-323.

Rapid Bioassessment Protocols for Use in Streams and Wadeable Rivers: Periphyton, Benthic Macroinvertebrates, and Fish, Second Edition

11-5

DeShon, J.E. 1995. Development and application of the invertebrate community index (ICI). Pages 217-243 *in* W.S. Davis and T.P. Simon (editors). *Biological assessment and criteria: Tools for water resource planning and decision making.* Lewis Publishers, Boca Raton, Florida.

Diamond, J.M., M.T. Barbour, and J.B. Stribling. 1996. Characterizing and comparing bioassessment methods and their results: A perspective. *Journal of the North American Benthological Society.* 15:713-727.

Dixit, S.S., J.P. Smol, J. C. Kingston, and D.F. Charles. 1992. Diatoms: Powerful indicators of environmental change. *Environmental Science and Technology* 26:23-33.

Dodds, W. K., J. R. Jones, and E. B. Welch. 1998. Suggested classification of stream trophic status: Distributions of temperate stream types by chlorophyll, total nitrogen, and phosphorus. *Water Research* 32:1455-1462.

Elliott, J.M. and P.A. Tullett. 1978. A bibliography of samplers for benthic invertebrates. *Freshwater Biological Association*, Publication No. 4.

Energy, Mines, and Resources Canada. 1986. *Canada Wetland Regions (Map scale 1:7,500,000).* MCR 4108. Canada Map Office, Energy, Mines, and Resources Canada, Ottawa, Ontario.

Ettinger, W. 1984. Variation between technicians sorting benthic macroinvertebrate samples. *Freshwater Invertebrate Biology* 3:147-149.

Faith, D.P., P.R. Minchin, and L. Belbin. 1987. Compositional dissimilarity as a robust measure of ecological distance. *Vegetation.* 69:57-68.

Fausch, D.D., J.R. Karr, and P.R. Yant. 1984. Regional application of an index of biotic integrity based on stream fish communities. *Transactions of the American Fisheries Society* 113:39-55.

Fausch, K.D. and L.H. Schrader. 1987. *Use of the index of biotic integrity to evaluate the effects of habitat, flow, and water quality on fish communities in three Colorado Front Range streams.* Final Report to the Kodak-Colorado Division and the Cities of Fort Collins, Loveland, Greeley, Longmont, and Windsor. Department of Fishery and Wildlife Biology, Colorado State University, Fort Collins.

Ferraro, S.P., F.A. Cole, W.A. DeBen, and R.C. Schwartz. 1989. Power-cost efficiency of eight macrobenthic sampling schemes in Puget Sound, Washington. *Canadian Journal of Fisheries and Aquatic Sciences* 46:2157-2165.

Florida Department of Environmental Protection (FL DEP). 1996. *Standard operating procedures for biological assessment.* Florida Department of Environmental Protection, Biology Section. July 1996.

Fore, L.S., J.R. Karr, and R.W. Wisseman. 1996. Assessing invertebrate responses to human activities: Evaluating alternative approaches. *Journal of the North American Benthological Society* 15(2):212-231.

Funk, J.L. 1957. Movement of stream fishes in Missouri. *Transactions of the American Fisheries Society* 85:39-57.

Gallant, A.L., T.R. Whittier, D.P. Larsen, J.M. Omernik, and R.M. Hughes. 1989. *Regionalization as a tool for managing environmental resources.* U. S. Environmental Protection Agency, Environmental Research Laboratory, Corvallis, Oregon. EPA 600/3-89/060.

Gammon, J.R. 1976. *The fish population of the middle 340km of the Wabash River.* Purdue University Water Resources Research Center, LaFayette, Indiana. Technical Report 86.

Gammon, J.R. 1980. *The use of community parameters derived from electrofishing catches of river fish as indicators of environmental quality, in Seminar on Water Quality Management Tradeoffs.* U.S. Environmental Protection Agency, Washington, D.C. EPA-905/9-80-009.

Gammon, J.R., A. Spacie, J.L. Hamelink, and R.L. Kaesler. 1981. Role of electrofishing in assessing environmental quality of the Wabash River. Pages 307-324 STP 730 *in* J.M. Bates and C.I. Weber (editors). *Ecological Assessments of Effluent Impacts on Communities of Indigenous Aquatic Organisms.* American Society for Testing and Materials, Philadelphia, Pennsylvania.

Gerking, S.D. 1959. The restricted movement of fish populations. *Biological Review* 34:221-242.

Gerritsen, J. 1995. Additive biological indices for resource management. *Journal of the North American Benthological Society* 14(3):451-457.

Gerritsen, J. 1996. *Biological criteria: technical guidance for survey design and statistical evaluation of survey data. Volume 2. Development of biological indices.* Prepared for Office of Science and Technology, U.S. Environmental Protection Agencey, Washington, D.C.

Gibson, G.R. 1992. *Procedures for initiating narrative biological criteria.* Office of Science and Technology, U. S. Environmental Protection Agency, Washington, D.C. EPA-822- B-92-002.

Gibson, G.R., M.T. Barbour, J.B. Stribling, J. Gerritsen, and J.R. Karr. 1996. *Biological criteria: Technical guidance for streams and small rivers (revised edition).* U.S. Environmental Protection Agency, Office of Water, Washington, D. C. EPA 822-B-96-001.

Gislason, J.C. 1985. Aquatic insect abundance in a regulated stream under fluctuating and stable diel flow patterns. *North American Journal of Fisheries Management* 5:39-46.

Gordon, N.D., T.A. McMahon, and B.L. Finlayson. 1992. *Stream hydrology: an introduction for ecologists.* John Wiley and Sons, Inc., West Sussex, England.

Gore, J.A. and R.D. Judy, Jr. 1981. Predictive models of benthic macroinvertebrate density for use in instream flow studies and regulated flow management. *Canadian Journal of Fisheries and Aquatic Sciences* 38:1363-1370.

Gorman, O.T. 1988. The dynamics of habitat use in a guild of Ozark minnows. *Ecological Monographs* 58(1):1-18.

Gregory, S.V., F.J. Swanson, W.A. McKee, and K.W. Cummins. 1991. An ecosystem perspective of riparian zones. *BioScience* 41(8):540-551.

Rapid Bioassessment Protocols for Use in Streams and Wadeable Rivers: Periphyton, Benthic Macroinvertebrates, and Fish, Second Edition

11-7

Gurtz, M.E. 1994. Design considerations for biological components of the National Water Quality Assessment (NAWQA) program. Pages 323-354 *in* S.L. Loeb and A. Spacie (editors). *Biological monitoring of aquatic systems.* Lewis Publishers, Boca Raton, Louisiana.

Gurtz, M.E. and T.A. Muir. 1994. *Report of the interagency biological methods workshop.* U.S. Geological Survey, Denver, Colorado. Open-file Report 94-490.

Hall, L.W., M.C. Scott, and W.D. Killen. 1996. *Development of biological indicators based on fish assemblages in Maryland coastal plain streams.* Maryland Department of Natural Resources, Chesapeake Bay and Watershed Programs, Annapolis, Maryland. CBWP-MANTA-EA-96-1.

Halliwell, D.B., R.W. Langdon, R.A. Daniels, J.P. Kurtenbach, and R.A. Jacobson. 1999. Classification of freshwater fish species of the northeastern United States for use in the development of IBIs. Pages 301-337 *in* T.P. Simon (editor). *Assessing the sustainability and biological integrity of water resources using fish communities.* CRC Press, Boca Raton, Florida.

Hannaford, M.J. and V.H. Resh. 1995. Variability in macroinvertebrate rapid-bioassessment surveys and habitat assessments in a northern California stream. *Journal of the North American Benthological Society* 14:430-439.

Hannaford, M.J., M.T. Barbour, and V.H. Resh. 1997. Training reduces observer variability in visual-based assessments of stream habitat. *Journal of the North American Benthological Society* 16(4):853-860.

Hawkins, C.P., and R.H. Norris. 1997. Abstract — Comparison of the ability of multimetric and multivariate assessment techniques to detect biological impairment in mountainous streams of California. *Bulletin of the North American Benthological Society* 14(1):96.

Hawkins, C.P., M.L. Murphy, and N.H. Anderson. 1982. Effects of canopy, substrate composition, and gradient on the structure of macroinvertebrate communities in Cascade Range streams of Oregon. *Ecology* 63(6):1840-1856.

Hawkins, C.P., J.L. Kershner, P.A. Bisson, M.D. Bryant, L.M. Decker, S.V. Gregory, D.A. McCullough, C.K. Overton, G.H. Reeves, R.J. Steedman, and M.K. Young. 1993. A hierarchical approach to classifying stream habitat features. *Fisheries* 18:3-12.

Hayslip, G.A. 1993. *EPA Region 10 in-stream biological monitoring handbook (for wadable streams in the Pacific Northwest).* U.S. Environmental Protection Agency-Region 10, Environmental Services Division, Seattle, Washington. EPA-910-9-92-013.

Hendricks, M.L., C.H. Hocutt, and J.R. Stauffer, Jr. 1980. Monitoring of fish in lotic habitats. *In* C.H. Hocutt and J.R. Stauffer, Jr. (editors). *Biological Monitoring of Fish.* D. C. Heath Co., Lexington, Massachusetts.

Hicks, B.J., R.L. Beschta, and R. D. Harr. 1991. Long-term changes in streamflow following logging in western Oregon and associated fisheries implications. *Water Resources Bulletin* 27(2):217-226.

Hill, B. H. 1997. The use of periphyton assemblage data in an index of biotic integrity. *Bulletin of the North American Benthological Society* 14, 158.

Hill, J. and G.D. Grossman. 1987. Home range estimates for three North American stream fishes. *Copeia* 1987:376-380.

Hilsenhoff, W.L. 1987. An improved biotic index of organic stream pollution. *Great Lakes Entomologist* 20: 31-39.

Hilsenhoff, W.L. 1988. Rapid field assessment of organic pollution with a family level biotic index. *Journal of the North American Benthological Society* 7(1):65-68.

Hornig, C.E., C.W. Bayer, S.R. Twidwell, J.R. Davis, R.J. Kleinsasser, G.W. Linam, and K.B. Mayes. 1995. Development of regionally based biological criteria in Texas. Pages 145-152 *in* W.S. Davis and T.P. Simon (editors). *Biological assessment and criteria: Tools for water resource planning and decision making.* Lewis Publishers, Ann Arbor, Michigan.

Hughes, R.M. 1985. Use of watershed characteristics to select control streams for estimating effects of metal mining wastes on extensively disturbed streams. *Environmental Management* 9:253-262.

Hughes, R.M. 1995. Defining acceptable biological status by comparing with reference conditions. Pages 31-47 *in* W.S. Davis and T.P. Simon (editors). *Biological assessment and criteria: Tools for water resource planning and decision making.* Lewis Publishers, Ann Arbor, Michigan.

Hughes, R.M. and J.M. Omernik. 1983. An alternative for characterizing stream size. Pages 87-101 *in* T.D. Fontaine, III and S.M. Bartell (editors). *Dynamics of Lotic Ecosystems.* Ann Arbor Science Publishers, Ann Arbor, Michigan.

Hughes, R.M. and J.R. Gammon. 1987. Longitudinal changes in fish assemblages and water quality in the Willamette River, Oregon. *Transactions of the American Fisheries Society* 116(2):196-209.

Hughes, R.M. and D.P. Larsen. 1988. Ecoregions: An approach to surface water protection. *Journal of the Water Pollution Control Federation* 60:486-493.

Hughes, R.M., J.H. Gakstatter, M.A. Shirazi, and J.M. Omernik. 1982. An approach for determining biological integrity in flowing waters. Pages 877-888 *in* T.B. Brann (editor). *Inplace Resource Inventories: Principles and Practices, Proceedings of a National Workshop.* Society of American Foresters, Bethesda, Maryland.

Hughes, R.M., D.P. Larsen, and J.M. Omernik. 1986. Regional reference sites: A method for assessing stream potentials. *Environmental Management* 10:629-635.

Hughes, R.M., E. Rexstad, and C.E. Bond. 1987. The relationship of aquatic ecoregions, river basins, and physiographic provinces to the ichthyogeographic regions of Oregon. *Copeia* 1987:423-432.

Hughes, R.M., P.R. Kaufmann, A.T. Herlihy, T.M. Kincaid, L. Reynolds, and D.P. Larsen. 1998. A process for developing and evaluating indices of fish assemblage integrity. *Canadian Journal of Fisheries and Aquatic Sciences* 55:1618-1631.

Hunsacker, C.T. and D.A. Levine. 1995. Hierarchical approaches to the study of water quality in rivers. *Bioscience* 45(3):193-203.

Rapid Bioassessment Protocols for Use in Streams and Wadeable Rivers: Periphyton, Benthic Macroinvertebrates, and Fish, Second Edition

11-9

Hupp, C.R. 1992. Riparian vegetation recovery patterns following stream channelization: A geomorphic perspective. *Ecology* 73(4):1209-1226.

Hupp, C.R. and A. Simon. 1986. Vegetation and bank-slope development. *Proceedings of the Fourth Federal Interagency Sedimentation Conference* 4:83-92.

Hupp, C.R. and A. Simon. 1991. Bank accretion and the development of vegetated depositional surfaces along modified alluvial channels. *Geomorphology* 4:111-124.

Hurlbert, S.H. 1971. The nonconcept of species diversity: A critique and alternative parameters. *Ecology* 52:577-586.

Intergovernmental Task Force on Monitoring Water Quality (ITFM). 1992. *Ambient water quality monitoring in the United States. First year review, evaluation, and recommendations.* ITFM, Interagency Advisory Committee on Water Data, Water Information Coordination Program, U. S. Geological Survey, Washington, D.C.

Intergovernmental Task Force on Monitoring Water Quality (ITFM). 1995a. *The strategy for improving water-quality monitoring in the United States: Final report of the Intergovernmental Task Force on Monitoring Water Quality.* U.S. Geological Survey, Reston, Virginia.

Intergovernmental Task Force on Monitoring Water Quality (ITFM). 1995b. *The strategy for improving water-quality monitoring in the United States: Final report of the Intergovernmental Task Force on Monitoring Water Quality.* Technical appendixes. U.S. Geological Survey, Reston, Virginia.

Jackson, S. 1992. Re-examining independent applicability: Agency policy and current issues. Pages 135-138 *in* K. Swetlow (editor). *Water quality standards for the 21st century, proceedings of the third national conference.* Office of Science and Technology, U.S. Environmental Protection Agency, Washington, D.C. EPA 823-R-92-009.

Johnson, R.A. and D.W. Wichern. 1992. Applied multivariate statistical analysis. Third Edition. Prentice Hall, Englewood Cliffs, NJ.

Jongman, R.H., C.J.F. terBrook, and O.F.R. vanTongeren. 1987. *Data analysis in community and landscape ecology.* Pudoc Wageningen Publishing, Netherlands.

Karr, J.R. 1981. Assessment of biotic integrity using fish communities. *Fisheries* 66:21-27.

Karr, J.R. 1987. Biological monitoring and environmental assessment: A conceptual framework. *Environmental Management.* 11:249-256.

Karr, J.R. 1991. Biological integrity: A long-neglected aspect of water resource management. *Ecological Applications* 1:66-84.

Karr, J.R. 1993. Defining and assessing ecological integrity beyond water quality. *Environmental Toxicology and Chemistry* 12:1521-1531.

Karr, J.R. 1998. Rivers as sentinels: Using the biology of rivers to guide landscape management. Pages 502-528 *in* R.J. Naiman and R.E. Bilby, editors. *River Ecology and Management: Lessons from the Pacific Coastal Ecosystem.* Springer, NY.

Chapter 11: Literature Cited

Karr, J.R. and D.R. Dudley. 1981. Ecological perspectives on water quality goals. *Environmental Management* 5:55-68.

Karr, J.R., and E.W. Chu. 1997. Biological monitoring: Essential foundation for ecological risk assessment. *Human and Ecological Risk Assessment.* 3:933-1004.

Karr, J.R., and E.W. Chu. 1999. Restoring life in running waters: Better biological monitoring. Island Press, Washington, D.C.

Karr, J.R., K.D. Fausch, P.L. Angermeier, P.R. Yant, and I.J. Schlosser. 1986. *Assessing biological integrity in running waters: A method and its rationale.* Special publication 5. Illinois Natural History Survey.

Kaufmann, P.R. 1993. Physical Habitat. Pages 59-69 *in* R.M. Huges, ed. *Stream Indicator and Design Workshop.* EPA/600/R-93/138. U.S. Environmental Protection Agency, Corvallis, Oregon.

Kaufmann, P.R. and E.G. Robison. 1997. Physical Habitat Assessment. Pages 6-1 to 6-38 *in* D.J. Klemm and J.M. Lazorchak (editors). *Environmental Monitoring and Assessment Program. 1997 Pilot Field Operations Manual for Streams.* EPA/620/R-94/004. Environmental Monitoring Systems Laboratory, Office of Research and Development, U.S. Environmental Protection Agency, Cincinnati, Ohio.

Keller, E.A. and F.J. Swanson. 1979. Effects of large organic material on channel form and fluvial processes. *Earth Surface Processes.* 4:361-380.

Kentucky Department of Environmental Protection (KDEP). 1993. *Methods for assessing biological integrity of surface waters.* Kentucky Department of Environmental Protection, Division of Water, Frankfort, Kentucky.

Kerans, B.L., J.R. Karr, and S.A. Ahlstedt. 1992. Aquatic invertebrate assemblages: Spatial and temporal differences among sampling protocols. *Journal of the North American Benthological Society* 11:377-390.

Kerans, B.L. and J.R. Karr. 1994. A benthic index of biotic integrity (B-IBI) for rivers of the Tennessee Valley. *Ecological Applications* 4:768-785.

Klemm, D.J., P.A. Lewis, F. Fulk, and J.M. Lazorchak. 1990. *Macroinvertebrate field and laboratory methods for evaluating the biological integrity of surface waters.* U.S. Environmental Protection Agency, Environmental Monitoring and Support Laboratory, Cincinnati, Ohio. EPA-600-4-90-030.

Klemm, D.J., Q.J. Stober, and J.M. Lazorchak. 1993. *Fish field and laboratory methods for evaluating the biological integrity of surface waters.* Environmental Monitoring and Support Laboratory, U. S. Environmental Protection Agency, Cincinnati, Ohio. EPA/600/R-92/111.

Klemm, D.J. and J.M. Lazorchak (editors). 1994. *Environmental monitoring and assessment program -- surface waters and Region 3 regional environmental monitoring and assessment program. 1994. Pilot*

Rapid Bioassessment Protocols for Use in Streams and Wadeable Rivers: Periphyton, Benthic Macroinvertebrates, and Fish, Second Edition

11-11

field operation and methods manual for streams. Environmental Monitoring Systems Lab. Office of Research and Development, U.S. Environmental Protection Agency, Cincinnati, Ohio. EPA/620/R-94/004.

Klemm, D.J. and J.M. Lazorchak. 1995. *Environmental monitoring and assessment program — surface waters: Field operations and methods for measuring the ecological conditions of wadeable streams.* Environmental Monitoring Systems Laboratory, Office of Research and Development, U.S. Environmental Protection Agency, Cincinnati, Ohio. EPA/620/R-94/004.

Kolkwitz, R. and M. Marsson. 1908. Ecology of plant saprobia. [Translated 1967]. Pages 47-52 *in* L.E. Keup, W.M. Ingram and K.M. MacKenthum (eds.). *Biology of Water Pollution.* Federal Water Pollution Control Administration, Washington, DC.

Kuehne, R.A. and R.W. Barbour. 1983. *The American Darters.* University Press of Kentucky, Lexington, Kentucky.

Lange-Bertalot, H. 1979. Pollution tolerance as a criterion for water quality estimation. *Nova Hedwigia* 64:285-304.

Larsen, D.P., J.M. Omernik, R.M. Hughes, C.M. Rohm, T.R. Whittier, A.J. Kinney, A.L. Gallant, and D.R. Dudley. 1986. The correspondence between spatial patterns in fish assemblages in Ohio streams and aquatic ecoregions. *Environmental Management* 10:815-828.

Lazorchak, J.M., Klemm, D.J., and D.V. Peck (editors). 1998. *Environmental Monitoring and Assessment Program - Surface Waters: Field Operations and Methods for Measuring the Ecological Condition of Wadeable Streams.* EPA/620/R-94/004F. U.S. Environmental Protection Agency, Washington, D.C.

Lenat, D.R. 1993. A biotic index for the southeastern United States: Derivation and list of tolerance values, with criteria for assigning water-quality ratings. *Journal of the North American Benthological Society* 12:279-290.

Leonard, P.M. and D.J. Orth. 1986. Application and testing of an index of biotic integrity in small, coolwater streams. *Transactions of the American Fisheries Society* 115:401-414.

Leopold, L.B., M.G. Wolman, and J.P. Miller. 1964. *Fluvial processes in geomorphology.* W. H. Freeman and Company, San Francisco, California.

Lowe, R.L. 1974. *Environmental requirements and pollution tolerance of freshwater diatoms.* U.S. Environmental Protection Agency, Environmental Monitoring Series, Cincinnati, Ohio.

Lowe, R. L., and Pan, Y. 1996. Benthic algal communities and biological monitors. In *Algal Ecology: Freshwater Benthic Ecosystems.* R. J. Stevenson, M. Bothwell, and R. L. Lowe. pp. 705-39. Academic Press, San Diego, California, USA.

Ludwig, J.A. and J.F. Reynolds. 1988. *Statistical ecology: A primer on methods and computing.* John Wiley and Sons, Inc., New York, New York.

Lyons, J. 1992a. *Using the index of biotic integrity (IBI) to measure environmental quality in warmwater streams of Wisconsin.* General Technical Report, NC-149. U.S. Department of Agriculture, Forest Service, St. Paul, Minnesota.

Lyons, J. 1992b. The length of stream to sample with a towed electrofishing unit when fish species richness is estimated. *North American Journal of Fisheries Management* 12:198-203.

Lyons, J., L. Wang, and T.D. Simonson. 1996. Development and Validation of an Index of Biotic Integrity for Coldwater Streams in Wisconsin. *North American Journal of Fisheries Management* 16:241-256.

MacDonald, L.H., A.W. Smart, and R.C. Wissmar. 1991. *Monitoring guidelines to evaluate effects of forestry activities on streams in the Pacific Northwest and Alaska.* Prepared for Region 10, U.S. Environmental Protection Agency, Seattle, Washington. EPA 910/9-91-001.

Massachusetts Department of Environmental Protection (MA DEP). 1995. *Massachusetts DEP preliminary biological monitoring and assessment protocols for wadable rivers and streams.* Massachusetts Department of Environmental Protection, North Grafton, Massachusetts.

Matthews, R.A., P.F. Kondratieff, and A. L. Buikema, Jr. 1980. A field verification of the use and of the autotrophic index in monitoring stress effects. *Bulletin of Environmental Contamination and Toxicology* 25:226-233.

Matthews, W.J. 1986. Fish faunal structure in an Ozark stream: Stability, persistence, and a catastrophic flood. *Copeia* 1986:388-397.

Maughan, J.T. 1993. *Ecological assessment of hazardous waste sites.* Van Nostrand Reinhold, New York, New York.

Maxted, J.R., M.T. Barbour, J. Gerritsen, V. Poretti, N. Primrose, A. Silvia, D. Penrose, and R. Renfrow. In Press. Assessment framework for mid-Atlantic coastal plain streams using benthic macroinvertebrates. Submitted to *Journal of North American Benthological Society.*

McFarland, B H., Hill, B. H., and Willingham, W. T. 1997. Abnormal *Fragilaria* spp. (Bacillariophyceae) in streams impacted by mine drainage. *Journal of Freshwater Ecology* 12, 141-9.

Meador, M.R., C.R. Hupp, T.F. Cuffney, and M.E. Gurtz. 1993. *Methods for characterizing stream habitat as part of the national water-quality assessment program.* U.S. Geological Survey Open-File Report, Raleigh, North Carolina. USGS/OFR 93-408.

Merritt, R.W., K.W. Cummins, and V.H. Resh. 1996. Collecting, sampling, and rearing methods for aquatic insects. Pages 12-28 *in* R.W. Merritt and K.W. Cummins (editors). *An introduction to the aquatic insects of North America.* 3rd edition. Kendall/Hunt Publishing, Dubuque, Iowa.

Mid-Atlantic Coastal Streams Workgroup (MACS). 1996. *Standard operating procedures and technical basis: Macroinvertebrate collection and habitat assessment for low-gradient nontidal streams.* Delaware Department of Natural Resources and Environmental Conservation, Dover, Delaware.

Miller, D.L., P.M. Leonard, R.M. Hughes, J.R. Karr, P.B. Moyle, L.H. Schrader, B.A. Thompson, R.A. Daniel, K.D. Fausch, G.A. Fitzhugh, J.R. Gammon, D.B. Halliwell, P.L. Angermeier, and D.J. Orth. 1988. Regional applications of an Index of Biotic Integrity for use in water resource management. *Fisheries* 13(5):12-20.

Rapid Bioassessment Protocols for Use in Streams and Wadeable Rivers: Periphyton, Benthic Macroinvertebrates, and Fish, Second Edition

11-13

Mount, D.I., N. Thomas, M. Barbour, T. Norberg, T. Roush, and R. Brandes. 1984. *Effluent and ambient toxicity testing and in-stream community response on the Ottawa River, Lima, Ohio.* Permits Division, Washington, D.C., and Office of Research and Development, Duluth, Minnesota. EPA 600/3-84-080.

Moyle, P.B. 1976. *Inland fishes of California.* University of California Press, Berkeley, California.

Moyle, P.B., L.R. Brown, and B. Herbold. 1986. *Final report on development and preliminary tests of indices of biotic integrity for California.* Final report to the U.S. Environmental Protection Agency, Environmental Research Laboratory, Corvallis, Oregon.

Myers, T.J. and S. Swanson. 1991. Aquatic habitat condition index, stream type, and livestock bank damage in northern Nevada. *Water Resources Bulletin* 27(4):667-677.

Naiman, R.J., H. Decamps, and M. Pollack. 1993. The role of riparian corridors in maintaining regional biodiversity. *Ecological Applications* 3(2):209-212.

Needham, P.R. and R.L. Usinger. 1956. Variability in the macrofauna of a single riffle in Prosser Creek, California, as indicated by the Surber sampler. *Hilgardia* 24:383-409.

Nielsen, L.A. and D.L. Johnson (editors). 1983. *Fisheries Techniques.* American Fisheries Society, Bethesda, Maryland.

Norris, R.H. 1995. Biological monitoring: The dilemma of data analysis. *Journal of North American Benthological Society* 14:440-450.

Norris, R.H., and A. Georges. 1993. Analysis and interpretation of benthic macroinvertebrate surveys. Pages 234-286 *in* D.M. Rosenberg and V.H. Resh (editors). *Freshwater Biomonitoring and Benthic Macroinvertebrates.* Chapman and Hall, New York, New York.

Ohio Environmental Protection Agency (Ohio EPA). 1987. *Biological criteria for the protection of aquatic life: volumes I-III.* Ohio Environmental Protection Agency, Columbus, Ohio.

Ohio EPA. 1992. *Ohio Water Resource Inventory. Volume I: Summary, Status, and Trends.* Ohio EPA, Columbus, Ohio.

Oklahoma Conservation Commission (OCC). 1993. *Development of rapid bioassessment protocols for Oklahoma utilizing characteristics of the diatom community.* Oklahoma Conservation Commission, Oklahoma City, Oklahoma.

Omernik, J. M. 1987. Ecoregions of the Conterminous United States. *Annals of the Association of American Geographers* 77(1):118-125.

Omernik, J.M. 1995. Ecoregions: A spatial framework for environmental management. Pages 49-62 *in* W.S. Davis and T.P Simon (editors). *Biological assessment and criteria: Tools for water resource planning and decision making.* Lewis Publishers, Boca Raton, Florida.

Osborne, L.L. and E.E. Hendricks. 1983. *Streamflow and Velocity as Determinants of Aquatic Insect Distribution and Benthic Community Structure in Illinois.* Water Resources Center, University of Illinois. U.S. Department of the Interior, Bureau of Reclamation. UILU-WRC-83-183.

Osborne, L.L., B. Dickson, M. Ebbers, R. Ford, J. Lyons, D. Kline, E. Rankin, D. Ross, R. Sauer, P. Seelbach, C. Speas, T. Stefanavage, J. Waite, and S. Walker. 1991. Stream habitat assessment programs in states of the AFS North Central Division. *Fisheries* 16(3):28-35.

Oswood, M.E. and W.E. Barber. 1982. Assessment of fish habitat in streams: Goals, constraints, and a new technique. *Fisheries* 7(3):8-11.

Overton, W.S., D. White, and D.L. Stevens, Jr. 1991. *Design report for EMAP, the environmental monitoring and assessment program.* U.S. Environmental Protection Agency, Office of Research and Development, Washington, D.C. EPA-600-3- 91-053.

Palmer, C.M. 1969. A composite rating of algae tolerating organic pollution. *Journal of Phycology* 5:78-82.

Palmer, C.M. 1977. *Algae and water pollution.* U.S. Environmental Protection Agency, Cincinnati, Ohio. EPA-600/9-77-036.

Pan, Y. and R.J. Stevenson. 1996. Gradient analysis of diatom assemblages in Western Kentucky wetlands. *Journal of Phycology* 32:222-232.

Pan, Y., R. J. Stevenson, B. H. Hill, A. T. Herlihy, and G. B. Collins. 1996. Using diatoms as indicators of ecological conditions in lotic systems: A regional assessment. *Journal of the North American Benthological Society* 15:481-495.

Patrick, R. 1973. Use of algae, especially diatoms, in the assessment of water quality. *In* J. Cairns, Jr. and K.L. Dickson (editors). *Biological methods for the assessment of water quality.* Special Technical Publication 528. American Society for Testing and Materials, Philadelphia, Pennsylvania.

Patrick, R. 1977. Ecology of freshwater diatoms. Pages 284-332 *in* D. Werner (editor). *The biology of diatoms.* Botanical monographs volume 13. University of California Press, Berkeley, California.

Patrick, R., M.H. Hohn, and J.H. Wallace. 1954. A new method for determining the pattern of the diatom flora. *Notulae Naturae* 259:1-12.

Patrick, R. and C.W. Reimer. 1966. *The diatoms of the United States, exclusive of Alaska and Hawaii.* Monograph No. 13. Academy of Natural Sciences, Philadelphia, Pennsylvania.

Patrick, R. and C.W. Reimer. 1975. *The Diatoms of the United States.* Vol. 2, Part 1. Monograph No. 13. Academy of Natural Sciences, Philadelphia, Pennsylvania.

Paulsen, S.G., D.P. Larsen, P.R. Kaufmann, T.R. Whittier, J.R. Baker, D. Peck, J. McGue, R.M. Hughes, D. McMullen, D. Stevens, J.L. Stoddard, J. Lazorchak, W. Kinney, A.R. Selle, and R. Hjort. 1991. *EMAP - surface waters monitoring and research strategy, fiscal year 1991.* EPA-600-3-91-002. U.S. Environmental Protection Agency, Office of Research and Development, Washington, D.C. and Environmental Research Laboratory, Corvallis, Oregon.

Pearsons, T.N., H.W. Li, and G.A. Lamberti. 1992. Influence of habitat complexity on resistance to flooding and resilience of stream fish assemblages. *Transactions of the American Fisheries Society* 121:427-436.

Peckarsky, B. 1984. Sampling the stream benthos. Pages 131-160 *in* J. Downing and F. Rigler (editors). *A manual of methods for the assessment of secondary productivity in freshwater.* 2nd edition. Oxford, Blackwell Scientific Publications, IBP Handbook 19.

Peterson, C.G. and R.J. Stevenson. 1990. Post-spate development of epilithic algal communities in different current environments. *Canadian Journal of Botany* 68:2092-2102.

Plafkin, J.L., M.T. Barbour, K.D. Porter, S.K. Gross, and R.M. Hughes. 1989. *Rapid bioassessment protocols for use in streams and rivers: Benthic macroinvertebrates and fish.* U.S. Environmental Protection Agency, Office of Water Regulations and Standards, Washington, D.C. EPA 440-4-89-001.

Platts, W.S., W.F. Megahan, and G.W. Minshall. 1983. *Methods for Evaluating Stream, Riparian, and Biotic Conditions.* U.S. Department of Agriculture, U.S. Forest Service, Ogden, Utah. General Technical Report INT-138.

Porter, S. D., T. F. Cuffney, M. E. Gurtz, and M. R. Meador. 1993. *Methods for Collecting Algal Samples as Part of the National Water-Quality Assessment Program.* U. S. Geological Survey, Report 93-409. Raleigh, North Carolina, USA.

Prescott, G.W. 1968. *The algae: A review.* Houghton Mifflin Company, Boston, Massachussets.

Rankin, E.T. 1991. The use of the qualitative habitat evaluation index for use attainability studies in streams and Rivers in Ohio. *In* George Gibson, editor. *Biological Criteria: Research and Regulation,* Office of Water, U.S. Environmental Protection Agency, Washington, D.C. EPA 440/5-91-005.

Rankin, E.T. 1995. Habitat indices in water resource quality assessments. Pages 181-208 *in* W.S. Davis and T.P Simon (editors). *Biological assessment and criteria: Tools for water resource planning and decision making.* Lewis Publishers, Boca Raton, Florida.

Raven, P.J., N.T.H. Holmes, F.H. Dawson, P.J.A. Fox, M. Everard, I.R. Fozzard, and K.J. Rowen. 1998. *River Habitat Quality: The physical character of rivers and streams in the UK and Isle of Man.* Environment Agency. ISBN1 873760 42 9. Bristol, England.

Reckhow, K.H. and W. Warren-Hicks. 1996. *Biological criteria: Technical guidance for survey design and statistical evaluation of biosurvey data.* Draft document prepared for U.S. EPA, Office of Science and Technology, Washington, DC.

Reice, S.R. 1980. The role of substratum in benthic macroinvertebrate microdistribution and litter decomposition in a woodland stream. *Ecology* 61:580-590.

Resh, V.H. 1979. Sampling variability and life history features: Basic consideration in the design of aquatic insect studies. *Journal of the Fisheries Research Board of Canada* 36:290-311.

Resh, V.H. and J.K. Jackson. 1993. Rapid assessment approaches to biomonitoring using benthic macroinvertebrates. Pages 195-233 *in* D.M. Rosenberg and V.H. Resh (editors). *Freshwater biomonitoring and benthic macroinvertebrates.* Chapman and Hall, New York.

Resh, V.H., J.W. Feminella, and E.P. McElravy. 1990. *Sampling aquatic insects.* Videotape. Office of Media Services, University of California, Berkeley, California.

Resh, V.H., R.H. Norris, and M.T. Barbour. 1995. Design and implementation of rapid assessment approaches for water resource monitoring using benthic macroinvertebrates. *Australian Journal of Ecology* 20:108-121.

Reynolds, J.B. 1983. Electrofishing. Pages 147-164 *in* L.A. Nielsen and D.L. Johnson (editors). Fisheries Techniques. *American Fisheries Society,* Bethesda, Maryland.

Reynoldson, T.B., R.C. Bailey, K.E. Day, and R.H. Norris. 1995. Biological guidelines for freshwater sediment based on **BE**nthic **A**ssessment of **S**edimen**T** (the BEAST) using a multivariate approach for predicting biological state. *Australian Journal of Ecology* (1995) 20:198-219.

Robins, C.R., R.M. Bailey, C.E. Bond, J.R. Brooker, E.A. Lachner, R.N. Lea, and W.B. Scott. 1991. *Common and scientific names of fishes from the United States and Canada.* American Fisheries Society Special Publication 20, Bethesda, Maryland.

Rodgers, J.H., Jr., K.L. Dickson, and J. Cairns, Jr. 1979. A review and analysis of some methods used to measure functional aspects of periphyton. *In* R.L. Weitzel (editor). *Methods and measurements of periphyton communities: A review.* Special Technical Publication 690. American Society for Testing and Materials.

Rohm, C.M., J.W. Giese, and C.C. Bennett. 1987. Evaluation of an aquatic ecoregion classification of streams in Arkansas. *Freshwater Ecology* 4:127-140.

Rosen, B.H. 1995. Use of periphyton in the development of biocriteria. Pages 209-215 *in* W.S. Davis and T.P. Simon (editors). *Biological assessment and criteria: Tools for water resource planning and decision making.* Lewis Publishers, Boca Raton, Florida.

Rosgen, D.L. 1985. A stream classification system. *In* Proceedings of the First North American Riparian Conference *Riparian Ecosystem and their Management: reconciling conflicting uses.* U.S. Department of Agriculture Forest Service, Tucson, Arizona. General Technical Report RM-120.

Rosgen, D.L. 1994. A classification of natural rivers. *Catena* 22:169-199.

Rosgen, D. 1996. *Applied river morphology.* Wildland Hydrology Books, Pagosa Springs, Colorado.

Ross, L.T. and D.A. Jones (editors). 1979. *Biological aspects of water quality in Florida.* Technical Series Volume 4, no. 3. Florida Department of Environmental Regulation, Tallahassee.

Ross, S.T., W.J. Matthews, and A.E. Echelle. 1985. Persistence of stream fish assemblages: Effects of environmental change. *American Naturalist* 126:24-40.

Roth, N.E., M.T. Southerland, J.C. Chaillou, J.H. Vølstad, S.B. Weisberg, H.T. Wilson, D.G. Heimbuch, J.C. Seibel. 1997. *Maryland Biological Stream Survey: Ecological status of non-tidal streams in six basins sampled in 1995.* Maryland Department of Natural Resources, Chesapeake Bay and Watershed Programs, Monitoring and Non-tidal Assessment, Annapolis, Maryland. CBWP-MANTA-EA-97-2.

Rott, E. 1991. Methodological aspects and perspectives in the use of periphyton for monitoring and protecting rivers. *In* B.A. Whitton, E. Rott, and G. Friedrich (editors). *Use of algae for monitoring rivers.* Institut fur Botanik, University of Innsbruck, Austria.

Rapid Bioassessment Protocols for Use in Streams and Wadeable Rivers: Periphyton, Benthic Macroinvertebrates, and Fish, Second Edition

11-17

Sabater, S., F. Sabater, and J. Armengol. 1988. Relationships between diatom assemblages and physico-chemical variables in the River Ter (NE Spain). *International Review of Ges. Hydrobiologia* 73:171-179.

Science Advisory Board (SAB). 1993. *Evaluation of draft technical guidance on biological criteria for streams and small rivers (prepared by the Biological Criteria Subcommittee of the Ecological Processes and Effects Committee).* An SAB Report. US Environmental Protection Agency, Washington, D.C. EPA-SAB-EPEC-94-003.

Shackleford, B. 1988. *Rapid Bioassessments of Lotic Macroinvertebrate Communities: Biocriteria Development.* Arkansas Department of Pollution Control and Ecology, Little Rock, Arkansas.

Shields, F.D., S.S. Knight Jr., and C.M. Cooper. 1995. Use of the index of biotic integrity to assess physical habitat degradation in warmwater streams. *Hydrobiologia* 312(3):191-208.

Shields, F.D., S.S. Knight Jr., and C.M. Cooper. 1998. Rehabilitation of aquatic habitats in warmwater streams damaged by channel incision in Mississippi. *Hydrobiologia* 382:63-86.

Simon, A. 1989a. The discharge of sediment in channelized alluvial streams. *Water Resources Bulletin* 25(6):1177-1187.

Simon, A. 1989b. A model of channel response in disturbed alluvial channels. *Earth Surface Processes and Landforms* 14:11-26.

Simon, T.P. 1991. *Development of ecoregion expectations for the index of biotic integrity (IBI) Central Corn Belt Plain.* U.S. Environmental Protection Agency, Region V, Chicago, Illinois. EPA 905/9-91/025.

Simon, T.P. (editor). 1999. *Assessing the sustainability and biological integrity of water resources using fish communities.* CRC Press, Boca Raton, Florida.

Simon, A. and C.R. Hupp. 1987. Geomorphic and vegetative recovery processes along modified Tennessee streams: An interdisciplinary approach to disturbed fluvial systems. *Proceedings of the Forest Hydrology and Watershed Management Symposium*, Vancouver, August 1987. Publication No. 167:251-261.

Simon, T.P. and J. Lyons. 1995. Application of the index of biotic integrity to evaluate water resource integrity in freshwater ecosystems. Pages 245-262 *in* W.S. Davis and T.P Simon (editors). *Biological assessment and criteria: Tools for water resource planning and decision making.* Lewis Publishers, Boca Raton, Florida.

Simonson, T.D., J. Lyons, and P.D. Kanehl. 1994. Quantifying fish habitat in streams: Transect spacing, sample size, and a proposed framework. *North American Journal of Fisheries Management* 14:607-615.

Simonson, T.D., and J. Lyons. 1995. Comparison of catch per effort and removal procedures for sampling stream fish assemblages. *North American Journal of Fisheries Management* 15:419-427.

Simpson, J., R. Norris, L. Barmuta, and P. Blackman. 1996. *Australian River assessment system: National river health program predictive model manual.* http.//ausrivas.canberra.au.

Smith, E.P., and J.R. Voshell, Jr. 1997. *Studies of Benthic Macroinvertebrates and Fish in Streams within EPA Region 3 for Development of Biological Indicators of Ecological Condition.* Virginia Polytechnic Institute and State University, Blacksburg, VA.

Snyder, B.D., J.B. Stribling, and M.T. Barbour. 1998. *Codorus Creek biological assessment in the vicinity of the P.H. Glatfelter Company Spring Grove, Pennsylvania.* Prepared for P.H. Glatfelter Company.

Southerland, M.T. and J.B. Stribling. 1995. Status of biological criteria development and implementation. Pages 81-96 *in* W.S. Davis and T.P. Simon (editors). *Biological assessment and criteria: Tools for water resource planning and decision making.* Lewis Publishers, Boca Raton, Florida.

Southwood, T.R.E. 1977. Habitat, the templet for ecological strategies? *Journal of Animal Ecology* 46:337-365.

Spindler, P. 1996. *Using ecoregions for explaining macroinvertebrate community distribution among reference sites in Arizona, 1992.* Arizona Department of Environmental Quality, Hydrologic Support and Assessment Section, Flagstaff, Arizona.

Statzner, B., J.A. Gore, and V.H. Resh. 1988. Hydraulic stream ecology: Observed patterns and potential applications. *Journal of the North American Benthological Society* 7(4):307-360.

Steedman, R.J. 1988. Modification and assessment of an index of biotic integrity to quantify stream quality in southern Ontario. *Canadian Journal of Fisheries and Aquatic Science* 45:492-501.

Stevenson, R.J. 1984. Epilithic and epipelic diatoms in the Sandusky River, with emphasis on species diversity and water pollution. *Hydrobiologia* 114:161-174.

Stevenson, R.J. 1990. Benthic algal community dynamics in a stream during and after a spate. *Journal of the North American Benthological Society* 9:277-288.

Stevenson, R.J. 1996. An introduction to algal ecology in freshwater benthic habitats. Pages 3-30 *in* R.J. Stevenson, M. Bothwell, R.L. Lowe, editors. *Algal Ecology: Freshwater Benthic Ecosystems.* Academic Press, San Diego, California.

Stevenson, R. J. 1998. Diatom indicators of stream and wetland stressors in a risk management framework. *Environmental Monitoring and Assessment* 51:107-118.

Stevenson, R. J. and Y. Pan. 1999. Assessing ecological conditions in rivers and streams with diatoms. Pages 11-40 *in* E. F. Stoermer and J. P. Smol, editors. *The Diatoms: Applications to the Environmental and Earth Sciences.* Cambridge University Press, Cambridge, UK.

Stevenson, R.J. and R.L. Lowe. 1986. Sampling and interpretation of algal patterns for water quality assessments. Pages 118-149 *in* B.G. Isom (editor). *Rationale for sampling and interpretation of ecological data in the assessment of freshwater ecosystems.* American Society of Testing and Materials. ASTM STP 894.

Stribling, J.B., B.D. Snyder, and W.S. Davis. 1996a. *Biological assessment methods, biocriteria, and biological indicators. Bibliography of selected technical, policy and regulatory literature.* U.S.

Environmental Protection Agency, Office of Policy, Planning, and Evaluation, Washington, D.C. EPA 230-B-96-001.

Stribling, J.B., C. Gerardi, and B.D. Snyder. 1996b. *Biological Assessment of the Mattaponi Creek and Brier Ditch Watersheds, Prince George's County, Maryland: 1996 Winter Index Period.* Prepared for Prince George's County, Department of Environmental Resources, Largo, Maryland.

Stribling, J.B., B.K. Jessup, and J. Gerritsen. 1999. Development of Biological and Habitat Criteria for Wyoming Streams and Their Use in the TMDL Process. Prepared by Tetra Tech, Inc., Owings Mills, MD, for U.S. EPA, Region 8, Denver, CO.

Suter, G.W., II, L.W. Barnthouse, S.M. Bartell, T. Mill, D. Mackay, and S. Paterson. 1993. *Ecological risk assessment.* Lewis Publishers, Ann Arbor, Michigan.

ter Braak, C. J. F., and van Dam, H. 1989. Inferring pH from diatoms: A comparison of old and new calibration methods. *Hydrobiologia* 178:209-23.

Underwood, A.J. 1994. On beyond Baci: Sampling designs that might reliably detect environmental disturbances. *Ecological Applications* 4:3-15.

U.S. Department of Agriculture (USDA), Soil Conservation Service. 1981. *Land resource regions and major land resource areas of the United States.* Agricultural handbook 296. U.S. Government Printing Office, Washington, D.C.

U.S. Environmental Protection Agency (U.S. EPA). 1980. *National accomplishments in pollution control 1970-1980: Some case histories.* U.S. Environmental Protection Agency, Office of Planning and Management, Program Evaluation Division, Washington, D.C.

U.S. Environmental Protection Agency (U.S. EPA). 1983. *Technical support manual: Waterbody surveys and assessments for conducting use attainability analyses.* U.S. Environmental Protection Agency, Office of Water Regulations and Standards, Washington, D.C. Volumes 1-3.

U.S. Environmental Protection Agency (U.S. EPA). 1984. *The development of data quality objectives. Prepared by the EPA quality assurance management staff and the DQO workgroup.* U.S. Environmental Protection Agency, Washington, D.C.

U.S. Environmental Protection Agency (U.S. EPA). 1986. *Development of data quality objectives. Descriptions of stages I and II.* Prepared by the EPA Quality Assurance Management staff. Office of Research and Development, U.S. Environmental Protection Agency, Washington, D.C.

U.S. Environmental Protection Agency (U.S. EPA). 1987. *Surface water monitoring: A framework for change.* U.S. Environmental Protection Agency, Office of Water, Office of Policy Planning and Evaluation, Washington, D.C.

U.S. Environmental Protection Agency (U.S. EPA). 1988. *Proceedings of the first national workshop on biological criteria*, Lincolnwood, Illinois, December 2-4. 1987. U.S. Environmental Protection Agency, Chicago, Illinois. 905/9-89/003.

U.S. Environmental Protection Agency (U.S. EPA). 1989. *Overview of selected EPA regulations and guidance affecting POTW management.* U.S. Environmental Protection Agency, Office of Water, Washington, D.C. EPA 440/69-89/008.

U.S. Environmental Protection Agency (U.S. EPA). 1990a. *Second national symposium on water quality assessment: Meeting summary* October 16-19, 1989, Fort Collins, Colorado. U.S. Environmental Protection Agency, Office of Water, Washington, D.C.

U.S. Environmental Protection Agency (U.S. EPA). 1990b. *Biological criteria: National program guidance for surface waters.* U.S. Environmental Protection Agency, Office of Water Regulations and Standards, Washington, D.C. EPA 440-5-90-004.

U.S. Environmental Protection Agency (U.S. EPA). 1990c. *Methods for measuring the acute toxicity of effluents and receiving waters to aquatic organisms.* 4th edition. Office of Research and Development, U.S. Environmental Protection Agency, Cincinnati, Ohio. EPA/600-4-90-027.

U. S. Environmental Protection Agency (U.S. EPA). 1991a. *Biological criteria: State development and implementation efforts.* U.S. Environmental Protection Agency, Office of Water, Washington, D.C. EPA-440/5-91-003.

U. S. Environmental Protection Agency (U.S. EPA). 1991b. *Biological criteria: Guide to technical literature.* U.S. Environmental Protection Agency, Office of Water, Washington, D.C. EPA-440-5-91-004.

U.S. Environmental Protection Agency (U.S. EPA). 1991c. *Technical support document for water quality based toxics control.* U.S. Environmental Protection Agency, Office of Water, Washington, D.C. EPA 505-2-90-001.

U. S. Environmental Protection Agency (U.S. EPA). 1991d. *Biological Criteria: Research and Regulation: Proceedings of a Symposium.* U.S. Environmental Protection Agency, Office of Water, Washington, D.C. EPA-440-5-91-005.

U. S. Environmental Protection Agency (U.S. EPA). 1991e. *Report of the ecoregions subcommittee of the ecological processes and effects committee. Evaluation of the ecoregion concept.* U.S. Environmental Protection Agency, Science Advisory Board, Washington, D.C. EPA-SAB-EPEC-91-003.

U.S. Environmental Protection Agency (U.S. EPA). 1991f. *Guidance for the implementation of water quality-based decisions: The TMDL process.* U.S. Environmental Protection Agency, Office of Water Regulations and Standards, Washington, D.C. EPA 440/4-91-001.

U.S. Environmental Protection Agency (U.S. EPA). 1992. *Framework for ecological risk assessment.* U.S. Environmental Protection Agency, Washington, D.C. EPA/630/R-92/001.

U.S. Environmental Protection Agency (U.S. EPA). 1994a. *Watershed protection: TMDL Note #2 bioassessment and TMDLs.* U.S. Environmental Protection Agency, Office of Water, Washington, D.C. EPA841-K-94-005a.

U.S. Environmental Protection Agency (U.S. EPA). 1994b. *National water quality inventory: 1992 report to Congress.* U.S. Environmental Protection Agency, Office of Water, Washington, D.C. EPA 841-R-94-001.

U.S. Environmental Protection Agency (U.S. EPA). 1994c. *The watershed protection approach 1993/94 supplement to 1992 annual report draft.* U.S. Environmental Protection Agency, Office of Water, Washington, D.C.

U.S. Environmental Protection Agency (U.S. EPA). 1994d. *Guidance on implementation of biological criteria.* Draft. U.S. Environmental Protection Agency, Office of Science and Technology, Washington, D.C. January 13.

U. S. Environmental Protection Agency (U.S. EPA). 1995a. *Generic quality assurance project plan guidance for programs using community-level biological assessment in streams and wadeable rivers.* U.S. Environmental Protection Agency, Office of Water, Washington, D.C. EPA 841-B-95-004.

U.S. Environmental Protection Agency (U.S. EPA). 1995b. *Guidelines for preparation of the 1996 State Water Quality Assessments (305[b] Reports).* Office of Water, U.S. Environmental Protection Agency, Washington, D.C. EPA 841-B-95-001.

U.S. Environmental Protection Agency (U.S. EPA). 1996a. *The volunteer monitor's guide to quality assurance project plans.* U.S. Environmental Protection Agency, Office of Wetlands, Oceans, and Watersheds, Washington, D.C. EPA 841-B-96-003.

U.S. Environmental Protection Agency (U.S. EPA). 1996b. *Nonpoint source monitoring and evaluation guide.* U.S. Environmental Protection Agency, Office of Water, Washington, D.C.

U.S. Environmental Protection Agency (U.S.EPA). 1996c. *Level III ecoregions of the continental United States.* U.S. Environmental Protection Agency, National Health and Environmental Effects Research Laboratory, Corvallis, Oregon.

U.S. Environmental Protection Agency (U.S. EPA). 1997a. *Estuarine and coastal marine waters bioassessment and biocriteria technical guidance.* U.S. Environmental Protection Agency, Office of Water, Washington, D.C. EPA-822-B-97-001.

U.S. Environmental Protection Agency (U.S. EPA). 1997b. *Volunteer stream monitoring: A methods manual.* U.S. Environmental Protection Agency, Office of Water, Washington, D.C. EPA 841-B-97-003.

U.S. Environmental Protection Agency (U.S. EPA). 1997c. *Guidelines for the preparation of the Comprehensive State Water Quality Assessments (305[b] reports).* U.S. Environmental Protection Agency, Office of Water, Washington, D.C. EPA-841-B-97-002A.

U.S. Environmental Protection Agency (U.S. EPA). 1998. *Lake and reservoir bioassessment and biocriteria technical guidance document.* U.S. Environmental Protection Agency, Office of Water, Washington, D.C. EPA-841-B-98-007.

VanLandingham, S. L. 1982. *Guide to the identification, environmental requirements and pollution tolerance of freshwater blue-green algae (Cyanophyta).* EPA-600/3-82-073.

Vinson, M.R. and C.P. Hawkins. 1996. Effects of sampling area and subsampling procedure on comparisons of taxa richness among streams. *Journal of the North American Benthological Society* 15(3):392-399.

Volstad, J.H., D.H. Heimbuch, M.T. Southerland, P.T. Jacobson, J.A. Chaillou, S.B. Weisberg, H.T. Wilson. 1995. *Maryland Biological Stream Survey: The 1993 pilot study.* Maryland Department of Natural Resources, Chesapeake Bay Research and Monitoring Division, Annapolis Maryland.

Wade, D.C. and S.B. Stalcup. 1987. *Assessment of the Sport Fishery Potential for the Bear Creek Floatway: Biological Integrity of Representative Sites, 1986.* Tennessee Valley Authority, Muscle Shoals, Alabama. Report No. TVA/ONRED/AWR-87/30.

Wallace, J.B., J.W. Grubaugh, and M.R. Whiles. 1996. Biotic indices and stream ecosystem processes: results from an experimental study. *Ecological Applications* (6):140-151.

Wallace, J.B., J.R. Webster, and W.R. Woodall. 1977. The role of filter feeders in flowing waters. *Archiv fur Hydrobiologie* 79:506-532.

Ward, G.M. and N.G. Aumen. 1986. Woody debris as a source of fine particulate organic matter in coniferous forest stream ecosystems. *Canadian Journal of Fisheries and Aquatic Sciences.* 43:1635-1642.

Warren, C.E. 1979. *Toward Classification and Rationale for Watershed Management and Stream Protection.* U.S. Environmental Protection Agency, Corvallis, Oregon. EPA-600/3-79-059.

Warren, M.L., Jr., and B.M. Burr. 1994. Status of freshwater fishes of the US: Overview of an imperiled fauna. *Fisheries* 19(1):6-18.

Weber, C.I. (editor). 1973. *Biological field and laboratory methods for measuring the quality of surface water and effluents.* U.S. Environmental Protection Agency, Office of Research and Development, Cincinnati, Ohio. EPA 670-4-73-001.

Weitzel, R. L. 1979. Periphyton measurements and applications. *In* R. L. Weitzel (editor). *Methods and measurements of periphyton communities: A review.* Special Technical Publication 690. American Society for Testing and Materials.

Wesche, T.A., C.M. Goertler, C. B. Frye. 1985. Importance and evaluation of instream and riparian cover in smaller trout streams. Pages 325-328 *in* The Proceedings of the First North American Riparian Conference *Riparian Ecosystems and their Management: Reconciling conflicting uses.* U.S. Department of Agriculture Forest Service, General Technical Report TM-120. Tucson, Arizona.

Whittaker, R.H. 1952. A study of summer foliage insect communities in the Great Smoky Mountains. *Ecological Monographs* 22:6.

Whittaker, R.H. and C.W. Fairbanks. 1958. A study of plankton copepod communities in the Columbia basin, Southeastern Washington. *Ecology* 39:46-65.

Rapid Bioassessment Protocols for Use in Streams and Wadeable Rivers: Periphyton, Benthic Macroinvertebrates, and Fish, Second Edition

11-23

Whittier, T.R., R.M. Hughes, and D.P. Larsen. 1988. Correspondence between ecoregions and spatial patterns in stream ecosystems in Oregon. *Canadian Journal of Fisheries and Aquatic Sciences* 45:1264-1278.

Whitton, B. A., and Kelly, M. G. 1995. Use of algae and other plants for monitoring rivers. *Australian Journal of Ecology* 20, 45-56.

Whitton, B. A. and E. Rott. 1996. *Use of algae for monitoring rivers II.* E. Rott, Publisher, Institut für Botanik, Universität Innsbruck, Innsbruck, Austria

Whitton, B. A., Rott, E., and Friedrich, G., ed. 1991. *Use of Algae for Monitoring Rivers.* E. Rott, Publisher, Institut für Botanik, Universität Innsbruck, Innsbruck, Austria

Wilhm, J.L. and T.C. Doris. 1968. Biological parameters for water quality criteria. *Bioscience* 18:477-481.

Winget, R.N. and F.A. Mangum. 1979. *Biotic condition index: Integrated biological, physical, and chemical stream parameters for management.* Intermountain Region, U.S. Department of Agriculture, Forest Service, Ogden, Utah.

Wright, J.F., M.T. Furse, and P.D. Armitage. 1993. RIVPACS: A technique for evaluating the biological quality of rivers in the UK. *European Water Pollution Control* 3(4):15-25.

Yoder, C.O. 1991.The integrated biosurvey as a tool for evaluation of aquatic life use attainment and impairment in Ohio surface waters. Pages 110-122 *in* George Gibson, editor. *Biological criteria: Research and regulation, proceedings of a symposium.* U.S. Environmental Protection Agency, Office of Water, Washington, D.C. EPA-440-5-91-005.

Yoder, C.O. 1995. Policy issues and management applications for biological criteria. Pages 327-343 *in* W.S. Davis and T.P. Simon (editors). *Biological assessment and criteria: Tools for water resource planning and decision making.* Lewis Publishers, Boca Raton, Florida.

Yoder, C.O. and E.T. Rankin. 1995a. Biological criteria program development and implementation in Ohio. Pages 109-144 *in* W.S. Davis and T.P Simon (editors). *Biological assessment and criteria: Tools for water resource planning and decision making.* Lewis Publishers, Boca Raton, Florida.

Yoder, C.O. and E.T. Rankin. 1995b. Biological response signatures and the area of degradation value: New tools for interpreting multimetric data. Pages 263-286 *in* W.S. Davis and T.P Simon (editors). *Biological assessment and criteria: Tools for water resource planning and decision making.* Lewis Publishers, Boca Raton, Florida.

APPENDIX A:

SAMPLE DATA FORMS FOR THE PROTOCOLS

Rapid Bioassessment Protocols For Use in Streams and Wadeable Rivers: Periphyton, Benthic Macroinvertebrates, and Fish, Second Edition

A-1

This Page Intentionally Left Blank

APPENDIX A-1:

Habitat Assessment and Physicochemical Characterization Field Data Sheets

Form 1: Physical Characterization/Water Quality Field Data Sheet
Form 2: Habitat Assessment Field Data Sheet - High Gradient Streams
Form 3: Habitat Assessment Field Data Sheet - Low Gradient Streams

Rapid Bioassessment Protocols For Use in Streams and Wadeable Rivers: Periphyton, Benthic Macroinvertebrates, and Fish, Second Edition

A-3

This Page Intentionally Left Blank

PHYSICAL CHARACTERIZATION/WATER QUALITY FIELD DATA SHEET
(FRONT)

STREAM NAME	LOCATION
STATION #_____ RIVERMILE_____	STREAM CLASS
LAT _____ LONG _____	RIVER BASIN
STORET #	AGENCY
INVESTIGATORS	
FORM COMPLETED BY	DATE _____ TIME _____ AM PM	REASON FOR SURVEY

WEATHER CONDITIONS	Now		Past 24 hours	Has there been a heavy rain in the last 7 days? ❑ Yes ❑ No
	❑	storm (heavy rain)	❑	
	❑	rain (steady rain)	❑	Air Temperature_____ ^{0} C
	❑	showers (intermittent)	❑	
	____%❑	%cloud cover	❑ ____ %	Other_____
	❑	clear/sunny	❑	

SITE LOCATION/MAP	Draw a map of the site and indicate the areas sampled (or attach a photograph)

STREAM CHARACTERIZATION	**Stream Subsystem** ❑ Perennial ❑ Intermittent ❑ Tidal	**Stream Type** ❑ Coldwater ❑ Warmwater
	Stream Origin ❑ Glacial ❑ Non-glacial montane ❑ Swamp and bog	❑ Spring-fed ❑ Mixture of origins ❑ Other_____	**Catchment Area**_____ km^2

PHYSICAL CHARACTERIZATION/WATER QUALITY FIELD DATA SHEET
(BACK)

WATERSHED FEATURES	**Predominant Surrounding Landuse** ❑ Forest ❑ Commercial ❑ Field/Pasture ❑ Industrial ❑ Agricultural ❑ Other _____ ❑ Residential	**Local Watershed NPS Pollution** ❑ No evidence ❑ Some potential sources ❑ Obvious sources **Local Watershed Erosion** ❑ None ❑ Moderate ❑ Heavy
RIPARIAN VEGETATION (18 meter buffer)		**Indicate the dominant type and record the dominant species present** ❑ Trees ❑ Shrubs ❑ Grasses ❑ Herbaceous dominant species present _____
INSTREAM FEATURES	**Estimated Reach Length** _____ m **Estimated Stream Width** _____ m **Sampling Reach Area** _____ m² **Area in km² (m²x1000)** _____ km² **Estimated Stream Depth** _____ m **Surface Velocity** _____ m/sec (at thalweg)	**Canopy Cover** ❑ Partly open ❑ Partly shaded ❑ Shaded **High Water Mark** _____ m **Proportion of Reach Represented by Stream Morphology Types** ❑ Riffle _____% ❑ Run _____% ❑ Pool _____% **Channelized** ❑ Yes ❑ No **Dam Present** ❑ Yes ❑ No
LARGE WOODY DEBRIS		**LWD** _____ m² **Density of LWD** _____ m²/km² **(LWD/ reach area)**
AQUATIC VEGETATION		**Indicate the dominant type and record the dominant species present** ❑ Rooted emergent ❑ Rooted submergent ❑ Rooted floating ❑ Free floating ❑ Floating Algae ❑ Attached Algae dominant species present _____ **Portion of the reach with aquatic vegetation** _____ %
WATER QUALITY	**Temperature** _____ ⁰C **Specific Conductance** _____ **Dissolved Oxygen** _____ **pH** _____ **Turbidity** _____ **WQ Instrument Used** _____	**Water Odors** ❑ Normal/None ❑ Sewage ❑ Petroleum ❑ Chemical ❑ Fishy ❑ Other _____ **Water Surface Oils** ❑ Slick ❑ Sheen ❑ Globs ❑ Flecks ❑ None ❑ Other _____ **Turbidity (if not measured)** ❑ Clear ❑ Slightly turbid ❑ Turbid ❑ Opaque ❑ Stained ❑ Other _____
SEDIMENT/ SUBSTRATE	**Odors** ❑ Normal ❑ Sewage ❑ Petroleum ❑ Chemical ❑ Anaerobic ❑ None ❑ Other _____ **Oils** ❑ Absent ❑ Slight ❑ Moderate ❑ Profuse	**Deposits** ❑ Sludge ❑ Sawdust ❑ Paper fiber ❑ Sand ❑ Relict shells ❑ Other _____ **Looking at stones which are not deeply embedded, are the undersides black in color?** ❑ Yes ❑ No

INORGANIC SUBSTRATE COMPONENTS (should add up to 100%)			ORGANIC SUBSTRATE COMPONENTS (does not necessarily add up to 100%)		
Substrate Type	Diameter	% Composition in Sampling Reach	Substrate Type	Characteristic	% Composition in Sampling Area
Bedrock			Detritus	sticks, wood, coarse plant materials (CPOM)	
Boulder	> 256 mm (10")				
Cobble	64-256 mm (2.5"-10")		Muck-Mud	black, very fine organic (FPOM)	
Gravel	2-64 mm (0.1"-2.5")				
Sand	0.06-2mm (gritty)		Marl	grey, shell fragments	
Silt	0.004-0.06 mm				
Clay	< 0.004 mm (slick)				

HABITAT ASSESSMENT FIELD DATA SHEET—HIGH GRADIENT STREAMS (FRONT)

STREAM NAME		LOCATION	
STATION #_____ RIVERMILE_____		STREAM CLASS	
LAT _____ LONG _____		RIVER BASIN	
STORET #		AGENCY	
INVESTIGATORS			
FORM COMPLETED BY		DATE _____ TIME _____ AM PM	REASON FOR SURVEY

	Habitat Parameter	Condition Category			
		Optimal	**Suboptimal**	**Marginal**	**Poor**
Parameters to be evaluated in sampling reach	**1. Epifaunal Substrate/ Available Cover**	Greater than 70% of substrate favorable for epifaunal colonization and fish cover; mix of snags, submerged logs, undercut banks, cobble or other stable habitat and at stage to allow full colonization potential (i.e., logs/snags that are <u>not</u> new fall and not transient).	40-70% mix of stable habitat; well-suited for full colonization potential; adequate habitat for maintenance of populations; presence of additional substrate in the form of newfall, but not yet prepared for colonization (may rate at high end of scale).	20-40% mix of stable habitat; habitat availability less than desirable; substrate frequently disturbed or removed.	Less than 20% stable habitat; lack of habitat is obvious; substrate unstable or lacking.
	SCORE	20 19 18 17 16	15 14 13 12 11	10 9 8 7 6	5 4 3 2 1 0
	2. Embeddedness	Gravel, cobble, and boulder particles are 0-25% surrounded by fine sediment. Layering of cobble provides diversity of niche space.	Gravel, cobble, and boulder particles are 25-50% surrounded by fine sediment.	Gravel, cobble, and boulder particles are 50-75% surrounded by fine sediment.	Gravel, cobble, and boulder particles are more than 75% surrounded by fine sediment.
	SCORE	20 19 18 17 16	15 14 13 12 11	10 9 8 7 6	5 4 3 2 1 0
	3. Velocity/Depth Regime	All four velocity/depth regimes present (slow-deep, slow-shallow, fast-deep, fast-shallow). (Slow is < 0.3 m/s, deep is > 0.5 m.)	Only 3 of the 4 regimes present (if fast-shallow is missing, score lower than if missing other regimes).	Only 2 of the 4 habitat regimes present (if fast-shallow or slow-shallow are missing, score low).	Dominated by 1 velocity/ depth regime (usually slow-deep).
	SCORE	20 19 18 17 16	15 14 13 12 11	10 9 8 7 6	5 4 3 2 1 0
	4. Sediment Deposition	Little or no enlargement of islands or point bars and less than 5% of the bottom affected by sediment deposition.	Some new increase in bar formation, mostly from gravel, sand or fine sediment; 5-30% of the bottom affected; slight deposition in pools.	Moderate deposition of new gravel, sand or fine sediment on old and new bars; 30-50% of the bottom affected; sediment deposits at obstructions, constrictions, and bends; moderate deposition of pools prevalent.	Heavy deposits of fine material, increased bar development; more than 50% of the bottom changing frequently; pools almost absent due to substantial sediment deposition.
	SCORE	20 19 18 17 16	15 14 13 12 11	10 9 8 7 6	5 4 3 2 1 0
	5. Channel Flow Status	Water reaches base of both lower banks, and minimal amount of channel substrate is exposed.	Water fills >75% of the available channel; or <25% of channel substrate is exposed.	Water fills 25-75% of the available channel, and/or riffle substrates are mostly exposed.	Very little water in channel and mostly present as standing pools.
	SCORE	20 19 18 17 16	15 14 13 12 11	10 9 8 7 6	5 4 3 2 1 0

Rapid Bioassessment Protocols For Use in Streams and Wadeable Rivers: Periphyton, Benthic Macroinvertebrates, and Fish, Second Edition - Form 2

A-7

HABITAT ASSESSMENT FIELD DATA SHEET—HIGH GRADIENT STREAMS (BACK)

Habitat Parameter	Condition Category			
	Optimal	**Suboptimal**	**Marginal**	**Poor**
6. Channel Alteration	Channelization or dredging absent or minimal; stream with normal pattern.	Some channelization present, usually in areas of bridge abutments; evidence of past channelization, i.e., dredging, (greater than past 20 yr) may be present, but recent channelization is not present.	Channelization may be extensive; embankments or shoring structures present on both banks; and 40 to 80% of stream reach channelized and disrupted.	Banks shored with gabion or cement; over 80% of the stream reach channelized and disrupted. Instream habitat greatly altered or removed entirely.
SCORE	20 19 18 17 16	15 14 13 12 11	10 9 8 7 6	5 4 3 2 1 0
7. Frequency of Riffles (or bends)	Occurrence of riffles relatively frequent; ratio of distance between riffles divided by width of the stream <7:1 (generally 5 to 7); variety of habitat is key. In streams where riffles are continuous, placement of boulders or other large, natural obstruction is important.	Occurrence of riffles infrequent; distance between riffles divided by the width of the stream is between 7 to 15.	Occasional riffle or bend; bottom contours provide some habitat; distance between riffles divided by the width of the stream is between 15 to 25.	Generally all flat water or shallow riffles; poor habitat; distance between riffles divided by the width of the stream is a ratio of >25.
SCORE	20 19 18 17 16	15 14 13 12 11	10 9 8 7 6	5 4 3 2 1 0
8. Bank Stability (score each bank) Note: determine left or right side by facing downstream.	Banks stable; evidence of erosion or bank failure absent or minimal; little potential for future problems. <5% of bank affected.	Moderately stable; infrequent, small areas of erosion mostly healed over. 5-30% of bank in reach has areas of erosion.	Moderately unstable; 30-60% of bank in reach has areas of erosion; high erosion potential during floods.	Unstable; many eroded areas; "raw" areas frequent along straight sections and bends; obvious bank sloughing; 60-100% of bank has erosional scars.
SCORE ___ (LB)	Left Bank 10 9	8 7 6	5 4 3	2 1 0
SCORE ___ (RB)	Right Bank 10 9	8 7 6	5 4 3	2 1 0
9. Vegetative Protection (score each bank)	More than 90% of the streambank surfaces and immediate riparian zone covered by native vegetation, including trees, understory shrubs, or nonwoody macrophytes; vegetative disruption through grazing or mowing minimal or not evident; almost all plants allowed to grow naturally.	70-90% of the streambank surfaces covered by native vegetation, but one class of plants is not well-represented; disruption evident but not affecting full plant growth potential to any great extent; more than one-half of the potential plant stubble height remaining.	50-70% of the streambank surfaces covered by vegetation; disruption obvious; patches of bare soil or closely cropped vegetation common; less than one-half of the potential plant stubble height remaining.	Less than 50% of the streambank surfaces covered by vegetation; disruption of streambank vegetation is very high; vegetation has been removed to 5 centimeters or less in average stubble height.
SCORE ___ (LB)	Left Bank 10 9	8 7 6	5 4 3	2 1 0
SCORE ___ (RB)	Right Bank 10 9	8 7 6	5 4 3	2 1 0
10. Riparian Vegetative Zone Width (score each bank riparian zone)	Width of riparian zone >18 meters; human activities (i.e., parking lots, roadbeds, clear-cuts, lawns, or crops) have not impacted zone.	Width of riparian zone 12-18 meters; human activities have impacted zone only minimally.	Width of riparian zone 6-12 meters; human activities have impacted zone a great deal.	Width of riparian zone <6 meters; little or no riparian vegetation due to human activities.
SCORE ___ (LB)	Left Bank 10 9	8 7 6	5 4 3	2 1 0
SCORE ___ (RB)	Right Bank 10 9	8 7 6	5 4 3	2 1 0

Parameters to be evaluated broader than sampling reach

Total Score _____

HABITAT ASSESSMENT FIELD DATA SHEET—LOW GRADIENT STREAMS (FRONT)

STREAM NAME	LOCATION
STATION #_____ RIVERMILE_____	STREAM CLASS
LAT _____ LONG _____	RIVER BASIN
STORET #	AGENCY
INVESTIGATORS	
FORM COMPLETED BY	DATE _____ TIME _____ AM PM	REASON FOR SURVEY

	Habitat Parameter	Condition Category			
		Optimal	**Suboptimal**	**Marginal**	**Poor**
Parameters to be evaluated in sampling reach	**1. Epifaunal Substrate/ Available Cover**	Greater than 50% of substrate favorable for epifaunal colonization and fish cover; mix of snags, submerged logs, undercut banks, cobble or other stable habitat and at stage to allow full colonization potential (i.e., logs/snags that are <u>not</u> new fall and <u>not</u> transient).	30-50% mix of stable habitat; well-suited for full colonization potential; adequate habitat for maintenance of populations; presence of additional substrate in the form of newfall, but not yet prepared for colonization (may rate at high end of scale).	10-30% mix of stable habitat; habitat availability less than desirable; substrate frequently disturbed or removed.	Less than 10% stable habitat; lack of habitat is obvious; substrate unstable or lacking.
	SCORE	20 19 18 17 16	15 14 13 12 11	10 9 8 7 6	5 4 3 2 1 0
	2. Pool Substrate Characterization	Mixture of substrate materials, with gravel and firm sand prevalent; root mats and submerged vegetation common.	Mixture of soft sand, mud, or clay; mud may be dominant; some root mats and submerged vegetation present.	All mud or clay or sand bottom; little or no root mat; no submerged vegetation.	Hard-pan clay or bedrock; no root mat or vegetation.
	SCORE	20 19 18 17 16	15 14 13 12 11	10 9 8 7 6	5 4 3 2 1 0
	3. Pool Variability	Even mix of large-shallow, large-deep, small-shallow, small-deep pools present.	Majority of pools large-deep; very few shallow.	Shallow pools much more prevalent than deep pools.	Majority of pools small-shallow or pools absent.
	SCORE	20 19 18 17 16	15 14 13 12 11	10 9 8 7 6	5 4 3 2 1 0
	4. Sediment Deposition	Little or no enlargement of islands or point bars and less than <20% of the bottom affected by sediment deposition.	Some new increase in bar formation, mostly from gravel, sand or fine sediment; 20-50% of the bottom affected; slight deposition in pools.	Moderate deposition of new gravel, sand or fine sediment on old and new bars; 50-80% of the bottom affected; sediment deposits at obstructions, constrictions, and bends; moderate deposition of pools prevalent.	Heavy deposits of fine material, increased bar development; more than 80% of the bottom changing frequently; pools almost absent due to substantial sediment deposition.
	SCORE	20 19 18 17 16	15 14 13 12 11	10 9 8 7 6	5 4 3 2 1 0
	5. Channel Flow Status	Water reaches base of both lower banks, and minimal amount of channel substrate is exposed.	Water fills >75% of the available channel; or <25% of channel substrate is exposed.	Water fills 25-75% of the available channel, and/or riffle substrates are mostly exposed.	Very little water in channel and mostly present as standing pools.
	SCORE	20 19 18 17 16	15 14 13 12 11	10 9 8 7 6	5 4 3 2 1 0

Rapid Bioassessment Protocols For Use in Streams and Wadeable Rivers: Periphyton, Benthic Macroinvertebrates, and Fish, Second Edition - Form 3

A-9

HABITAT ASSESSMENT FIELD DATA SHEET—LOW GRADIENT STREAMS (BACK)

Parameters to be evaluated broader than sampling reach

Habitat Parameter	Condition Category			
	Optimal	Suboptimal	Marginal	Poor
6. Channel Alteration	Channelization or dredging absent or minimal; stream with normal pattern.	Some channelization present, usually in areas of bridge abutments; evidence of past channelization. i.e., dredging, (greater than past 20 yr) may be present, but recent channelization is not present.	Channelization may be extensive; embankments or shoring structures present on both banks; and 40 to 80% of stream reach channelized and disrupted.	Banks shored with gabion or cement; over 80% of the stream reach channelized and disrupted. Instream habitat greatly altered or removed entirely.
SCORE	20 19 18 17 16	15 14 13 12 11	10 9 8 7 6	5 4 3 2 1 0
7. Channel Sinuosity	The bends in the stream increase the stream length 3 to 4 times longer than if it was in a straight line. (Note - channel braiding is considered normal in coastal plains and other low-lying areas. This parameter is not easily rated in these areas.)	The bends in the stream increase the stream length 1 to 2 times longer than if it was in a straight line.	The bends in the stream increase the stream length 1 to 2 times longer than if it was in a straight line.	Channel straight; waterway has been channelized for a long distance.
SCORE	20 19 18 17 16	15 14 13 12 11	10 9 8 7 6	5 4 3 2 1 0
8. Bank Stability (score each bank)	Banks stable; evidence of erosion or bank failure absent or minimal; little potential for future problems. <5% of bank affected.	Moderately stable; infrequent, small areas of erosion mostly healed over. 5-30% of bank in reach has areas of erosion.	Moderately unstable; 30-60% of bank in reach has areas of erosion; high erosion potential during floods.	Unstable; many eroded areas; "raw" areas frequent along straight sections and bends; obvious bank sloughing; 60-100% of bank has erosional scars.
SCORE ___ (LB)	Left Bank 10 9	8 7 6	5 4 3	2 1 0
SCORE ___ (RB)	Right Bank 10 9	8 7 6	5 4 3	2 1 0
9. Vegetative Protection (score each bank) Note: determine left or right side by facing downstream.	More than 90% of the streambank surfaces and immediate riparian zone covered by native vegetation, including trees, understory shrubs, or nonwoody macrophytes; vegetative disruption through grazing or mowing minimal or not evident; almost all plants allowed to grow naturally.	70-90% of the streambank surfaces covered by native vegetation, but one class of plants is not well-represented; disruption evident but not affecting full plant growth potential to any great extent; more than one-half of the potential plant stubble height remaining.	50-70% of the streambank surfaces covered by vegetation; disruption obvious; patches of bare soil or closely cropped vegetation common; less than one-half of the potential plant stubble height remaining.	Less than 50% of the streambank surfaces covered by vegetation; disruption of streambank vegetation is very high; vegetation has been removed to 5 centimeters or less in average stubble height.
SCORE ___ (LB)	Left Bank 10 9	8 7 6	5 4 3	2 1 0
SCORE ___ (RB)	Right Bank 10 9	8 7 6	5 4 3	2 1 0
10. Riparian Vegetative Zone Width (score each bank riparian zone)	Width of riparian zone >18 meters; human activities (i.e.. parking lots, roadbeds, clear-cuts, lawns, or crops) have not impacted zone.	Width of riparian zone 12-18 meters; human activities have impacted zone only minimally.	Width of riparian zone 6-12 meters; human activities have impacted zone a great deal.	Width of riparian zone <6 meters; little or no riparian vegetation due to human activities.
SCORE ___ (LB)	Left Bank 10 9	8 7 6	5 4 3	2 1 0
SCORE ___ (RB)	Right Bank 10 9	8 7 6	5 4 3	2 1 0

Total Score _____

APPENDIX A-2:

Periphyton Field and Laboratory Data Sheets

Form 1: Periphyton Field Data Sheet
Form 2: Periphyton Sample Log-In Sheet
Form 3: Periphyton Soft Algae Laboratory Bench Sheet (front and back)
Form 4: Periphyton Diatom Laboratory Bench Sheet (front and back)
Form 5: Rapid Periphyton Survey Field Sheet

Rapid Bioassessment Protocols For Use in Streams and Wadeable Rivers: Periphyton, Benthic Macroinvertebrates, and Fish, Second Edition

A-11

This Page Intentionally Left Blank

PERIPHYTON FIELD DATA SHEET

STREAM NAME	LOCATION	
STATION #_____ RIVERMILE_____	STREAM CLASS	
LAT _____ LONG _____	RIVER BASIN	
STORET #	AGENCY	
INVESTIGATORS		LOT NUMBER
FORM COMPLETED BY	DATE _____ TIME _____ AM PM	REASON FOR SURVEY

HABITAT TYPES	**Indicate the percentage of each habitat type present** ❑ Sand-Silt-Mud-Muck_____% ❑ Gravel-Cobble_____% ❑ Bedrock_____% ❑ Small Woody Debris_____% ❑ Large Woody Debris_____% ❑ Plants, Roots_____% ❑ Riffle_____% ❑Run_____% ❑ Pool_____% ❑ Canopy_____%
SAMPLE COLLECTION	**Gear used** ❑ suction device ❑ bar clamp sample ❑ scraping ❑ Other_____ **How were the samples collected?** ❑ wading ❑ from bank ❑ from boat **If natural habitat collections, indicate the number of samples taken in each habitat type.** ❑ Sand-Silt-Mud-Muck_____% ❑ Gravel-Cobble_____% ❑ Bedrock_____% ❑ Small Woody Debris_____% ❑ Large Woody Debris_____% ❑ Plants, Roots_____%
GENERAL COMMENTS	

QUALITATIVE LISTING OF AQUATIC BIOTA

Indicate estimated abundance: 0 = Absent/Not Observed, 1 = Rare (<5%), 2 = Common (5% - 30%), 3= Abundant (30% - 70%), 4 = Dominant (>70%)

Periphyton	0	1	2	3	4	Slimes	0	1	2	3	4
Filamentous Algae	0	1	2	3	4	Macroinvertebrates	0	1	2	3	4
Macrophytes	0	1	2	3	4	Fish	0	1	2	3	4

Rapid Bioassessment Protocols For Use in Streams and Wadeable Rivers: Periphyton, Benthic Macroinvertebrates, and Fish, Second Edition - Form 1

A-13

This Page Intentionally Left Blank

PERIPHYTON SAMPLE LOG-IN SHEET

Date Collected	Collected By	Number of Containers	Preservation	Station #	Stream Name and Location	Date Received by Lab	Lot Number	Date of Completion — sorting	Date of Completion — mounting	Date of Completion — identification

Serial Code Example: P0754001(1)
P = Periphyton (B = Benthos, F = Fish # 0754 = project number # 001 = sample number # (1) = lot number (e.g., winter 1996 =1; summer 1996 = 2)

Rapid Bioassessment Protocols For Use in Streams and Wadeable Rivers: Periphyton, Benthic Macroinvertebrates, and Fish, Second Edition - Form 2

A-15

This Page Intentionally Left Blank

PERIPHYTON SOFT ALGAE LABORATORY BENCH SHEET (FRONT)

page _____ of _____

STREAM NAME		LOCATION	
STATION # RIVERMILE		STREAM CLASS	
LAT LONG		RIVER BASIN	
STORET # LOT #		AGENCY	
COLLECTORS INITIALS DATE		TAXONOMISTS INITIALS DATE	
SUBSAMPLE TARGET FOR SOFT ALGAE ❏ 300 ❏ 400 ❏ 500 ❏ Other ____			

TAXA NAME	TALLY	CODE	# OF CELLS	TCR

Taxonomic certainty ratings (TCR) can be determined for each taxa or for the laboratory as a whole. The TCR scale is 1-5, with: 1 = most certain and 5 = least certain. If rating is 3-5, give reason. The number of cells for filamentous algae is an estimate of relative biomass.

Total No. Algal cells _____   **Total No. Taxa** _____

Rapid Bioassessment Protocols For Use in Streams and Wadeable Rivers: Periphyton, Benthic Macroinvertebrates, and Fish, Second Edition - Form 3

A-17

PERIPHYTON SOFT ALGAE LABORATORY BENCH SHEET (BACK)

STREAM IDENTIFICATION CODE	DATE COUNTED
COUNTED TRANSECT LENGTH	COUNTED TRANSECT WIDTH
SIZE OF COVERGLASS	TOTAL SAMPLE VOLUME
VOLUME OF SAMPLE ON COVERGLASS	SAMPLE DILUTION FACTOR
PROPORTION OF SAMPLE COUNTED	AREA OF SUBSTRATE SAMPLED
TOTAL NUMBER OF CELLS COUNTED	TOTAL ASSEMBLAGE CELL DENSITY

TAXONOMY

ID _____

Date _____

Explain TCR ratings of 3-5:

Other Comments (e.g. condition of algae):

_____

QC: ❏ YES   ❏ NO   **QC Checker**
_____

Algal recognition   ❏ pass   ❏ fail
Verification complete   ❏ YES   ❏ NO

General Comments (use this space to add additional comments):

_____
_____
_____
_____
_____
_____
_____
_____
_____
_____
_____
_____

PERIPHYTON DIATOM LABORATORY BENCH SHEET (FRONT)

page _____ of _____

STREAM NAME		LOCATION	
STATION # RIVERMILE		STREAM CLASS	
LAT LONG		RIVER BASIN	
STORET # LOT #		AGENCY	
COLLECTORS INITIALS DATE		TAXONOMISTS INITIALS DATE	
SUBSAMPLE TARGET FOR DIATOM ❏ 300 ❏ 400 ❏ 600 ❏ Other ____			

TAXA NAME	TALLY (# of valves)	CODE	# OF CELLS	TCR

Taxonomic certainty ratings (TCR) can be determined for each taxa or for the laboratory as a whole. The TCR scale is 1-5, with 1 = most certain and 5 = least certain. If rating is 3-5, give reason. The number of cells for filamentous algae is an estimate of relative biomass.

Total No. Algal cells _____ **Total No. Taxa** _____

Rapid Bioassessment Protocols For Use in Streams and Wadeable Rivers: Periphyton, Benthic Macroinvertebrates, and Fish, Second Edition - Form 4

A-19

PERIPHYTON DIATOM LABORATORY BENCH SHEET (BACK)

TAXONOMY	Explain TCR ratings of 3-5:
ID _____	
Date _____	Other Comments (e.g. condition of algae):
	QC: ❏ YES ❏ NO **QC Checker** _____
	Algal recognition ❏ pass ❏ fail
	Verification complete ❏ YES ❏ NO

General Comments (use this space to add additional comments):

_____

_____

_____

_____

_____

_____

_____

_____

_____

_____

_____

_____

_____

_____

_____

_____

_____

_____

RAPID PERIPHYTON SURVEY FIELD SHEET

STREAM NAME		LOCATION	
STATION #	RIVERMILE	STREAM CLASS	
LAT	LONG	RIVER BASIN	
STORET #	LOT #	AGENCY	
COLLECTORS INITIALS	DATE	TAXONOMISTS INITIALS	DATE

ASSESSED BY
GRID AREA
ID MACROALGA #1
ID MACROALGA #2
ID MICROALGA #1
ID MICROALGA #2

Macroalga #1 Maximum Length _____
Macroalga #2 Maximum Length _____

TRANSECT/ VIEW #	# DOTS IN GRID AREA	MACROALGA #1 DOTS COVERED	MACROALGA #2 DOTS COVERED	# DOTS MICROALGA SUBSTRATE	MICROALGA #1 DOTS COVERED BY THICKNESS RANK							MICROALGA #2 DOTS COVERED BY THICKNESS RANK						
					0	0.5	1	2	3	4	5	0	0.5	1	2	3	4	5
TOTAL # DOTS AT SITE																		

General Comments:

This Page Intentionally Left Blank

APPENDIX A-3:

Benthic Macroinvertebrate Field and Laboratory Data Sheets

Form 1: Benthic Macroinvertebrate Field Data Sheet
Form 2: Benthic Macroinvertebrate Sample Log-In Sheet
Form 3: Benthic Macroinvertebrate Laboratory Bench Sheet
Form 4: Preliminary Assessment Score Sheet (Pass)

Rapid Bioassessment Protocols For Use in Streams and Wadeable Rivers: Periphyton, Benthic Macroinvertebrates, and Fish, Second Edition

A-23

This Page Intentionally Left Blank

BENTHIC MACROINVERTEBRATE FIELD DATA SHEET

STREAM NAME	LOCATION	
STATION #_____ RIVERMILE_____	STREAM CLASS	
LAT _____ LONG _____	RIVER BASIN	
STORET #	AGENCY	
INVESTIGATORS		LOT NUMBER
FORM COMPLETED BY	DATE _____ TIME _____ AM PM	REASON FOR SURVEY

HABITAT TYPES	**Indicate the percentage of each habitat type present** ❑ Cobble_____% ❑ Snags_____% ❑ Vegetated Banks_____% ❑ Sand_____% ❑ Submerged Macrophytes_____% ❑ Other ( )_____%
SAMPLE COLLECTION	**Gear used** ❑ D-frame ❑ kick-net ❑ Other _____ **How were the samples collected?** ❑ wading ❑ from bank ❑ from boat **Indicate the number of jabs/kicks taken in each habitat type.** ❑ Cobble_____ ❑ Snags_____ ❑ Vegetated Banks_____ ❑ Sand_____ ❑ Submerged Macrophytes_____ ❑ Other ( )_____
GENERAL COMMENTS	

QUALITATIVE LISTING OF AQUATIC BIOTA
Indicate estimated abundance: 0 = Absent/Not Observed, 1 = Rare, 2 = Common, 3= Abundant, 4 = Dominant

Periphyton	0 1 2 3 4	Slimes	0 1 2 3 4
Filamentous Algae	0 1 2 3 4	Macroinvertebrates	0 1 2 3 4
Macrophytes	0 1 2 3 4	Fish	0 1 2 3 4

FIELD OBSERVATIONS OF MACROBENTHOS
Indicate estimated abundance: 0 = Absent/Not Observed, 1 = Rare (1-3 organisms), 2 = Common (3-9 organisms), 3= Abundant (>10 organisms), 4 = Dominant (>50 organisms)

Porifera	0 1 2 3 4	Anisoptera	0 1 2 3 4	Chironomidae	0 1 2 3 4
Hydrozoa	0 1 2 3 4	Zygoptera	0 1 2 3 4	Ephemeroptera	0 1 2 3 4
Platyhelminthes	0 1 2 3 4	Hemiptera	0 1 2 3 4	Trichoptera	0 1 2 3 4
Turbellaria	0 1 2 3 4	Coleoptera	0 1 2 3 4	Other	0 1 2 3 4
Hirudinea	0 1 2 3 4	Lepidoptera	0 1 2 3 4		
Oligochaeta	0 1 2 3 4	Sialidae	0 1 2 3 4		
Isopoda	0 1 2 3 4	Corydalidae	0 1 2 3 4		
Amphipoda	0 1 2 3 4	Tipulidae	0 1 2 3 4		
Decapoda	0 1 2 3 4	Empididae	0 1 2 3 4		
Gastropoda	0 1 2 3 4	Simuliidae	0 1 2 3 4		
Bivalvia	0 1 2 3 4	Tabinidae	0 1 2 3 4		
		Culcidae	0 1 2 3 4		

Rapid Bioassessment Protocols For Use in Streams and Wadeable Rivers: Periphyton, Benthic Macroinvertebrates, and Fish, Second Edition - Form 1

A-25

This Page Intentionally Left Blank

BENTHIC MACROINVERTEBRATE SAMPLE LOG-IN SHEET

Date Collected	Collected By	Number of Containers	Preservation	Station #	Stream Name and Location	Date Received by Lab	Lot Number	Date of Completion		
								sorting	mounting	identification

Serial Code Example: B0754001(1)
B = Benthos (F = Fish; P = Periphyton # 0754 = project number # 001 = sample number # (1) = lot number (e.g., winter 1996 =1; summer 1996 = 2)

This Page Intentionally Left Blank

BENTHIC MACROINVERTEBRATE LABORATORY BENCH SHEET (FRONT)

STREAM NAME	LOCATION
STATION #_____ RIVERMILE_____	STREAM CLASS
LAT _____ LONG _____	RIVER BASIN
STORET #	AGENCY
COLLECTED BY DATE_____	LOT #
TAXONOMIST DATE_____	SUBSAMPLE TARGET ❏ 100 ❏ 200 ❏ 300 ❏ Other ____

Enter Family and/or Genus and Species name on blank line.

Organisms	No.	LS	TI	TCR	Organisms	No.	LS	TI	TCR
Oligochaeta					Megaloptera				
Hirudinea					Coleoptera				
Isopoda									
Amphipoda					Diptera				
Decapoda									
Ephemeroptera									
					Gastropoda				
					Pelecypoda				
Plecoptera									
					Other				
Trichoptera									
Hemiptera									

Taxonomic certainty rating (TCR) 1-5:1=most certain, 5=least certain. If rating is 3-5, give reason (e.g., missing gills). LS= life stage: I = immature; P = pupa; A = adult TI = Taxonomists initials

Total No. Organisms _____   Total No. Taxa _____

BENTHIC MACROINVERTEBRATE LABORATORY BENCH SHEET (BACK)

SUBSAMPLING/SORTING INFORMATION	Number of grids picked: _____
Sorter _____ Date _____	Time expenditure _____ No. of organisms _____ Indicate the presence of large or obviously abundant organisms: QC: ❏ YES ❏ NO **QC Checker** _____ $$\frac{\text{\# organisms originally sorted}}{\left(\begin{array}{c}\text{\# organisms recovered by checker}\end{array} + \text{\# organisms originally sorted}\right)} = \text{\% sorting efficiency}$$ ≥90%, sample passes _____ <90%, sample fails, action taken _____ _____
TAXONOMY ID _____ Date _____	Explain TCR ratings of 3-5: Other Comments (e.g. condition of specimens): QC: ❏ YES ❏ NO **QC Checker** _____ Organism recognition ❏ pass ❏ fail Verification complete ❏ YES ❏ NO

PRELIMINARY ASSESSMENT SCORE SHEET
(PASS)

STREAM NAME		LOCATION	
STATION #_____ RIVERMILE_____		STREAM CLASS	
LAT _____ LONG _____		RIVER BASIN	
STORET #		AGENCY	
COLLECTED BY DATE_____		LOT # _____ NUMBER OF SWEEPS _____	
HABITATS: ❑ COBBLE ❑ SHOREZONE ❑ SNAGS ❑ VEGETATION			

Enter Family and/or Genus and Species name on blank line.

Organisms		No.	LS	TI	TCR	Organisms		No.	LS	TI	TCR
Oligochaeta						Megaloptera					
Hirudinea						Coleoptera					
Isopoda											
Amphipoda						Diptera					
Decapoda											
Ephemeroptera											
						Gastropoda					
						Pelecypoda					
Plecoptera											
						Other					
Trichoptera											
						Taxonomic certainty rating (TCR) 1-5:1=most certain, 5=least certain. If rating is 3-5, give reason (e.g., missing gills). LS= life stage: I = immature; P = pupa; A = adult TI = Taxonomists initials					
Hemiptera											

	Site Value	Target Threshold	If 2 or more metrics are ≥ target threshold, site is
Total No. Taxa			**HEALTHY**
EPT Taxa			If less than 2 metrics are within target range, site is
Tolerance Index			**SUSPECTED IMPAIRED**

This Page Intentionally Left Blank

APPENDIX A-4:

Fish Field and Laboratory Data Sheets

Form 1: Fish Sampling Field Data Sheet
Form 2: Fish Sample Log-In Sheet

Rapid Bioassessment Protocols For Use in Streams and Wadeable Rivers: Periphyton, Benthic Macroinvertebrates, and Fish, Second Edition

A-33

This Page Intentionally Left Blank

FISH SAMPLING FIELD DATA SHEET (FRONT)

STREAM NAME		LOCATION	
STATION #_____ RIVERMILE_____		STREAM CLASS	
LAT _____ LONG _____		RIVER BASIN	
STORET #		AGENCY	
GEAR		INVESTIGATORS	
FORM COMPLETED BY		DATE _____ TIME _____ AM PM	REASON FOR SURVEY

SAMPLE COLLECTION	**How were the fish captured?** ❏ back pack ❏ tote barge ❏ other _____ **Block nets used?** ❏ YES ❏ NO **Sampling Duration** Start time _____ End time _____ Duration _____ **Stream width (in meters)** Max_____ Mean_____
HABITAT TYPES	**Indicate the percentage of each habitat type present** ❏ Riffles_____% ❏ Pools_____% ❏ Runs_____% ❏ Snags_____% ❏ Submerged Macrophytes_____% ❏ Other (_____)_____%
GENERAL COMMENTS	

SPECIES	TOTAL (COUNT)	OPTIONAL: LENGTH (mm)/WEIGHT (g) (25 SPECIMEN MAX SUBSAMPLE)					ANOMALIES*							
							D	E	F	L	M	S	T	Z

FISH SAMPLING FIELD DATA SHEET (BACK)

SPECIES	TOTAL (COUNT)	OPTIONAL: LENGTH (mm)/WEIGHT (g) (25 SPECIMEN MAX SUBSAMPLE)					ANOMALIES*							
							D	E	F	L	M	S	T	Z

*
ANOMALY CODES: D = deformities; E = eroded fins; F = fungus; L = lesions; M = multiple DELT anomalies; S = emaciated; Z = other

FISH SAMPLE LOG-IN SHEET

Date Collected	Collected By	Number of Containers	Preservation	Station #	Stream Name and Location	Date Received by Lab	Lot Number	Date of Completion		
								sorting	mounting	identification

Serial Code Example: F0754001(1)

F = Fish (B = Benthos; P = Periphyton)# 0754 = project number # 001 = sample number # (1) = lot number (e.g., winter 1996 =1; summer 1996 = 2)

Rapid Bioassessment Protocols For Use in Streams and Wadeable Rivers: Periphyton, Benthic Macroinvertebrates, and Fish, Second Edition - Form 2

A-37

This Page Intentionally Left Blank

APPENDIX B:

REGIONAL TOLERANCE VALUES, FUNCTIONAL FEEDING GROUPS AND HABIT/BEHAVIOR ASSIGNMENTS FOR BENTHIC MACROINVERTEBRATES

Rapid Bioassessment Protocols for Use in Streams and Wadeable Rivers: Periphyton, Benthic Macroinvertebrates, and Fish, Second Edition

B-1

This Page Intentionally Left Blank

APPENDIX B

Appendix B is a list of selected benthic macroinvertebrates of the United States in phylogenetic order. Included are the Taxonomic Serial Number (TSN) and the Parent Taxonomic Serial Number for each of the taxa listed according to the Integrated Taxonomic Information System (ITIS). The ITIS generates a national taxonomic list that is constantly updated and currently posted on the World Wide Web at <www.itis.usda.gov>. If you are viewing this document electronically, this page is linked to the ITIS web site.

This Appendix displays regional tolerance values, primary and secondary functional feeding group information, and primary and secondary habit designations for selected benthic macroinvertebrates. In an effort to provide regionally accurate tolerance information, lists included in this Appendix were taken from the following states (and workgroup): Idaho (Northwest), Ohio[1] (Midwest), North Carolina (Southeast), Wisconsin (Upper Midwest), and the MACS workgroup (Mid-Atlantic Coastal Streams). Tolerance values are on a 0 to 10 scale, 0 representing the tolerance value of an extremely sensitive organism and 10 for a tolerant organism. For functional feeding group and habit/behavior assignments, primary and secondary designations are listed, if both are known. Each characterization is based on the organisms' larval qualities, except a group of beetles (listed as 'adult') that are aquatic as adults. The following are lists of the abbreviations used in this appendix.

FUNCTIONAL FEEDING DESIGNATIONS

PA=parasite
PR=predator
OM=omnivore
GC=gatherer/collector

FC=filter/collector
SC=scraper
SH=shredder
PI=piercer

HABIT/BEHAVIOR DESIGNATIONS

cn=clinger
cb=climber
sp=sprawler
bu=burrower

sw=swimmer
dv=diver
sk=skater

Sources For Benthic Tolerance, Functional Feeding Group, and Habit/Behavior Designations [a]

ID= Idaho DEP (Northwest)

OH= Ohio EPA (Midwest)

NC = North Carolina DEM (Southeast)

WI = Wisconsin DNR (Upper Midwest)

MACS= Mid-Atlantic Coastal Streams Workgroup (NJ DEP, DE DNREC, MD DNR, VA DEC, NC DEM, SC DHES)

[a] Habit/Behavior information is primarily based on Merritt and Cummins (1996) and pertains to insect larval forms (except for Dryopidae adults) and is mostly at genus level.

---

[1]Ohio traditionally uses an inverted 60-point scale compared to the other states in this list. In order to be comparable to the other listed states, the Ohio values were converted to a 0-10 scale as discussed above.

Rapid Bioassessment Protocols for Use in Streams and Wadeable Rivers: Periphyton, Benthic Macroinvertebrates, and Fish, Second Edition

B-3

This Page Intentionally Left Blank

Regional Tolerance Values, Functional Feeding Groups, and Habit/Behavior Assignments for Benthic Macroinvertebrates

Parent TSN	TSN	Scientific Name	Southeast (NC)	Upper Midwest (WI)	Midwest (OH)	Northwest (ID)	Mid-Atlantic (MACS)	primary	secondary	primary	secondary
			Regional Tolerance Values					Functional Feeding Group		Habit/ Behavior	
202423	59490	Nematoda				5		PA			
202423	64183	Nematomorpha						PA			
202423	57411	Nemertea				8		PR			
	57412	Rhynchocoela									
57577	57578	Prostoma graecense	6.6					PR			
57577	193496	Prostoma rubrum									
202423	53963	Platyhelminthes									
53963	53964	Turbellaria				4		PR			
53965	54468	Tricladida				4		GC			
54552	54553	Cura									
54468	54502	Planariidae				1		OM			
54502	54503	Dugesia				4		OM			
54503	54504	Dugesia tigrina	7.5					PR			
54502	54510	Polycelis				6		GC			
54510	54512	Polycelis coronata				1		OM			
202423	46861	Porifera						FC			
47690	47691	Spongillidae						FC			
47691	47692	Spongilla						FC			
47692	47696	Spongilla aspinosa						FC			
	155470	Ectoprocta									
156691	156692	Plumatella repens									
174619	174662	Hydrobates									
202423	48738	Cnidaria									
50844	50845	Hydra				5		PR			
50845	50846	Hydra americana									
156753	156754	Urnatella gracilis									
69458	79118	Bivalvia						FC			
	79119	Pelecypoda				8		FC			
79517	79519	Brachidontes exustus						FC			
79912	79913	Unionidae				8		FC			
79913	79930	Anodonta				8		FC			
79930	79946	Anodonta couperiana						FC			
		Anodonta nuttalliana idahoensis				8		FC			
79913	79951	Elliptio						FC			
79951	79975	Elliptio buckleyi						FC			
79951	79952	Elliptio complanata	5.4					FC			
79951	79964	Elliptio lanceolata	1.9								
79913	80032	Gonidea				4		FC			
80032	80033	Gonidea angulata				8		FC			
79986	80006	Lampsilis teres						FC			
79913	80370	Margaritifera				4		FC			
80370	80371	Margaritifera margaritifera				8		FC			
80059	80067	Quadrula cylindrica						FC			
81381	81385	Corbicula						FC			
81385	81387	Corbicula fluminea	6.3		3.2			FC			

Rapid Bioassessment Protocols for Use in Streams and Wadeable Rivers: Periphyton, Benthic Macroinvertebrates, and Fish, Second Edition

B-5

Parent TSN	TSN	Scientific Name	Southeast (NC)	Upper Midwest (WI)	Midwest (OH)	Northwest (ID)	Mid-Atlantic (MACS)	Functional Feeding Group primary	Functional Feeding Group secondary	Habit/ Behavior primary	Habit/ Behavior secondary
81385	81386	Corbicula manilensis						FC			
81333	81335	Mytilopsis leucophaeata						FC			
80384	81388	Pisidiidae				8		GC			
	81389	Sphaeriidae				8	8	FC			
81388	81436	Eupera									
	205642	Byssanodonta cubensis (= Eupera)						FC			
81436	81438	Eupera cubensis						FC			
81388	81427	Musculium					5	FC			
81427	81430	Musculium lacustre					5	FC			
		Byssanodonta (= Eupera)						FC			
81427	81434	Musculium securis					5	FC			
81427	81428	Musculium transversum									
81388	81400	Pisidium	6.8		4.6	8	8	FC			
81400	81405	Pisidium casertanum				8		SC			
81400		Pisidium lilljborgi					8	FC			
81400	81406	Pisidium compressum				8		FC			
81400	81402	Pisidium dubium						FC			
81400	81408	Pisidium fallax				8		FC			
81400	81403	Pisidium idahoense				8		FC			
81400	81424	Pisidium punctatum				8		FC			
81400	81425	Pisidium punctiferum						FC			
81400	81420	Pisidium walkeri				8		FC			
81388	81391	Sphaerium	7.7		4.7	6		GC	FC		
81391	81395	Sphaerium patella				8		FC			
81391	81398	Sphaerium striatinum						FC			
69458	69459	Gastropoda				7		SC			
76437	76568	Ancylidae				6		SC			
76568	76569	Ferrissia	6.9		5.2	6	7	SC			
76569	76573	Ferrissia hendersoni						SC			
76569	76572	Ferrissia rivularis						SC			
76569	76575	Ferrissia walkeri					7	SC			
76585	76586	Hebetancylus excentricus						SC			
76568	76576	Laevapex						SC			
76576	76578	Laevapex diaphanus						SC			
76576	76577	Laevapex fuscus	7.3		6.7			SC			
76576	76579	Laevapex peninsulae						SC			
76476	76477	Lanx				6		GC			
76437	76483	Lymnaeidae			6.9	6	6	SC			
76483	76497	Fossaria			2.6	8		SC			
76483	76484	Lymnaea				8		SC			
76483	76528	Pseudosuccinea						SC			
76528	76529	Pseudosuccinea columella	7.2					SC			
76483	76525	Radix									
76483	76534	Stagnicola	8			10	7	SC			
76437	76676	Physidae				8		SC			
76676	76677	Physa				8		SC			
76676	76698	Physella	9.1		7.6	8	8	SC			
76698	76707	Physella cubensis						SC			

Parent TSN	TSN	Scientific Name	Regional Tolerance Values					Functional Feeding Group		Habit/ Behavior	
			Southeast (NC)	Upper Midwest (WI)	Midwest (OH)	Northwest (ID)	Mid-Atlantic (MACS)	primary	secondary	primary	secondary
76698	76724	Physella hendersoni						SC			
76698	76736	Physella heterostropha						SC			
76437	76591	Planorbidae				7		SC			
76591	76592	Gyraulus				8		SC			
76592	76593	Gyraulus circumstriatus					7	SC			
76592	76595	Gyraulus parvus			5.5			SC			
76591	76599	Helisoma						SC			
76599	76600	Helisoma anceps	6.5		6		7	SC			
76591	76626	Menetus									
76626	205210	Menetus dilatatus	8.4		8.1			SC			
76591	76643	Micromenetus						SC			
76643	76648	Micromenetus dilatatus						SC			
76643	76646	Micromenetus floridensis						SC			
76591	76654	Planorbella				6		SC			
76654	76662	Planorbella duryi						SC			
76654	76667	Planorbella pilsbryi			7.4						
76654	76668	Planorbella scalaris						SC			
76671	205212	Planorbella trivolvis			9.5			SC			
76591	76621	Promenetus						GC			
76591	76673	Vorticifex				8		SC			
76673		Vorticifex effusa				6		SC			
77064	77300	Limacidae									
70160	70163	Neritina reclivata						SC			
70745	70747	Amnicola	4.8			5		SC			
70747	70764	Amnicola dalli						SC			
70747		Amnicola grana					8	SC			
70764	205008	Amnicola dalli johnsoni						SC			
70747	70748	Amnicola limosa					8	SC			
70745	70778	Fluminicola				5		SC			
70778	70782	Fluminicola hindsi				5		SC			
	71549	Pleurocera			3.7						
70298	70493	Hydrobiidae			7			SC			
		Pyrgulopsis idahoensis				8		SC			
70493	70509	Cincinnatia						SC			
70509	70513	Cincinnatia floridana						SC			
70493	70643	Fontelicella				8		SC			
70493	70527	Littoridinops						SC			
70527	70530	Littoridinops monroensis						SC			
70633	70634	Notogillia wetherbyi						SC			
70493	205005	Potamopyrgus				10		SC			
205005	205006	Potamopyrgus antipodarum				8		SC			
70699	70700	Pyrgophorus platyrachis						SC			
70712	70713	Rhapinema dacryon						SC			
	70548	Somatogyrus	6.5								
70548	70582	Somatogyrus walkerianus						SC			
70493	70702	Spilochlamys						SC			
70702	70703	Spilochlamys conica						SC			
71541	71654	Elimia	2.5		3.6		2	SC			

Rapid Bioassessment Protocols for Use in Streams and Wadeable Rivers: Periphyton, Benthic Macroinvertebrates, and Fish, Second Edition

B-7

Parent TSN	TSN	Scientific Name	Southeast (NC)	Upper Midwest (WI)	Midwest (OH)	Northwest (ID)	Mid-Atlantic (MACS)	Functional Feeding Group		Habit/ Behavior	
								primary	secondary	primary	secondary
71654	71858	Elimia athearni						SC			
71654	71746	Elimia curvicostata						SC			
71654	71761	Elimia floridensis						SC			
71541	71542	Goniobasis									
71541	71570	Juga				7		SC			
71541	71601	Leptoxis	1.6								
70298	71531	Thiaridae						SC			
71531	71532	Melanoides						SC			
71532	71533	Melanoides tuberculata						SC			
70298	70345	Valvatidae						SC			
70345	70346	Valvata				8		SC			
73194	73195	Marisa cornuarietis									
70342	70343	Pomacea paludosa						SC			
331584	70304	Viviparidae				6		SC			
331600	70311	Campeloma						SC			
70311	70312	Campeloma decisum	6.7			6		SC			
70311	70322	Campeloma floridense						SC			
70311	70315	Campeloma geniculum						SC			
70311	70317	Campeloma limum						SC			
70333	70336	Lioplax pilsbryi						SC			
331585	70305	Viviparus						SC			
70305	70307	Viviparus georgianus						SC			
202423	64357	Annelida						GC			
64357	68422	Oligochaeta				5		GC			
68498	69069	Lumbricina				8		GC			
68422	69168	Branchiobdellida									
69168	69169	Branchiobdellidae				6		GC			
69069	69080	Glossoscolecidae					10	GC			
69069	69165	Lumbricidae					10	GC			
68498	68499	Sparganophilidae									
68509	68510	Enchytraeidae	10			10	10	GC			
68509	68854	Naididae						GC			
68423	68424	Aeolosoma									
68854	68967	Allonais						GC			
68967	68971	Allonais inequalis						GC			
68854	69021	Bratislavia						GC			
69021	69022	Bratislavia bilongata						GC			
69021	69023	Bratislavia unidentata						GC			
68934	68935	Chaetogaster diaphanus									
68854	68898	Dero	10				10	GC			
68898	555636	Dero botrytis						GC			
68898	68904	Dero digitata						GC			
68898	68902	Dero flabelliger						GC			
68898	68912	Dero furcata						GC			
68898	68924	Dero lodeni						GC			
68898	68900	Dero nivea						GC			
68898	68907	Dero obtusa						GC			
68898	68923	Dero pectinata						GC			

Appendix B: Regional Tolerance Values, Functional Feeding Groups, and Habit/Behavior Assignments for Benthic Macroinvertebrates

Parent TSN	TSN	Scientific Name	Regional Tolerance Values					Functional Feeding Group		Habit/ Behavior	
			Southeast (NC)	Upper Midwest (WI)	Midwest (OH)	Northwest (ID)	Mid-Atlantic (MACS)	primary	secondary	primary	secondary
68898	68903	Dero trifida						GC			
68898	68915	Dero vaga						GC			
69003	69004	Haemonais waldvogeli						GC			
	68946	Nais	9.1								
68946	68949	Nais behningi						GC			
68946	68950	Nais communis						GC			
68946	68952	Nais elinguis						GC			
68946	68954	Nais pardalis						GC			
68946	68956	Nais pseudobtusa						GC			
68946	68957	Nais simplex						GC			
68946	68959	Nais variabilis						GC			
68862	68863	Paranais litoralis						GC			
68854	68876	Pristina	9.9					GC			
68876	68879	Pristina aequiseta						GC			
68876	68880	Pristina breviseta						GC			
68876	68881	Pristina foreli						GC			
68876	68894	Pristina leidyi						GC			
68876	68893	Pristina longisoma						GC			
68876	68887	Pristina osborni						GC			
68876	68891	Pristina plumaseta						GC			
68876	68878	Pristina sima						GC			
68876	68895	Pristina synclites						GC			
68854	69024	Pristinella						GC			
69024	69030	Pristinella jenkinae						GC			
69024	69025	Pristinella longisoma						GC			
69024	69026	Pristinella osborni						GC			
68854	68855	Slavina						GC			
68855	68856	Slavina appendiculata	7.1					GC			
68984	68985	Specaria josinae						GC			
69017	69018	Stephensoniana trivandrana						GC			
68871	68873	Stylaria fossularis					8	GC			
68871	68872	Stylaria lacustris	8.5					GC			
68854	69009	Vejdovskyella						GC			
69009	69010	Vejdovskyella comata						GC			
68509	69041	Opistocystidae									
68509	68585	Tubificidae				10	10	GC			
	68588	Peloscolex	8.8								
68679	68683	Aulodrilus americanus						GC			
68679	68682	Aulodrilus limnobius	5.2					GC			
68679	68680	Aulodrilus pigueti	4.7					GC			
68679	68684	Aulodrilus pluriseta					8	GC			
68619	68621	Branchiura sowerbyi	8.4					GC			
68585	68745	Haber									
68745	68746	Haber speciosus	2.8								
68660	68662	Ilyodrilus templetoni	9.4					GC			
68808	68809	Isochaetides curvisetosus	7.2					GC			
68808	68810	Isochaetides freyi	7.6								
68585	68638	Limnodrilus	9.6					GC			

Rapid Bioassessment Protocols for Use in Streams and Wadeable Rivers: Periphyton, Benthic Macroinvertebrates, and Fish, Second Edition

B-9

Parent TSN	TSN	Scientific Name	Southeast (NC)	Upper Midwest (WI)	Midwest (OH)	Northwest (ID)	Mid-Atlantic (MACS)	Functional Feeding Group primary	secondary	Habit/Behavior primary	secondary
68638	68653	Limnodrilus angustipenis						GC			
68638	68652	Limnodrilus cervix	10								
68638	68639	Limnodrilus hoffmeisteri	9.8					GC			
68638	68649	Limnodrilus profundicola						GC			
68638	68644	Limnodrilus udekemianus	9.7					GC			
68780	68610	Spirosperma ferox						GC			
68780	68781	Spirosperma nikolskyi	7.7								
68585	68751	Psammoryctides									
68751	68752	Psammoryctides convolutus						GC			
68793	68794	Quistradrilus multisetosus					10	GC			
68839	68844	Rhyacodrilus sodalis				10		GC			
68585	68780	Spirosperma						GC			
68780	68782	Spirosperma carolinensis				10		GC			
68585	68622	Tubifex				10		GC			
68622	68623	Tubifex tubifex	10					GC			
68439	68440	Lumbriculidae	7.3			8		GC			
68440	68473	Eclipidrilus				8					
68473	68476	Eclipidrilus palustris						GC			
68440	68441	Lumbriculus						GC			
68441	68447	Lumbriculus inconstans						GC			
68441	68444	Lumbriculus variegata						GC			
68422	69290	Hirudinea				10		PR			
69406	69407	Hirudinidae				7		PR			
69407	69408	Haemopsis				10		PR			
69408	69412	Haemopsis marmorata						PR			
69418	69421	Macrobdella ditetra									
69407	69430	Percymoorensis				10		PR			
69407	69423	Philobdella									
69437	69438	Erpobdellidae				8		PR			
69438	69439	Dina				8		PR			
69438	69449	Mooreobdella	7.8					PR			
69449	69454	Mooreobdella tetragon	9.7					PR			
69455	69456	Nephelopsis obscura						PR			
69295	69357	Glossiphoniidae				8		PR			
69388	69389	Alboglossiphonia heteroclita						PR			
69380	69390	Glossiphonia heteroclita									
69357	69358	Batracobdella						PA			
69358	69359	Batracobdella paludosa						PA			
69357	69380	Glossiphonia						PR			
555637	555638	Desserobdella phalera						PR			
69380	69381	Glossiphonia complanata						PR			
69357	69396	Helobdella				6		PA	PR		
	204822	Gloiobdella elongata						PR			
69396	69397	Helobdella elongata	9.9					PR			
69396	69401	Helobdella fusca						PA			
69396	69398	Helobdella stagnalis	6.7					PR			
69396	69399	Helobdella triserialis	8.9					PA			
69357	69363	Placobdella				6		PR			

Parent TSN	TSN	Scientific Name	Southeast (NC)	Upper Midwest (WI)	Midwest (OH)	Northwest (ID)	Mid-Atlantic (MACS)	primary	secondary	primary	secondary
			Regional Tolerance Values					Functional Feeding Group		Habit/ Behavior	
69363	69367	Placobdella multilineata						PR			
69363	69364	Placobdella papillifera	9					PA			
69363	69365	Placobdella parasitica	6.6					PA			
	69374	Batracobdella phalera	7.1								
69363	69372	Placobdella translucens						PA			
69357	69375	Theromyzon				10		PR			
69315	69316	Myzobdella lugubris						PR			
69296	69304	Piscicola				10		PR			
69304	69309	Piscicola salmositica				7		PR			
		Acari						PR			
		Acariformes						PR			
		Corticacarus delicatus				8		PR			
83538	83544	Oribatei									
		Parasitengona									
		Protzia californensis				8		PR			
82754	82769	Trombidiformes									
82862	82864	Arrenurus						PR			
82864	82907	Arrenurus apetiolatus						PR			
82864	82953	Arrenurus bicaudatus						PR			
82864	205790	Arrenurus hovus						PR			
82864	205791	Arrenurus problecornis						PR			
82864	205792	Arrenurus zapus						PR			
83434	83435	Albia						PR			
83176	83177	Clathrosperchon						PR			
82770	82771	Halacaridae									
82770	83122	Hydrachnidae									
83122	83123	Hydrachna						PR			
83224	83225	Hydrodroma						PR			
82770	83281	Hygrobatidae				8		PR			
83281	83282	Atractides						PR			
83281	83297	Hygrobates						PR			
83297	83310	Hygrobates occidentalis				8		PR			
83499	83500	Geayia									
83499	83502	Krendowskia									
82770	83033	Lebertiidae				8		PR			
83033	83034	Lebertia				8		PR			
83050	205794	Centrolimnesia						PR			
83050	83051	Limnesia						PR			
83145	83146	Limnochares						PR			
83476	83479	Mideopsis						PR			
83239	83240	Frontipoda						PR			
83239	83244	Oxus						PR			
82770	83159	Piersigiidae				8		PR			
83330	83350	Piona						PR			
83164	83172	Wandesia									
82770	83005	Sperchonidae				8		PR			
83005	83006	Sperchon						PR			
83006		Sperchon pseudoplumifer				8		PR			

Rapid Bioassessment Protocols for Use in Streams and Wadeable Rivers: Periphyton, Benthic Macroinvertebrates, and Fish, Second Edition

B-11

Parent TSN	TSN	Scientific Name	Southeast (NC)	Upper Midwest (WI)	Midwest (OH)	Northwest (ID)	Mid-Atlantic (MACS)	primary	secondary	primary	secondary
			Regional Tolerance Values					Functional Feeding Group		Habit/ Behavior	
83005	83029	Sperchonopsis						PR			
83249	83254	Torrenticola						PR			
83072	83093	Koenikea									
83093	205798	Koenikea angulata									
83093	193512	Koenikea aphrasta									
83093	193513	Koenikea elaphra									
83099	205797	Koenikea spinipes carella									
83072	83103	Neumania						PR			
83103	83106	Neumania distincta						PR			
83072	83073	Unionicola						PR			
82697	83677	Crustacea				8		GC			
95495	95599	Decapoda				8		SH			
98789	98790	Rhithropanopeus harrisii									
97250	97251	Potimirim potimirim									
96106	96213	Palaemonidae									
96213	96220	Macrobrachium									
96220	96225	Macrobrachium acanthurus									
96220	96221	Macrobrachium ohione									
96213	96383	Palaemonetes									
96383	96396	Palaemonetes kadiakensis					4	OM			
96383	96385	Palaemonetes paludosus					4				
97306	97324	Astacidae	7.2			8		SC			
97324	97325	Pacifastacus				6		OM			
97325		Pacifastacus cambilii				6		SH			
97325	97328	Pacifastacus connectens				6		SH			
97325	97326	Pacifastacus leniusculus				6		SH			
97306	97336	Cambaridae					6	GC			
97336	97337	Cambarus	8.1								
97336	97421	Orconectes	2.7								
97421	97423	Orconectes limosus					6	SH			
97336	97490	Procambarus	9.5								
97490	97492	Procambarus acutus					9	SH			
97490	97498	Procambarus alleni									
97490	97514	Procambarus fallax									
97490	97555	Procambarus pygmaeus									
97490	97566	Procambarus spiculifer									
89802	93294	Amphipoda				4		GC			
93584	93589	Corophium						FC			
93589	93594	Corophium lacustre						FC			
93641	93642	Grandidierella bonnieroides						GC			
95080	95081	Crangonyx	8				4	GC			
95081	95088	Crangonyx richmondensis						OM			
95081	193517	Crangonyx serratus	8.1					GC			
93295	93745	Gammaridae						GC			
93745	93747	Anisogammarus				4		GC			
	97160	Argis	8.7	8							
93745	93773	Gammarus				4		OM			
93773	93780	Gammarus fasciatus	6.9				6	GC			

Appendix B: Regional Tolerance Values, Functional Feeding Groups, and Habit/Behavior Assignments for Benthic Macroinvertebrates

| Parent TSN | TSN | Scientific Name | Regional Tolerance Values | | | | | Functional Feeding Group | | Habit/ Behavior | |
			Southeast (NC)	Upper Midwest (WI)	Midwest (OH)	Northwest (ID)	Mid-Atlantic (MACS)	primary	secondary	primary	secondary
93773	93789	Gammarus lacustris						OM			
93773	93781	Gammarus tigrinus						GC			
	93862	Stygonectes									
93947	93949	Synurella chamberlaini						GC			
94022	94025	Hyalella				8		GC			
94025	94026	Hyalella azteca	7.9	8			8	GC			
93295	95032	Talitridae				8		GC			
89802	92120	Isopoda				8		GC			
92148	92149	Cyathura polita						GC			
92650	92657	Asellidae						GC			
92657	92658	Asellus	9.4	8		8		GC			
92658	92659	Asellus occidentalis				8		GC			
92657	92686	Caecidotea				8	6	GC			
92686		Caecidotea attenuatus					6				
92686		Caecidotea communis					6	GC			
92686	92701	Caecidotea forbesi					6				
92686	92692	Caecidotea racovitzai					6				
92692	92695	Caecidotea racovitzai australis						GC			
92657	92666	Lirceus	7.7				8	GC			
	92977	Munna reynoldsi						GC			
92973	92976	Uromunna reynoldsi						GC			
93207	93209	Probopyris floridensis						GC			
93132	93133	Probopyus pandalicola						GC			
92224	92225	Cirolanidae						GC			
92225	541967	Anopsilana						GC			
92345	92348	Cassidinidea ovalis						GC			
92283	92301	Exosphaeroma						GC			
92283	92337	Sphaeroma						GC			
92337	92338	Sphaeroma destructor						GC			
92337	92342	Sphaeroma terebrans						GC			
206378	206379	Oniscus asellus									
92623	92624	Edotea montosa						GC			
92564	92588	Idotea						GC			
89802	89807	Mysidacea									
89856	90138	Mysidopsis						FC			
89856	90041	Mysis									
90275	90277	Taphromysis bowmani						FC			
89802	91061	Tanaidacea						FG			
	92068	Hargeria rapax						FC			
92026	92067	Leptochelia rapax									
	91502	Tanais cavolinii (part)									
	91396	Tanais cavolinii (part)									
	91400	Tanais cavolinii (part)									
	91519	Tanais cavolinii (part)									
83677	85257	Copepoda				8		GC			
83677	84195	Ostracoda				8		GC			
83767	83832	Cladocera				8		FC			
83872	83873	Daphnia				8		FC			

Rapid Bioassessment Protocols for Use in Streams and Wadeable Rivers: Periphyton, Benthic Macroinvertebrates, and Fish, Second Edition

B-13

Parent TSN	TSN	Scientific Name	Southeast (NC)	Upper Midwest (WI)	Midwest (OH)	Northwest (ID)	Mid-Atlantic (MACS)	Functional Feeding Group primary	secondary	Habit/Behavior primary	secondary
89599	89600	Balanus						FC			
89600	89621	Balanus eburneus						FC			
85780	85801	Diaptomus pribilofensis									
85257	88530	Cyclopoida				8		FC			
84409	84763	Entocytheridae									
82697	99208	Insecta									
99209	99237	Collembola				10		GC			
99239	99240	Podura						GC			
99240	99241	Podura aquatica									
99917	99918	Hypogastrura						GC			
99238	99245	Isotomidae						OM			
99245	99246	Isotomurus						GC			
99246	99247	Isotomurus palustris						GC			
99238	99643	Entomobryidae						GC			
100257	100258	Sminthuridae									
100258	100402	Bourletiella						GC			
100402	100436	Bourletiella spinata									
100500	100502	Ephemeroptera						GC			
		Polymitarcidae				2		GC			
101569	101570	Ephoron				2		GC		bu	
101570	101572	Ephoron leukon	1.5	2							
101459	101467	Caenidae				7		GC			
101467	101468	Brachycercus	3.5	3				GC			
101468	101475	Brachycercus maculatus						GC			
101468	101477	Brachycercus prudens				3		GC			
101467	101478	Caenis	7.6	7	3.1	7	7	GC		sp	cb
101478	101480	Caenis amica						OM			
101478	101488	Caenis latipennis				7		GC	SC		
101478		Caenis macafferti					7	GC			
101478	101483	Caenis diminuta						OM			
101478	101486	Caenis hilaris						OM			
101478	101489	Caenis punctata					7	GC			
101508	101525	Ephemeridae				4		GC			
101525	101526	Ephemera	2.2	1	3.1	4		GC		bu	
101526		Ephemera guttalata	0								
101525	101537	Hexagenia	4.7	6	3.6	6	6	GC		bu	
101537	101538	Hexagenia bilineata						GC			
101537	101552	Hexagenia limbata			2.6			GC			
101540	101549	Hexagenia munda orlando						GC			
101566	101567	Litobrancha recurvata	0	6							
100503	100755	Baetidae				4	4	GC			
	100801	Acentrella				4	4	GC		sw	cn
100801		Acentrella amplus	3.6								
100801		Acentrella insignificans				4		GC			
100801		Acentrella turbida				4		GC			
		Acerpenna					4	SH		sw	cn
		Acerpenna macdunnoughi			1.1		4	SH			
	206620	Acerpenna pygmaeus	3.7	4	2.3			OM			

Appendix B: Regional Tolerance Values, Functional Feeding Groups, and Habit/Behavior Assignments for Benthic Macroinvertebrates

Parent TSN	TSN	Scientific Name	Regional Tolerance Values					Functional Feeding Group		Habit/Behavior	
			Southeast (NC)	Upper Midwest (WI)	Midwest (OH)	Northwest (ID)	Mid-Atlantic (MACS)	primary	secondary	primary	secondary
100755	100800	Baetis			3.1	5	6	GC		sw	cb
100800		Baetis diphetorhageni									
100800	206621	Baetis alachua						OM			
100800	100803	Baetis alius				1		GC	SC		
100800	100821	Baetis australis						OM			
100800	100823	Baetis bicaudatus						GC			
100800	100833	Baetis ephippiatus	3.9					OM			
100800	100835	Baetis flavistriga	7.2	4	2.9	4		GC			
100800	100838	Baetis frondalis	8	5				OM			
100800	100807	Baetis insignificans						GC			
100800	100808	Baetis intercalaris	5.8	6	2.7	5	6	OM	GC		
100800	100810	Baetis intermedius						GC			
100800		Baetis notos				4		GC	SC		
100800	100858	Baetis pluto	4.8								
100800	100860	Baetis propinquus	6.2	6				OM			
100800	100861	Baetis pygmaeus						OM			
100800	100817	Baetis tricaudatus	1.8					GC			
100800	206618	Baetis armillatus			1.5			OM			
100800	206619	Baetis punctiventris						OM			
		Barbaetis						GC		sw	cn
		Plauditus									
		Plauditus cestus					4	GC			
100755	100903	Callibaetis	9.3	9	5.6	9	9	GC		sw	cn
100903	100919	Callibaetis floridanus						GC			
100903	100928	Callibaetis pretiosus						GC			
		Camelobaetidius								sw	cn
100755	100873	Centroptilum	6.3	2	2.7	2	2	GC			
100873	100884	Centroptilum hobbsi						OM			
100873	100897	Centroptilum viridocularis						OM			
100755	100756	Cloeon	7.4	4	3.5			OM		sw	cn
100756	100758	Cloeon rubropictum						OM			
		Diphetor				5		GC		sw	cn
		Diphetor hageni			2.3	5		GC			
		Fallceon quilleri						GC			
	100794	Heterocloeon	3.6					SC		sw	cn
		Labiobaetis				6		GC		sw	cn
		Labiobaetis frondalis									
		Labiobaetis propinquus				6		GC			
	100899	Paracloeodes	8.7					SC			
	206622	Procloeon						OM	GC	sw	cn
206622	206617	Procloeon rubropictum						OM			
206622	206623	Procloeon viridocularis						OM			
100755	100771	Pseudocloeon	4.4	4	1.7	4		SC			
100771	100776	Pseudocloeon bimaculatum						OM			
100771	100783	Pseudocloeon parvulum						OM			
100771	100784	Pseudocloeon punctiventris						OM			
		Ametropodidae									
101073	101074	Ametropus						GC		bu	

Rapid Bioassessment Protocols for Use in Streams and Wadeable Rivers: Periphyton, Benthic Macroinvertebrates, and Fish, Second Edition

B-15

Parent TSN	TSN	Scientific Name	Regional Tolerance Values					Functional Feeding Group		Habit/ Behavior	
			Southeast (NC)	Upper Midwest (WI)	Midwest (OH)	Northwest (ID)	Mid-Atlantic (MACS)	primary	secondary	primary	secondary
100503	100504	Heptageniidae				4		SC			
100504	100598	Cinygma				4		SC		cn	
100598	100600	Cinygma integrum						SC			
100504	100557	Cinygmula				4		SC		cn	
100557	100570	Cinygmula subaequalis	0								
100504	100626	Epeorus	1.2	0		0		SC		cn	
100626		Epeorus iron				0		SC			
100626		Epeorus ironopis				1		SC			
100626	100629	Epeorus albertae				0		SC			
100626	100632	Epeorus deceptivus				0		SC			
100626	100651	Epeorus dispar	1								
100626	100635	Epeorus grandis				0		SC			
100626	100637	Epeorus longimanus				0		SC			
100626	100642	Epeorus pleuralis	2								
100626	100645	Epeorus rubidus	1.4								
100627	100636	Ironopsis grandis				3		SC			
100504	100602	Heptagenia	2.8	3		4		SC		cn	sw
100602	100694	Heptagenia criddlei						SC			
100602	100608	Heptagenia diabasia			1.9						
100602	100604	Heptagenia elegantula				4		SC			
100602	100610	Heptagenia flavescens						OM			
100602	100612	Heptagenia julia	0.5								
100602	100616	Heptagenia marginalis	2.5								
100602	100619	Heptagenia pulla	2.3								
100602	100620	Heptagenia simpliciodes						SC			
100504	100666	Ironodes				4		SC		cn	
100504	100676	Leucrocuta	0	1	2.4	1		SC	GC	cn	
100676		Leucrocuta aphrodite	2.5	1							
100676	100677	Leucrocuta hebe			2.7						
100676	100679	Leucrocuta maculipennis			2.1						
100504	100692	Nixe				4		SC	GC	cn	
100692		Nixe simplicioides				2		SH			
100692	100693	Nixe criddlei				2		SH			
100692	100705	Nixe perfida			5.1						
100504	100572	Rhithrogena	0.4	0		0		SC		cn	
100572	100577	Rhithrogena amica	0								
100572	100579	Rhithrogena exilis	0								
100572	100595	Rhithrogena fuscifrons	0								
100572	100583	Rhithrogena hageni						GC			
100572	100575	Rhithrogena morrisoni						SC			
100572	100589	Rhithrogena robusta						GC			
100504	100713	Stenacron			3.1		4	SC		cn	
100713	100735	Stenacron carolina	1.7								
100713	100739	Stenacron floridense						OM			
100713	100714	Stenacron interpunctatum	7.1	7				OM			
100713	100736	Stenacron pallidum	2.9								
100504	100507	Stenonema				2	4	SC		cn	
100507	100513	Stenonema carlsoni	2.1								

Parent TSN	TSN	Scientific Name	Southeast (NC)	Upper Midwest (WI)	Midwest (OH)	Northwest (ID)	Mid-Atlantic (MACS)	Functional Feeding Group primary	secondary	Habit/ Behavior primary	secondary
100507	100514	Stenonema exiguum			1.9			OM			
100507	100516	Stenonema femoratum	7.5	5	3.1						
100507	100521	Stenonema integrum	5.5	4				OM			
100507	100527	Stenonema ithaca	4.1								
100507		Stenonema lenati	2.3								
100507	100530	Stenonema mediopunctatum	1.7	3	1.9						
100507	100531	Stenonema meririvulanum	0.3								
100507	206616	Stenonema mexicanum integrum				2.6		OM			
100507	100532	Stenonema modestum	5.8	1				SC			
100507	100536	Stenonema pudicum	2.1								
100507	100509	Stenonema pulchellum			2.3						
100507	100541	Stenonema smithae						OM			
100507	100542	Stenonema terminatum	4.5	4	2.3						
100507	100548	Stenonema vicarium	1	2	2.3						
100503	100951	Siphlonuridae				7		GC			
	100953	Siphlonurus	2.6	7		7		GC		sw	cb
100953	100955	Siphlonurus occidentalis				7		GC	SC		
		Acanthametropodidae									
100951	100996	Ameletus				0		GC		sw	cb
100996	101019	Ameletus celer				0		GC	SC		
100996	101009	Ameletus lineatus	2.1	0							
100996	101012	Ameletus similior						GC			
100996	101005	Ameletus connectus						GC			
100996	101006	Ameletus cooki				0		GC			
100996	101013	Ameletus sparsatus						GC			
100996	101002	Ameletus validus						GC			
100996	101003	Ameletus velox				0		GC			
101094	101232	Ephemerellidae				1		GC			
101232	101338	Attenella				3		GC			
101338	101340	Attenella attenuata	2.6	3							
101338	101345	Attenella delantala				3		GC			
101338	101343	Attenella margarita						GC			
101232	101347	Caudatella				1		GC		cn	
101347		Caudatella cascadia				1		GC			
101347		Caudatella edmundsi						SC			
101347	101351	Caudatella heterocaudata						GC			
101347	101348	Caudatella hystrix						SC			
		Caurinella				0		GC			
		Caurinella idahoensis				0		GC			
101232	101365	Drunella				0		PR		cn	sp
101365		Drunella allegheniensis	1.3								
101365	101389	Drunella coloradensis						PR			
101365		Drunella conestee	0								
101365	101366	Drunella cornutella	0								
101365	101368	Drunella doddsi						SC			
101365	101392	Drunella flavilinea						SC			
101365	101370	Drunella grandis						GC			
101365	185972	Drunella lata	0.1								

Rapid Bioassessment Protocols for Use in Streams and Wadeable Rivers: Periphyton, Benthic Macroinvertebrates, and Fish, Second Edition

B-17

Parent TSN	TSN	Scientific Name	Southeast (NC)	Upper Midwest (WI)	Midwest (OH)	Northwest (ID)	Mid-Atlantic (MACS)	Functional Feeding Group primary	secondary	Habit/Behavior primary	secondary
101365		Drunella pelosa						SC			
101365	101385	Drunella spinifera						PR			
101365	185974	Drunella tuberculata	0.2								
101365	185973	Drunella walkeri	1								
101365		Drunella wayah	0								
101232	101233	Ephemerella			2.9	1		GC		cn	sw
101233	101251	Ephemerella alleni						GC			
101233	101255	Ephemerella aurivillii						GC			
101233	101259	Ephemerella berneri	0								
101233	101262	Ephemerella catawba	4	1							
101233	101280	Ephemerella hispida	0.6								
101233	101239	Ephemerella inermis						SH			
101233	101240	Ephemerella infrequens						GC			
101233	101282	Ephemerella invaria	2.2	1							
101233	101285	Ephemerella lacustris				1		GC			
101233	101291	Ephemerella needhami	0	2							
101233	101296	Ephemerella rotunda	2.8					OM			
101233	101299	Ephemerella septentrionalis	2								
101233	101305	Ephemerella trilineata						OM			
101232	101324	Eurylophella			2.1		4	SC		cn	sp
101324	101334	Eurylophella bicolor	5.1	1							
101324		Eurylophella coxalis	2.6								
101324		Eurylophella doris						GC			
101324	101332	Eurylophella funeralis	2.3								
101324	101326	Eurylophella temporalis	4.6	5				GC			
101324	193519	Eurylophella trilineata						GC			
101324		Eurylophella verisimilis	0.3								
101232	101395	Serratella			0.6	2	2	GC		cn	
101395		Serratella carolina	0								
101395	101396	Serratella deficiens	2.7	2	2.1		2				
101395		Serratella micheneri				1		GC			
101395	185976	Serratella serrata	2.7			1		GC			
101395	185975	Serratella serratoides	1.5								
101395		Serratella teresa						GC			
101395	101399	Serratella tibialis						GC			
	101317	Timpanoga				7		GC			
101317	101318	Timpanoga hecuba				7		GC			
101360	101361	Dannella lita	0	4							
101360	101363	Dannella simplex	3.9	2	1.2						
101094	101095	Leptophlebiidae				2		GC			
101095	101108	Choroterpes			4			GC		cn	sp
101108	101114	Choroterpes hubbelli						OM			
101095	101183	Habrophlebia								sw	cn
101183	101184	Habrophlebia vibrans	0					OM			
101095	101122	Habrophlebiodes								sw	cn
101122	101124	Habrophlebiodes brunneipennis									
101095	101148	Leptophlebia	6.4	4		2		GC		sw	cn
101148		Leptophlebia bradleyi						OM			

*Appendix B: Regional Tolerance Values, Functional Feeding Groups,
and Habit/Behavior Assignments for Benthic Macroinvertebrates*

| Parent TSN | TSN | Scientific Name | Regional Tolerance Values | | | | | Functional Feeding Group | | Habit/ Behavior | |
			Southeast (NC)	Upper Midwest (WI)	Midwest (OH)	Northwest (ID)	Mid-Atlantic (MACS)	primary	secondary	primary	secondary
101148	101161	Leptophlebia intermedia						OM			
101095	101187	Paraleptophlebia	1.2	1	2.8	1	1	GC		sw	cn
101187	101206	Paraleptophlebia bicornuta				4		GC			
101187	101193	Paraleptophlebia debilis						GC			
101187	101195	Paraleptophlebia gregalis				4		GC			
101187	101212	Paraleptophlebia heteronea				2		GC			
101187	101214	Paraleptophlebia memorialis				4		GC			
101187	101227	Paraleptophlebia vaciva				4		GC			
101187	101199	Paraleptophlebia volitans						OM			
101094	101404	Tricorythidae				4		GC			
101404	101405	Tricorythodes	5.4	4	2.7	5	4	GC		sp	cn
101405	101406	Tricorythodes albilineatus						GC			
101405	101413	Tricorythodes minutus				4		GC			
	101429	Leptohyphes	2							cn	
101429	101432	Leptohyphes dolani									
		Baetiscidae									
101493	101494	Baetisca					4	GC		sp	
101494	101497	Baetisca becki						OM			
101494		Baetisca berneri	0.6								
101494	101499	Baetisca carolina	3.6	5							
101494	101503	Baetisca gibbera	1.4								
101494	101495	Baetisca obesa						OM			
101494	101506	Baetisca rogersi						OM			
		Metretopodidae									
		Siphloplecton	3.1	2			2	PR		sw	cn
		Isonychiidae									
101029	101041	Isonychia	3.8	2	1.9		2	FC		sw	cn
101041	101069	Isonychia arida									
101041	101060	Isonychia sayi									
101041	101062	Isonychia sicca									
		Neoephemeridae									
101460	101461	Neoephemera						GC		sp	cn
101461	101463	Neoephemera compressa						GC			
101461	101464	Neoephemera purpurea	2.1								
101461	101465	Neoephemera youngi						GC			
101523	101524	Dolania americana								bu	
		Anthopotamus			3.2						
	101510	Potamanthus	1.6	4							
109215	109216	Coleoptera						PR			
111952	111953	Amphizoa				1		PR		cn	
109226	109234	Carabidae				4		PR			
109234	111436	Chlaenius									
109226	111963	Dytiscidae				5		PR			
112072	112073	Agabetes acuductus						PR			
111963	111966	Agabus				8	5	PR		sw	dv
111963	112319	Bidessonotus								sw	cb
111963	112322	Bidessus									
111963	112362	Brachyvatus								sw	cb

Rapid Bioassessment Protocols for Use in Streams and Wadeable Rivers: Periphyton, Benthic Macroinvertebrates, and Fish, Second Edition

B-19

Parent TSN	TSN	Scientific Name	Southeast (NC)	Upper Midwest (WI)	Midwest (OH)	Northwest (ID)	Mid-Atlantic (MACS)	primary	secondary	primary	secondary
								Functional Feeding Group		Habit/ Behavior	
111963	112136	Celina					5	PR		sw	dv
112136	112142	Celina contiger						PR			
	112379	Colymbetes				5		PR		sw	dv
111963	112561	Copelatus	9.1				5	PR		sw	dv
112561	112567	Copelatus caelatipennis						PR			
111963	112371	Coptotomus	9					PR		sw	dv
112371	112375	Coptotomus interrogatus						PR			
111963	112364	Cybister						PR		sw	dv
111963	112153	Deronectes				5		PR		sw	
112153		Deronectes striatellus						PR			
111963	112159	Derovatellus								sw	cb
111963	112145	Desmopachria				5		PR		sw	cb
	112118	Dytiscus				5		PR		sw	dv
111963	112172	Hydaticus				5		PR		sw	dv
111963	112390	Hydroporus	8.9		4.1	5	5	PR		sw	cb
112390	112423	Hydroporus mellitus	1.8								
112390	112418	Hydroporus pilatei						PR			
111963	112257	Hydrovatus						PR			
112257	112259	Hydrovatus pustulatus						PR			
112257	112259	Hydrovatus pustulatus						PR			
112259	112261	Hydrovatus pustulatus compressus						PR		sw	cb
111963	112200	Hygrotus						PR		sw	dv
111963	112181	Ilybius					5	PR			
111963	112268	Laccodytes						PR		sw	dv
111963	112278	Laccophilus	10		7.9	5	5	PR			
112278	112281	Laccophilus fasciatus						PR			
112281	112283	Laccophilus fasciatus rufus						PR			
112278	112299	Laccophilus gentilis						PR			
112278	112285	Laccophilus proximus						PR			
112278	112298	Laccophilus schwarzi						PR			
112270	112276	Laccornis difformis								sw	cb
111963	112580	Liodessus						PR		sw	cb
111963	112595	Neoclypeodytes						PR		sw	cb
111963	112314	Oreodytes				5		PR			
112314		Oreodytes congruus				5		PR		sw	dv
111963	112086	Rhantus									
112109	112113	Thermonectus basillaris						PR		sw	cb
111963	112575	Uvarus									
109226	112653	Gyrinidae				5		PR		sw	dv
112653	112711	Dineutus	5.5		3.7	4	4	PR			
112711	112718	Dineutus carolinus									
112711	112715	Dineutus ciliatus									
112711	112713	Dineutus discolor									
112711	112727	Dineutus emarginatus									
112711	112719	Dineutus nigrior					4	PR			
112711	112717	Dineutus serrulatus								sw	dv
112653	112706	Cyretes									
112706	112707	Cyretes iricolor								sw	dv

Appendix B: Regional Tolerance Values, Functional Feeding Groups,
and Habit/Behavior Assignments for Benthic Macroinvertebrates

Parent TSN	TSN	Scientific Name	Regional Tolerance Values					Functional Feeding Group		Habit/ Behavior	
			Southeast (NC)	Upper Midwest (WI)	Midwest (OH)	Northwest (ID)	Mid-Atlantic (MACS)	primary	secondary	primary	secondary
112653	112654	Gyrinus	6.3		3.6	5	4	PR			
112654	112661	Gyrinus aeneolus					4	PR			
112654	112704	Gyrinus lugens									
112654	112701	Gyrinus pachysomus									
109226	111857	Haliplidae				7				cn	
111857	111947	Brychius						SC			
111947	111948	Brychius hornii								cb	
111857	111858	Haliplus									
111858	111872	Haliplus fasciatus					5	SH		cb	cn
111857	111923	Peltodytes	8.5		7		5	SH			
111923	111926	Peltodytes duodecimpuntatus									
111923	111927	Peltodytes floridensis									
111923	111928	Peltodytes lengi									
111923	111929	Peltodytes muticus									
111923	111930	Peltodytes oppositus									
111923	111932	Peltodytes sexmaculatus									
109226	112606	Noteridae						PR		cb	
112606	112623	Hydrocanthus	6.9								
112623	112626	Hydrocanthus iricolor						OM			
112623	112624	Hydrocanthus oblongus						OM		bu	
112606	112621	Notomicrus									
112636	193587	Suphis inflatus								cb	
112606	112607	Suphisellus						OM			
112607	112614	Suphisellus floridanus						OM			
112607	112613	Suphisellus gibbulus									
112607	193586	Suphisellus insularis						OM			
112607	112610	Suphisellus puncticollis						OM			
	112745	Hydroscapha				7		SC			
112736	112737	Sphaeriidae				8	8	FC			
114496	114509	Chrysomelidae						SH		cn	
114509	114613	Agasicles									
114613	114614	Agasicles hygrophila						SH		cn	
114509	114615	Disonycha						SH		cn	
114509	114510	Donacia						SH		cn	
114509	114546	Pyrrhalta									
113844	113869	Melyridae						PR			
114654	114666	Curculionidae						SH		cn	cb
114666	114667	Anchytarsus						SH			
114667	114668	Anchytarsus bicolor	3.8					SH		sp	cn
	114037	Lutrochus									
114037	114038	Lutrochus laticeps			2.9					cn	
114666	114779	Bagous						SH			
114779		Bagous carinatus						SH		cn	cb
114666	114676	Phytobius						SH			
	114679	Stenopelmus						SH			
206639	206640	Tyloderma capitale									
113918	113923	Helodidae (= Scirtidae)									
	113924	Scirtidae								cb	

Rapid Bioassessment Protocols for Use in Streams and Wadeable Rivers: Periphyton, Benthic Macroinvertebrates, and Fish, Second Edition

B-21

Parent TSN	TSN	Scientific Name	Regional Tolerance Values					Functional Feeding Group		Habit/ Behavior	
			Southeast (NC)	Upper Midwest (WI)	Midwest (OH)	Northwest (ID)	Mid-Atlantic (MACS)	primary	secondary	primary	secondary
113923	113948	Cyphon					7	SC		cb	sp
113923	113969	Elodes								cb	sp
113923	113925	Prionocyphon								cb	
113923	113929	Scirtes									
113998	114278	Chelonariidae									
114278	114279	Chelonarium lecontei									
113998	113999	Dryopidae (adult)						SH		cb	
113999	114025	Dryops (adult)								cn	
113999	114006	Helichus (adult)	5.4	5	3.2		5	SH			
114006	114011	Helichus basalis (adult)									
114006	114013	Helichus fastigiatus (adult)									
114006	114009	Helichus lithophilus (adult)									
114006	114017	Helichus striatus (adult)				5		SH			
114017	114019	Helichus striatus foveatus (adult)				5		SH		cb	
113999	114001	Pelonomus (adult)									
114001	114004	Pelonomus obscurus (adult)									
113998	114093	Elmidae				4		GC		cn	bu
	114196	Ampumixis				4		GC	SC	cn	bu
114196	114197	Ampumixis dispar				4		GC		cn	sp
114093	114193	Ancyronyx						OM			
114193	114194	Ancyronyx variegatus	6.9	6	4			OM		cn	
114093	114251	Atractelmis				4		GC		cn	
114093	114164	Cleptelmis				4		GC			
114164	114166	Cleptelmis addenda				4		GC	SC	cn	
114164	114165	Cleptelmis ornata				4		GC		cn	
114093	114208	Cylloepus				4		GC	SC	cn	cb
114093	114126	Dubiraphia	6.4	6	4.7	4	6	GC	SC		
114126	114129	Dubiraphia bivittata			3.1			OM			
114126		Dubiraphia giullianii				6		SC			
114126	114130	Dubiraphia quadrinotata			3.2			OM			
114126	114131	Dubiraphia vittata						OM		cn	cb
114093	114216	Gonielmis				5		GC			
114216	114217	Gonielmis dietrichi						OM		cn	
114093	114237	Heterelmis				4		GC		cn	
114093	114167	Heterlimnius				4		GC			
114167	114169	Heterlimnius corpulentus				4		GC		cn	bu
114167	114168	Heterlimnius koebelei				4		GC	SC	cn	
114093	114137	Lara				4		SH			
114137	114139	Lara avara				4		SH		cn	
114093	114212	Macronychus						OM			
114212	114213	Macronychus glabratus	4.7	4	2.9			OM		cn	cb
114093	114146	Microcylloepus				4		GC	SC		
114146	114147	Microcylloepus pusillus	2.1	3		2		GC			
114147	114151	Microcylloepus pusillus lodingi						OM			
114146	114160	Microcylloepus similis				2		GC		cn	
114093	114142	Narpus				4		GC			
114142	114144	Narpus concolor				4		GC		cn	
114093	114177	Optioservus	2.7	4	3.6	4	4	SC			

Appendix B: Regional Tolerance Values, Functional Feeding Groups, and Habit/Behavior Assignments for Benthic Macroinvertebrates

Parent TSN	TSN	Scientific Name	Southeast (NC)	Upper Midwest (WI)	Midwest (OH)	Northwest (ID)	Mid-Atlantic (MACS)	primary	secondary	primary	secondary
			Regional Tolerance Values					Functional Feeding Group		Habit/Behavior	
114177	193732	Optioservus castanipennis				4		SC			
114177	114178	Optioservus divergens				4		SC			
114177	114190	Optioservus fastiditus			1.9	4	4	SC			
114177	114180	Optioservus quadrimaculatus				4		SC			
114177	114181	Optioservus seriatus				4		SC		cn	
114093	114235	Ordobrevia				4		SC			
114235		Ordobrevia nubrifera				4		GC		cn	
114093	114244	Oulimnius				4		SC			
114244	114245	Oulimnius latiusculus	1.8							cn	
114093	114229	Promoresia					2	SC			
114229	114230	Promoresia elegans	2.2					OM			
114229	114231	Promoresia tardella	0				2	SC		cn	
114093	114198	Rhizelmis				1		SC		cn	
114093	114095	Stenelmis	5.4	5	3	7	5	SC			
114095	114117	Stenelmis antennalis						OM			
114095	114118	Stenelmis convexula						OM			
114095	114102	Stenelmis crenata						OM			
114095	114104	Stenelmis decorata					5	SC			
114095	114121	Stenelmis fuscata						OM			
114095	114105	Stenelmis humerosa						OM			
114095	114106	Stenelmis hungerfordi						SC			
114095	114108	Stenelmis markeli					5	SC			
114095	114114	Stenelmis sinuata						OM			
114095	114115	Stenelmis vittipennis						OM		cn	
114093	114205	Zaitzevia				4		GC			
114205	114207	Zaitzevia milleri				4		GC			
114205		Zaitzevia parvula				4		GC			
113998	114069	Psephenidae				4		SC		cn	
114069	114087	Ectopria				4	5	SC			
114087	114088	Ectopria nervosa	4.3	5	4			SC		cn	
114069	114085	Eubrianax				4		SC			
114085	114086	Eubrianax edwardsi				4		SC		cn	
114069	114070	Psephenus				4		SC			
114070	114074	Psephenus falli				4		SC		bu	
114070	114072	Psephenus herricki	2.5	4	3.5						
114265	114266	Anchycteis									
114266	114267	Anchycteis velutina									
114265	114273	Ptilodactyla					5	SH			
112752	112756	Hydraenidae					5	PR		cn	cb
112756	112757	Hydraena					5	PR			
112757	112758	Hydraena pennsylvanica								cn	
112756	112777	Ochthebius									
112777	112793	Ochthebius sculptus					5	PR			
112752	112811	Hydrophilidae					5	PR		sw	dv
	112890	Ametor					5				
112811	112812	Berosus	8.6		6.7		5	PR	PI		
112812	112824	Berosus peregrinus									
112812	112821	Berosus striatus								cb	

Rapid Bioassessment Protocols for Use in Streams and Wadeable Rivers: Periphyton, Benthic Macroinvertebrates, and Fish, Second Edition

B-23

| Parent TSN | TSN | Scientific Name | Regional Tolerance Values | | | | | Functional Feeding Group | | Habit/Behavior | |
			Southeast (NC)	Upper Midwest (WI)	Midwest (OH)	Northwest (ID)	Mid-Atlantic (MACS)	primary	secondary	primary	secondary
112811	112845	Chaetarthria				5				bu	
112811	113220	Crenitis				5		PR		bu	
112811	113017	Cymbiodyta								sw	dv
112811	113087	Derallus						OM			
113087	113088	Derallus altus						OM			
113085	113086	Dibolocelus ovatus								bu	sp
112811	112973	Enochrus	8.5			5		GC			
112973	112990	Enochrus ochraceus									
112811	113162	Helobata						OM			
113162	113165	Helobata striata						OM			
112811	113150	Helochares						OM			
112811	113106	Helophorus	7.9					SH		sw	dv
112811	113244	Hydrobiomorpha									
113244	113245	Hydrobiomorpha castus								cb	cn
112811	113196	Hydrobius				8		PR			
113196	113200	Hydrobius tumidus						OM		cb	
112811	113166	Hydrochus						SH		sw	dv
112811	113204	Hydrophilus									
112811	112858	Laccobius	8		1.9			PR			
112811	112909	Paracymus				5		PR	OM	cn	
112811	112931	Sperchopsis				5	5	PR	CG		
112931	112932	Sperchopsis tessellatus	6.5					OM		cb	
112811	112938	Tropisternus	9.8			5	10	PR			
112938	112951	Tropisternus blatchleyi									
112938	112944	Tropisternus lateralis									
112944	112946	Tropisternus lateralis nimbatus									
112938	193660	Tropisternus striolatus									
113264	113805	Ptiliidae									
113264	113265	Staphylinidae				8		PR		cn	
113265	113304	Bledius						PR		sk	
113265	113576	Stenus								bu	
113265	113440	Thinopinus									
114413	114429	Salpingidae									
109215	152741	Hymenoptera				8		PA			
109215	117232	Lepidoptera				6		SH	SC		
117294	117318	Noctuidae						SH		bu	
117915	117952	Pyroderces				5					
117639	117641	Pyralidae				5		SH		cb	
117641	117741	Acentria				1		SH		cb	
117641	117672	Munroessa						SH			
117672	117677	Munroessa gyralis						SH		cb	
117641	117756	Neargyractis						SH		cb	sw
117641	117642	Paraponyx					5	SH		cn	
117641	117682	Petrophila			2.7	5		SC		cb	sw
117654	117656	Synclita obliteralis						SH			
117906	117909	Prionoxystus				5					
117854	117856	Tortricidae									
109215	115000	Megaloptera									

Appendix B: Regional Tolerance Values, Functional Feeding Groups, and Habit/Behavior Assignments for Benthic Macroinvertebrates

Parent TSN	TSN	Scientific Name	Southeast (NC)	Upper Midwest (WI)	Midwest (OH)	Northwest (ID)	Mid-Atlantic (MACS)	primary	secondary	primary	secondary
			Regional Tolerance Values					Functional Feeding Group		Habit/ Behavior	
115000	115023	Corydalidae				0		PR		cn	cb
115023	115024	Chauliodes						PR			
115024	115027	Chauliodes pectinicornis						PR			
115024	115025	Chauliodes rastricornis						PR		cn	cb
115023	115033	Corydalus						PR			
115033	115034	Corydalus cornutus	5.6	6	2.4			PR		cn	cb
115023	115048	Neohermes								cn	cb
115023	115028	Nigronia						PR			
115028	115029	Nigronia fasciatus	6.2		1.8			PR			
115028	115031	Nigronia serricornis	5.5	0	3.6			PR		cn	cb
115023	115044	Orohermes				0		PR		cb	cn
115085	115086	Climacia									
115086	115087	Climacia areolaris	6.5								
115085	115090	Sisyra						PI			
115000	115001	Sialidae								bu	cb
115001	115002	Sialis	7.4	4	4.9	4	4	PR			
115002	193739	Sialis americana						PR			
115002	115017	Sialis iola						PR			
115002	115010	Sialis mohri						PR			
109215	115095	Trichoptera								sp	
		Beraeidae									
116489	116490	Beraea									
115095	116905	Brachycentridae				1		FC		cn	cb
116905	116933	Amiocentrus				1		GC			
116933	116934	Amiocentrus aspilus				2		GC		cn	
116905	116906	Brachycentrus	2.2			1		FC			
116906	116912	Brachycentrus americanus				1		FC			
116906	116921	Brachycentrus appalachia	1.1								
116906	116922	Brachycentrus chelatus	0								
116906	116914	Brachycentrus lateralis	0.4	1							
116906	116916	Brachycentrus nigrosoma	2.2								
116906	116910	Brachycentrus numerosus	1.8	1							
116906	116918	Brachycentrus occidentalis				1		FC		cn	sp
116906	116924	Brachycentrus spinae	0								
116905	116958	Micrasema				1	2	SH			
116958	116967	Micrasema bactro				1					
116958		Micrasema bennetti	0								
116958	116966	Micrasema burksi	0								
116958	116959	Micrasema charonis	0.3								
116958		Micrasema rickeri	0								
116958	116961	Micrasema rusticum	0					OM			
116958	116960	Micrasema wataga	3.2	2				OM		cn	
116905	116973	Oligoplectrum				1		GC			
115095	116529	Calamoceratidae								sp	
116529	116530	Anisocentropus						SH			
116530	116531	Anisocentropus pyraloides	0.8					SH		sp	
116537	553090	Heteroplectron americanum	2.9				3	SH			
116537	116538	Heteroplectron californicum				1		SH			

Rapid Bioassessment Protocols for Use in Streams and Wadeable Rivers: Periphyton, Benthic Macroinvertebrates, and Fish, Second Edition

B-25

| Parent TSN | TSN | Scientific Name | Regional Tolerance Values | | | | | Functional Feeding Group | | Habit/ Behavior | |
			Southeast (NC)	Upper Midwest (WI)	Midwest (OH)	Northwest (ID)	Mid-Atlantic (MACS)	primary	secondary	primary	secondary
		Uenoidae				0		SC			
115933	116331	Farula						SC			
115933	116046	Neophylax	1.6	3		3		SC			
116046	116047	Neophlax concinnus	1.2								
116046	116050	Neophlax mitchelli	0								
116046	116065	Neophylax occidentalis				3		SC			
116046	116057	Neophlax oligius	2.6								
116046	116052	Neophlax ornatus	1.6								
116046	116054	Neophylax rickeri				3		SC		cn	
116046	116063	Neophylax splendens				3		SC			
115933	116388	Neothremma				0		SC		cn	
116388	116389	Neothremma alicia				0		SC		sp	
115933	116039	Oligophlebodes				1		SC			
		Sericostriata				0		SC			
		Sericostriata surdickae				0		SC			
115095	117120	Glossosomatidae				0		SC		cn	
117120	117121	Agapetus	0			0		SC			
117120	117154	Anagapetus				0		SC		cn	
115236	115238	Culoptila cantha				0		SC		cn	
117120	117159	Glossosoma	1.5			0		SC			
117159	117165	Glossosoma penitus						SC			
117159	117167	Glossosoma alascense						SC			
117159	117162	Glossosoma intermedium				0		SC			
117159	117160	Glossosoma montana						SC			
117159	117202	Glossosoma oregonense						SC			
117159	117220	Glossosoma wenatchee						SC			
115246	115247	Matrioptila jeanae	0								
115096	115221	Protoptila	2.8	1		1		SC			
115221	183768	Protoptila coloma				1		SC			
115221	115232	Protoptila tenebrosa				1		SC		sp	
115095	117015	Helicopsychidae				3		SC		cn	
117015	117016	Helicopsyche				3		SC			
117016	117020	Helicopsyche borealis	0	3	1.8	3		SC			
115095	115398	Hydropsychidae				4	4	FC			
		Hydropsychidae									
		Arctopsychinae				2		FC		cn	
115398	115529	Arctopsyche				1		FC			
115529	115538	Arctopsyche californica				2		FC	OM		
115529	115530	Arctopsyche grandis				2		FC		cn	
115529	115533	Arctopsyche irrorata	0								
		Hydropsychinae						FC			
115398	115570	Ceratopsyche						FC		cn	
115570	115596	Ceratopsyche alhedra	0	3							
115570		Ceratopsyche bifida	1								
115570	115577	Ceratopsyche bronta	2.7	5							
115570		Ceratopsyche macleodi	0.9								
115570	115580	Ceratopsyche morosa	3.2	2	1.8						
115570	115586	Ceratopsyche slossonae	0	4	2						

Parent TSN	TSN	Scientific Name	Southeast (NC)	Upper Midwest (WI)	Midwest (OH)	Northwest (ID)	Mid-Atlantic (MACS)	Functional Feeding Group primary	secondary	Habit/ Behavior primary	secondary
115570	115589	Ceratopsyche sparna	3.2	1	3.2						
115570		Ceratopsyche ventura	0								
115398	115408	Cheumatopsyche	6.6	5	2.9	5	5	FC			
115408	115409	Cheumatopsyche campyla				6		FC			
115408	115441	Cheumatopsyche enonis				6		FC			
115408	115426	Cheumatopsyche pettiti				6		FC		cn	
115398	115399	Diplectrona				0		FC			
115399	115402	Diplectrona modesta	2.2					FC		cn	
115398	115618	Homoplectra								cn	
115398	115453	Hydropsyche				4	4	FC			
115453	115456	Hydropsyche aerata			2.6						
115453	115454	Hydropsyche betteni	8.1	6			4	FC			
115453	115458	Hydropsyche bidens			2.5						
115453	115455	Hydropsyche californica				4		FC			
115453	115462	Hydropsyche decalda	4.1					FC			
115453	115463	Hydropsyche demora	1.8								
115453	115465	Hydropsyche dicantha			3.5						
115453	115488	Hydropsyche elissoma						FC			
115453	115468	Hydropsyche frisoni			1.8						
115453	115469	Hydropsyche hageni	0								
115453	115471	Hydropsyche incommoda	5	7							
115453	115474	Hydropsyche mississippiensis						FC			
115453	115513	Hydropsyche occidentalis				4		FC			
115453	115485	Hydropsyche orris			2.6						
115453	115490	Hydropsyche oslari				4		FC			
115453	115477	Hydropsyche phalerata	3.7	1							
115453	206641	Hydropsyche rossi	4.9								
115453	115480	Hydropsyche scalaris	3	2							
115453	115481	Hydropsyche simulans			2.4						
115453	115527	Hydropsyche sparna					4	FC		cn	
115453	115484	Hydropsyche venularis	5.3		2.9						
115453	115482	Hydropsyche valanis			3						
115398	115603	Macrostemum	3.6	3			3	FC			
115603	115608	Macrostemum carolina						FC		cn	
115603	115606	Macrostemum zebratum			1.8						
115398	115556	Parapsyche				1		PR			
115556	115563	Parapsyche almota				3		PR			
115556	115559	Parapsyche cardis	0								
115556	115560	Parapsyche elsis				1		PR		cn	
115398	115551	Potamyia						FC			
115551	115552	Potamyia flava			2.5			FC			
115095	115629	Hydroptilidae				4				cb	
115629	115635	Agraylea			5.7	8				cn	
115629	115826	Dibusa								cn	
115826	115827	Dibusa angata			2.6						
115629	115641	Hydroptila	6.2	6	3.2	6	6	SC	PR		
115641	115643	Hydroptila ajax				6		SC			
115641	115695	Hydroptila arctia				6		SC			

Rapid Bioassessment Protocols for Use in Streams and Wadeable Rivers: Periphyton, Benthic Macroinvertebrates, and Fish, Second Edition

B-27

Parent TSN	TSN	Scientific Name	Regional Tolerance Values					Functional Feeding Group		Habit/ Behavior	
			Southeast (NC)	Upper Midwest (WI)	Midwest (OH)	Northwest (ID)	Mid-Atlantic (MACS)	primary	secondary	primary	secondary
115641	115696	Hydroptila argosa				6		SC		cn	
115629	115630	Leucotrichia				6		SC		cn	
115630	115631	Leucotrichia pictipes	4.3	2							
115629	115811	Mayatrichia				6		SC			
115811	115812	Mayatrichia ayama						SC		cn	
115629	115833	Neotrichia			3.6			SC			
115833		Neotrichia halia				4		SH		cn	
115629	115714	Ochrotrichia	7.2			4		GC		cn	
115629	115714	Ochrotrichia				4		GC		cb	
115629	115828	Orthotrichia				6		SC		cn	
115629	115779	Oxyethira			5.2						
115629	115817	Stactobiella				2		SH		cb	sp
		Limnephiloidea									
115095	116793	Lepidostomatidae				3		SH			
116793	116794	Lepidostoma	1	1		1	1	SH			
116794	116888	Lepidostoma cinereum				3		SH			
116794	116870	Lepidostoma quercinum				1		SH		sp	cb
115095	116547	Leptoceridae				4		GC		cb	sw
116547	116684	Ceraclea			2.6	5	3	GC		cn	sp
116684	116696	Ceraclea ancylus	2.5	3							
116684		Ceraclea flava	0								
116684	116725	Ceraclea maculata	6.4		3.6						
116684		Ceraclea transversa	2.7								
116547	116598	Mystacides				4	4	GC			
116598	116599	Mystacides sepulchralis	3.5	4							
116547	116651	Nectopsyche			2.4	3	3	SH			
116651	116661	Nectopsyche candida	3.8					OM			
116651	116663	Nectopsyche diarina			3.2						
116651	116659	Nectopsyche exquisita	4.2	3				OM			
116651	116662	Nectopsyche gracilis				3		SC			
116651	116660	Nectopsyche pavida	4.2		2.1			OM			
116651		Nectopsyche halia				3		SC			
116651		Nectopsyche lahontanensis				3		SC		sp	cb
116651		Nectopsyche stigmatica				3		SC		sp	cb
116547	116607	Oecetis	5.7	8	3	8	8	PR			
116607		Oecetis parva									
116607	116608	Oecetis avara									
116607	116609	Oecetis cinerascens									
116607	116643	Oecetis georgia					8				
116607	116613	Oecetis inconspicua					8				
116607	116631	Oecetis nocturna								sp	cn
116607	116636	Oecetis persimilis					8			sw	cb
116547	116548	Setodes	0.9	2				OM			
116547	116565	Triaenodes				6	6				
116565	206642	Triaenodes abus	4.3					SH			
116565	116569	Triaenodes flavescens						SH			
116565	206643	Triaenodes florida						SH			
116565	116571	Triaenodes ignitus						SH			

Appendix B: Regional Tolerance Values, Functional Feeding Groups,
and Habit/Behavior Assignments for Benthic Macroinvertebrates

Parent TSN	TSN	Scientific Name	Regional Tolerance Values					Functional Feeding Group		Habit/ Behavior	
			Southeast (NC)	Upper Midwest (WI)	Midwest (OH)	Northwest (ID)	Mid-Atlantic (MACS)	primary	secondary	primary	secondary
116565	116574	Triaenodes injusta	2.2								
116565	116575	Triaenodes marginatus				6	6	sh			
116565	116577	Triaenodes ochraceus						SH			
116565	206644	Triaenodes perna						SH			
116565	116580	Triaenodes tardus	4.7	6				SH			
115095	115933	Limnephilidae				4	4	SH			
115969	115970	Allocosmoecus partitus				0		SC		cn	cb
115867	115907	Cryptochia				0		SH			
	116438	Allomyia				0		SC			
115933	116253	Amphicosmoecus						SH		sp	
	115956	Anabolia						SH			
115933	115935	Apatania	0.6			1		SC			
		Apataniinae				1		SC			
	116247	Arctopora									
115933	116017	Chyranda				1		SH		sp	
116017	116018	Chyranda centralis				1		SH		sp	bu
115933	116013	Clostoeca						SH		sp	
115933	116023	Desmona				1		SH			
		Dicosmoecinae				1		SC			
115933	116265	Dicosmoecus				1		SH			
116265	116266	Dicosmoecus atripes				1		PR		bu	
116265	116268	Dicosmoecus gilvipes				2		SC		cn	
116340	116342	Ecclisocosmoecus scylla				0		SH			
115933	116025	Ecclisomyia				2		GC			
		Eocosmoecus						SH		sp	
		Eocosmoecus schmidi						SH			
115933	116030	Glyphopsyche				1				cn	
115933	116309	Grammotaulius				4		SH		sp	
115933	116295	Grensia				6		SH			
115933	116001	Hesperophylax				5		SH		sp	cb
115933	116286	Homophylax				0		SH			
115933	115995	Hydatophylax				1		SH			
115995	115997	Hydatophylax argus	2.3	2				SH		sp	
115933	116381	Imania						SC		cb	sp
115933	116382	Ironoquia								cn	
116382	116385	Ironoquia punctatissima	7.3	3							
		Limnephilinae				4		SH		sp	
115933	116069	Limnephilus				5		SH		sp	
115933	116344	Manophylax						SC		cn	
115933	116379	Moselyana				4		GC		cn	
115933	116315	Onocosmoecus				1		SH			
116315	116318	Onocosmoecus unicolor				2		SH		cb	
115972	115973	Pedomoecus sierra				0		SC		sp	
115933	116407	Platycentropus									
	115989	Pseudostenophylax				1		SH			
115933	115974	Psychoglypha				1		GC			
115974	115977	Psychoglypha bella				2		GC		sp	cb
115974	115981	Psychoglypha subborealis				2		GC			

Rapid Bioassessment Protocols for Use in Streams and Wadeable Rivers: Periphyton, Benthic Macroinvertebrates, and Fish, Second Edition

B-29

Parent TSN	TSN	Scientific Name	Regional Tolerance Values					Functional Feeding Group		Habit/ Behavior	
			Southeast (NC)	Upper Midwest (WI)	Midwest (OH)	Northwest (ID)	Mid-Atlantic (MACS)	primary	secondary	primary	secondary
115933	116409	Pycnopsyche	2.3	4	3.3		4	SH			
116409	116413	Pycnopsyche gentilis	0.8								
116409	116414	Pycnopsyche guttifer	2.7					SH		sp	cn
116409	116416	Pycnopsyche lepida	2.5								
116409	116417	Pycnopsyche scabripennis	4					SH			
	116473	Molannidae									
116473	116474	Molanna					6	SC		sp	
116474	116478	Molanna blenda	3.9		4						
116474	116479	Molanna tryphena								sp	
	116496	Odontoceridae									
116496	116520	Namamyia				0		OM	GC		
116496	116522	Nerophilus				0		OM		sp	
116522	116523	Nerophilus californicus				0		OM		sp	
116496	116527	Pseudogoera				0		OM	PR		
116496	116497	Psilotreta	0	0			0	SC			
116497	116498	Psilotreta frontalis								cn	
115095	115257	Philopotamidae				3	3	FC		cn	
115257	115273	Chimarra	2.8	4			4	FC		cn	
	115278	Chimarra aterrima			1.9						
	115276	Chimarra obscura			3.4						
115257	115319	Dolophilodes	1			1		GC			
115257	115258	Wormaldia	0.4			3		FC			
115258	115261	Wormaldia gabriella						SC			
115095	115867	Phryganeidae						SH		cb	
	115892	Phryganea				4		OM			
115867	115868	Ptilostomis	6.7	5			5	SH		cn	
		Goerinae					1	SC			
115933	116423	Goera	0.3							sn	
116423	116431	Goera archaon				1		SC		sb	
115933	116298	Goeracea				0		SC		sp	
		Goereilla						SH			
115095	117043	Polycentropodidae						FC		cn	
115334	115373	Cernotina						PR		cn	
115373	115375	Cernotina spicata						PR			
117043	117091	Cyrnellus						FC		cn	
117091	117092	Cyrnellus fraternus	7.4	8	4			FC			
117043	117095	Neureclipsis	4.4	7	2.7		7	FC		cn	
117095	117098	Neureclipsis crepuscularis									
117043	117104	Nyctiophylax	0.9	5	2.5	5		FC		cn	
	117112	Nyctiophylax moestus	2.6			5	5	PR			
		Paranyctiophylax									
117043	117044	Polycentropus	3.5	6	3.4	6	5	PR	FC	cn	
115334	115361	Phylocentropus	5.6	4			5	FC		cn	
115334	115395	Polyplectropus									
115095	115334	Psychomyiidae						GC			
115334	115391	Lype						SC		bu	
115391	115392	Lype diversa	4.3	2	2.8			SC			
115334	115335	Psychomyia				2		SC			

Appendix B: Regional Tolerance Values, Functional Feeding Groups, and Habit/Behavior Assignments for Benthic Macroinvertebrates

Parent TSN	TSN	Scientific Name	Southeast (NC)	Upper Midwest (WI)	Midwest (OH)	Northwest (ID)	Mid-Atlantic (MACS)	Functional Feeding Group primary	secondary	Habit/Behavior primary	secondary
115335	115341	Psychomyia flavida	3.3	2	1.9						
115335	115346	Psychomyia lumina				2		SC			
115335	115344	Psychomyia nomada	2								
115334	115350	Tinodes				2		SC			
115095	115096	Rhyacophilidae				0		PR		cn	
115096	115243	Himalopsyche						PR			
115096	115097	Rhyacophila				0		PR			
115097	115098	Rhyacophila acropedes				1		PR			
115097	115160	Rhyacophila acutiloba	0								
115097	115163	Rhyacophila alberta						PR			
115097	115099	Rhyacophila angelita						PR			
115097	115165	Rhyacophila arnaudi						PR			
115097	115146	Rhyacophila atrata	0								
115097	115101	Rhyacophila betteni						PR			
115097	115102	Rhyacophila bifila						PR			
115097	115153	Rhyacophila blarina						PR			
115097	115151	Rhyacophila brunnea						PR			
115097	115131	Rhyacophila carolina	0								
115097	115156	Rhyacophila coloradensis						PR			
115097	115133	Rhyacophila fuscula	2	0							
115097	115105	Rhyacophila grandis				1		PR			
115097	115159	Rhyacophila hyalinata						PR			
115097	115177	Rhyacophila iranda				0		PR			
115097	115134	Rhyacophila ledra	3.4								
115097	115147	Rhyacophila minor	0								
115097	115155	Rhyacophila narvae						PR			
115097	115111	Rhyacophila nevadensis				1		PR			
115097	115138	Rhyacophila nigrita	0								
115097	115208	Rhyacophila oreia						PR			
115097	115114	Rhyacophila pellisa				0		PR			
115097	115116	Rhyacophila rayneri				0		PR			
115097	115187	Rhyacophila robusta									
115097	115117	Rhyacophila rotunda						PR			
115097		Rhyacophila sibirica				0		PR			
115097	115144	Rhyacophila torva	1.8								
115097		Rhyacophila trissemani				1		PR			
115097	115189	Rhyacophila tucula									
115097	115120	Rhyacophila vaccua						PR			
115097	115191	Rhyacophila vaefes				1		PR			
115097		Rhyacophila vaeter				1		PR			
115097	115152	Rhyacophila vagrita						PR			
115097	115121	Rhyacophila valuma				1		PR			
115097	115123	Rhyacophila velora				1		PR			
115097	115124	Rhyacophila vepulsa									
115097	115125	Rhyacophila verrula									
115097	115195	Rhyacophila visor				1		PR		cn	
115097	115197	Rhyacophila vofixa				0		PR			
115097	115148	Rhyacophila vuphipes	0								

Rapid Bioassessment Protocols for Use in Streams and Wadeable Rivers: Periphyton, Benthic Macroinvertebrates, and Fish, Second Edition

B-31

Parent TSN	TSN	Scientific Name	Southeast (NC)	Upper Midwest (WI)	Midwest (OH)	Northwest (ID)	Mid-Atlantic (MACS)	primary	secondary	primary	secondary
			Regional Tolerance Values					Functional Feeding Group		Habit/ Behavior	
115095	116982	Sericostomatidae						SH			
116982	116983	Agarodes								sp	
116983	116991	Agarodes libalis	0	3							
117012	117013	Fattigia pele	1.1								
116982	117003	Gumaga				3		SH			
100900	103358	Hemiptera						PR		cb	sw
103358	103683	Belostomatidae						PR			
103683	103717	Abedus						PR		cb	sw
103717	103739	Abedus immaculatus						PR			
103683	103684	Belostoma	9.8					PR			
103684	103689	Belostoma flumineum						PR			
103684	103687	Belostoma lutarium						PR		cb	sw
103684	103688	Belostoma testaceum						PR			
103683	103699	Lethocerus						PR		sw	
103358	103364	Corixidae	9			10	5	PR		sw	
103364	103514	Callicorixa						PR			
103364	103501	Cenocorixa						PR		sw	
103501	103504	Cenocorixa bifida				8		PR		sw	
103364	103484	Corisella						PR		sw	
103364	103525	Cymatia				8		PI		sw	cb
103364	103547	Graptocorixa						PR		sw	
103364	103444	Hesperocorixa								sw	
103364	103491	Palmacorixa					5	PR		sw	cb
103364	103365	Ramphocorixa									
103364	103369	Sigara					9	PR			
103369	103370	Sigara alternata								sw	
103369	103398	Sigara washingtonensis				8		GC		sw	cb
103364	181192	Tenagobia				8					
103364	103423	Trichocorixa					5	PR			
103423	103424	Trichocorixa calva									
103423	103429	Trichocorixa sexcincta								sp	
103358	103768	Gelastocoridae						PR			
103768	103769	Gelastocoris						PR		sk	
103358	103801	Gerridae				5		PR			
103801	103829	Gerris						PR			
103829	103842	Gerris buenoi				5		PR		sk	
103829	103841	Gerris remigis				5		PR		sk	
103801	103872	Limnoporus						PR			
103801	103857	Metrobates						PR		sk	
103857	103859	Metrobates hesperius						PR			
103801	103881	Neogerris						PR		sk	
103881	103882	Neogerris hesione						PR			
103801	103802	Rheumatobates						PR			
103802	103807	Rheumatobates palosi								sk	
103802	103804	Rheumatobates tenuipes									
103801	103811	Trepobates				10		PR		cb	bu
103811	103815	Trepobates pictus						PR		cb	bu
103964	103965	Hebrus						PR		sk	cb

Appendix B: Regional Tolerance Values, Functional Feeding Groups,
and Habit/Behavior Assignments for Benthic Macroinvertebrates

Parent TSN	TSN	Scientific Name	Southeast (NC)	Upper Midwest (WI)	Midwest (OH)	Northwest (ID)	Mid-Atlantic (MACS)	Functional Feeding Group primary	secondary	Habit/Behavior primary	secondary
103964	103986	Lipogomphus						PR			
103964	103983	Merragata						PR			
103983	103984	Merragata brunnea						PR		sk	
103983	103985	Merragata hebroides						PR			
103938	103939	Hydrometra						PR			
103939	103944	Hydrometra wileyae						PR		sk	cb
103358	103953	Mesoveliidae						PR			
103953	103954	Mesovelia						PR			
103954	103955	Mesovelia cryptophila						PR			
103954	103956	Mesovelia mulsanti						PR		cn	sw
103358	103613	Naucoridae				5		PR		cb	sw
103613	103614	Ambrysus						PR			
103613	103665	Pelocoris					7	PR			
103665	103667	Pelocoris femoratus						PR		cb	
103358	103747	Nepidae						PR			
103747	103748	Ranatra	7.5					PR			
103748	103749	Ranatra australis						PR			
103748	103750	Ranatra buenoi						PR			
103748	103761	Ranatra drakei						PR			
103748	103755	Ranatra fusca						PR			
103748	103751	Ranatra kirkaldyi						PR			
103748	103754	Ranatra nigra						PR		sw	cb
103358	103557	Notonectidae						PR			
103557	103558	Notonecta						PR			
103558	103573	Notonecta irrorata						PR			
103558	103575	Notonecta uhleri						PR		sw	cb
103358	103602	Pleidae						PR			
103602	103603	Neoplea						PI			
103603	103604	Neoplea striola						PI		cb	
103358	104063	Saldidae				10		PR			
104063	104069	Pentacora						PR			
104063	104140	Saldula				10		PR		sk	
103358	103885	Veliidae						PR			
103885	103900	Microvelia					6	PR			
103900	103908	Microvelia hinei						PR			
103900	103910	Microvelia pulchella						PR			
103885	103923	Paravelia						PR		sk	
103923	103924	Paravelia brachialis						PR			
103885	103886	Rhagovelia					6	PR			
103886	103894	Rhagovelia choreutes						PR			
103886	103895	Rhagovelia disticta						PR		sk	
103886	103887	Rhagovelia obesa						PR			
	103935	Trochopus						PR			
100500	102467	Plecoptera						PR		cn	
102468	102643	Capniidae				1	1	SH		sp	cn
102643	102644	Allocapnia	2.8	3			3	SH		sp	cn
102643	102688	Capnia				1		SH			
102785	102786	Eucapnopsis brevicauda				1		SH		sp	cn

Rapid Bioassessment Protocols for Use in Streams and Wadeable Rivers: Periphyton, Benthic Macroinvertebrates, and Fish, Second Edition

B-33

Parent TSN	TSN	Scientific Name	Southeast (NC)	Upper Midwest (WI)	Midwest (OH)	Northwest (ID)	Mid-Atlantic (MACS)	Functional Feeding Group primary	secondary	Habit/ Behavior primary	secondary
102788	102804	Paracapnia				1		SH		sp	cn
102804	102805	Paracapnia angulata	0.2	1							
102468	102840	Leuctridae				0		SH			
102840	102841	Despaxia				0		SH		cn	
102841	102842	Despaxia augusta				0		SH		sp	cn
102840	102844	Leuctra	0.7				0	SH		sp	cn
102840	102877	Megaleuctra				0		SH		sp	cn
102909	102910	Moselia infuscata				0		SH			
102840	102887	Paraleuctra				0		SH		sp	cn
102887	102890	Paraleuctra occidentalis				0		SH			
103202	103239	Perlomyia				0		SH		sp	cn
102468	102517	Nemouridae				2		SH			
102517	102540	Amphinemura	3.4	3		2		SH			
102540	102541	Amphinemura delosa								sp	cn
102540	102542	Amphinemura nigritta								sp	cn
102517	102567	Malenka				2		SH		sp	cn
102517	102526	Nemoura								sp	cn
102517	102632	Ostrocera								sp	cn
102517	102622	Ostrocerca								sp	cn
102517	102605	Podmosta				2		SH			
102517	102584	Prostoia	6.1	2		2		SH		sp	cn
102584	102585	Prostoia besametsa				2		SH		sp	cn
102517	102640	Shipsa								sp	cn
102640	102641	Shipsa rotunda	0.3	2							
102517	102556	Soyedina				2		SH			
102517	102614	Visoka						SC		sp	cn
102614	102615	Visoka cataractae				1		SH			
102517	102591	Zapada				2		SH			
102591	102594	Zapada cinctipes				2		SH			
102591	102596	Zapada columbiana				2		SH			
102591	102601	Zapada frigida				2		SH			
102591	102597	Zapada oregonensis				2		SH		cn	sp
102468	102488	Peltoperlidae				2		SH		cn	sp
102488	102489	Peltoperla								cn	sp
102994	103142	Soliperla				2		SH			
102488	102500	Tallaperla	1.4							cn	sp
102500	102505	Tallaperla cornelia									
102488	102510	Yoraperla				2		SH			
102510		Yoraperla mariana				2		SH			
102510	102512	Yoraperla brevis				2		SH		cn	sp
102468	102470	Pteronarcidae						SH			
102470	102485	Pteronarcella				0		SH			
102485	102486	Pteronarcella badia				0		SH		cn	sp
102485	102487	Pteronarcella regularis				0		SH			
102470	102471	Pteronarcys	1.7		2.2	0		SH			
102471	102473	Pteronarcys californica				0		SH			
102471	102478	Pteronarcys dorsata	1.8					SH			
102471	102484	Pteronarcys princeps				0		SH		sp	cn

Appendix B: Regional Tolerance Values, Functional Feeding Groups, and Habit/Behavior Assignments for Benthic Macroinvertebrates

Parent TSN	TSN	Scientific Name	Southeast (NC)	Upper Midwest (WI)	Midwest (OH)	Northwest (ID)	Mid-Atlantic (MACS)	Functional Feeding Group primary	secondary	Habit/ Behavior primary	secondary
102468	102788	Taeniopterygidae				2		SH		sp	cn
102838	102839	Doddsia occidentalis				2		SC		sp	cn
102788	102830	Oemopteryx								sp	cn
102788	102808	Strophopteryx	2.5	3							
102788	102816	Taenionema				2		SC		sp	cn
102816	102827	Taenionema pallidum				2		SC			
102788	102789	Taeniopteryx	6.3	2			2	SH			
102789	102791	Taeniopteryx burksi	5.8					OM			
102789	102792	Taeniopteryx lita						OM		cn	
102789	102795	Taeniopteryx metequi	1.4								
102912	103202	Chloroperlidae				1		PR		cn	
	103236	Kathroperla				0		PR			
103236	103237	Kathroperla perdita				1		GC		cn	
		Chloroperlinae				1		PR			
103202	103203	Alloperla	1.4			1		PR		cn	
103202	103260	Haploperla								cn	
103260	103263	Haploperla brevis	1.3	1							
103202	103303	Neaviperla						PR		cn	
103303	103304	Neaviperla forcipata				1		PR		cn	
103202	103233	Paraperla				1		PR		cn	
103233	103234	Paraperla frontalis						PR			
103202	103305	Plumiperla						PR		cn	
103202	103254	Suwallia	0			1		PR		cn	
103202	103273	Sweltsa	0			1		PR			
103202	103308	Triznaka				1		PR		cn	
102912	102914	Perlidae				1	1	PR			
102914	102917	Acroneuria					0	PR			
102917	102919	Acroneuria abnormis	2.2	0				PR			
102917	102920	Acroneuria arenosa	2.2					PR			
102917	102922	Acroneuria carolinensis	0		2.3						
102917	102923	Acroneuria evoluta			2.8						
102917	102925	Acroneuria internata			2.2						
102917	102918	Acroneuria lycorias	1.5		2.4			PR			
102917	102926	Acroneuria mela	0.9					PR		cn	
102917	102927	Acroneuria perplexa						PR		cn	
102914	102975	Agnetina			1.8		2	PR		cn	
102975	102983	Agnetina annulipes	0				2			cn	
102975	102979	Agnetina capitata						PR		cn	
102975	102984	Agnetina flavescens	0								
102954	102955	Attaneuria ruralis						PR		cn	
102914	102934	Beloneuria	0			3		PR		cn	
102914	102985	Calineuria				3		PR		cn	
102985	102986	Calineuria californica				1		PR		cn	
102994	103121	Doroneuria				1		PR		cn	
103121	103123	Doroneuria baumanni				1		PR		cn	
103121	103122	Doroneuria theodora				1		PR		cn	
102914	102930	Claassenia				3		PR		cn	
102930	102932	Claassenia sabulosa				3		PR		cn	

Rapid Bioassessment Protocols for Use in Streams and Wadeable Rivers: Periphyton, Benthic Macroinvertebrates, and Fish, Second Edition

B-35

Parent TSN	TSN	Scientific Name	Southeast (NC)	Upper Midwest (WI)	Midwest (OH)	Northwest (ID)	Mid-Atlantic (MACS)	Functional Feeding Group primary	secondary	Habit/ Behavior primary	secondary
102914	102939	Eccoptura								cn	
102939	102940	Eccoptura xanthenes	4.1							cn	
102914	102971	Hesperoperla						PR		cn	
102971	102972	Hesperoperla pacifica				1		PR		cn	
102914	102942	Neoperla	1.6	1	3.1			PR		cn	
102942	102944	Neoperla clymene						PR			
102914	102962	Paragnetina						PR			
102962	102965	Paragnetina fumosa	3.5					PR			
102962	102970	Paragnetina ichusa	0								
102962	102966	Paragnetina immarginata	1.7								
102962	102967	Paragnetina kansensis	2					PR		cn	sp
102962	102968	Paragnetina media			2.1						
103202	103251	Perlesta	0		4.5	5		PR		cn	
103251	103253	Perlesta placida	4.9	5				OM			
103202	103244	Perlinella						PR			
103244	103246	Perlinella drymo	0	1				PR		cn	
103244	103248	Perlinella ephyre						PR		cn	
102912	102994	Perlodidae				2	2	PR		cn	sp
102994	103155	Calliperla				2		PR		cn	sp
102994	103157	Cascadoperla				2		PR			
102994	103118	Clioperla								cn	
103118	103119	Clioperla clio	4.8	1						cn	
102994	103137	Cultus				2		PR		cn	
103137	103139	Cultus decisus	1.6								
102994	103166	Diploperla	2							cn	
103166	103167	Diploperla duplicata	2.7								
103166	103169	Diploperla morgani	1.5								
	103094	Diura				2		PR			
103094	103096	Diura knowltoni				2		SC		cn	
103171	103172	Frisonia picticeps				2		PR		cn	
102994	103084	Helopicus								cn	
103084	103087	Helopicus bogaloosa	0							cn	
103084	103085	Helopicus subvarians	0.8								
	103124	Isogenoides				2		PR			
103124		Isogenoides hansoni	0								
102994	103070	Isogenus				2		PR			
102994	102995	Isoperla				2	2	PR			
102995	103012	Isoperla bilineata	5.5								
102995	103021	Isoperla dicala	2.2	2							
102995	103004	Isoperla fulva				2		PR			
102995	103029	Isoperla fusca				2		PR			
102995	103020	Isoperla holochlora	0								
102995	103007	Isoperla mormona				2		PR			
102995	103017	Isoperla namata	1.8								
102995	103018	Isoperla orata	0					OM			
102995	103009	Isoperla pinta				2		PR			
102995	103019	Isoperla similis	0.7								
102995	103035	Isoperla slossonae	2.6								

Appendix B: Regional Tolerance Values, Functional Feeding Groups, and Habit/Behavior Assignments for Benthic Macroinvertebrates

Parent TSN	TSN	Scientific Name	Southeast (NC)	Upper Midwest (WI)	Midwest (OH)	Northwest (ID)	Mid-Atlantic (MACS)	primary	secondary	primary	secondary
			Regional Tolerance Values					Functional Feeding Group		Habit/Behavior	

Parent TSN	TSN	Scientific Name	Southeast (NC)	Upper Midwest (WI)	Midwest (OH)	Northwest (ID)	Mid-Atlantic (MACS)	primary	secondary	primary	secondary
102995	103036	Isoperla transmarina	5.6								
102994	103149	Kogotus				2		PR		cn	
103174	103175	Malirekus hastatus	1.4								
102994	103110	Megarcys				2		PR		cn	
102994	103180	Oroperla				2		PR		cn	
102994	103134	Perlinodes						PR		cn	
103134	103135	Perlinodes aureus				2		PR		cn	
102994	103186	Pictetiella				2		PR		cn	
103186	103188	Pictetiella expansa				2		PR		cn	
103099	103100	Remenus bilobatus	0.3								
102994	103189	Rickera						PR		cn	
103189	103190	Rickera sorpta				2		PR		cn	
102994	103193	Setvena				2		PR		cn	
103193	103194	Setvena bradleyi				2		PR		cn	
102994	103102	Skwala				2		PR			
102994	103197	Yugus				2		PR		cn	sp
103197	103200	Yugus arinus	0								
103197	103198	Yugus bulbosus	0								
100500	101593	Odonata						PR		cb	
101595	101596	Aeshnidae				3		PR			
	101602	Aeshna				5		PR			
101596	101597	Anax				8	5	PR			
101597	101598	Anax junius						PR		cb	sp
101597	101599	Anax longipes						PR		cb	sp
101596	101648	Basiaeschna								cb	sp
101648	101649	Basiaeschna janata	7.7	6				PR			
101596	101645	Boyeria						PR		cb	
101645	101646	Boyeria grafiana	6.3								
101645	101647	Boyeria vinosa	6.3	2	3.5			PR		cb	sp
101639	101640	Coryphaeschna ingens						PR		cb	cn
101637	101638	Epiaeschna heros						PR		cb	cn
101634	101635	Gomphaeschna furcillata						PR			
101653	101654	Nasiaeschna pentacantha	8					PR		bu	
101595	101664	Gomphidae				1		PR		bu	
101715	101716	Aphylla williamsoni						PR			
101664	101770	Arigomphus								bu	
101770	101771	Arigomphus pallidus						PR			
101664	101730	Dromogomphus	6.3					PR			
101730	101731	Dromogomphus armatus						PR			
101730	101732	Dromogomphus spinosus						PR		bu	
	101725	Erpetogomphus				4		PR			
101777	101780	Gomphurus dilatatus	6.2	5	2.5			PR			
101664	101665	Gomphus					5	PR			
101665	101677	Gomphus dilatatus						PR			
101665	101668	Gomphus geminatus						PR			
101665	101685	Gomphus lividus					5	PR			
101665	101686	Gomphus minutus						PR			
101665	101689	Gomphus pallidus						PR		sp	

Rapid Bioassessment Protocols for Use in Streams and Wadeable Rivers: Periphyton, Benthic Macroinvertebrates, and Fish, Second Edition

B-37

Parent TSN	TSN	Scientific Name	Regional Tolerance Values					Functional Feeding Group		Habit/ Behavior	
			Southeast (NC)	Upper Midwest (WI)	Midwest (OH)	Northwest (ID)	Mid-Atlantic (MACS)	primary	secondary	primary	secondary
101665	101694	Gomphus spiniceps	4.9								
101734	101735	Hagenius brevistylus	4	1				PR		bu	
101791	206625	Hylogomphus geminatus						PR		bu	
101664	101766	Lanthus	2.7							bu	
101664	101736	Octogomphus				1		PR		bu	
101664	101738	Ophiogomphus	6.2	1		1		PR		bu	
101664	101718	Progomphus						PR		bu	
101718	101720	Progomphus obscurus	8.7					PR		bu	
101664	101761	Stylogomphus								bu	
101761	101762	Stylogomphus albistylus	4.8								
101664	206626	Stylurus						PR		sp	
206626	206627	Stylurus ivae						PR			
	101594	Anisoptera						PR			
101659	101660	Tachopteryx				10		PR		bu	
102025	102026	Cordulegastridae						PR		bu	
102026	102027	Cordulegaster	6.1	3		0	3	PR			
102027	102031	Cordulegaster maculata						PR		sp	
101796	102020	Corduliidae				2	5	PR		cb	sp
101851	101852	Didymops transversa						PR		cb	sp
	101862	Epicordulia	5.6								
101862	101863	Epicordulia princeps						PR		sp	
101862	101864	Epicordulia regina						PR		sp	
101797	101918	Macromia	6.7	2			2	PR		sp	
101918	101920	Macromia georgiana						PR		sp	
101918	101924	Macromia georgina						PR		cb	cn
101918	101922	Macromia taeniolata						PR			
101797	101934	Neurocordulia	5.8					PR			
101934	101938	Neurocordulia alabamensis						PR			
101934	101936	Neurocordulia molesta	3.3	5				PR			
101934	101939	Neurocordulia obsoleta	5.4	0				PR		sp	
101934	101935	Neurocordulia virginiensis	1.6					PR		sp	
101797	101947	Somatochlora	8.9	1		9	1	PR		cb	sp
101947	101949	Somatochlora linearis						PR			
102026	102035	Epitheca					4	PR			
102035	206629	Epitheca princeps						PR			
102035		Epitheca sepia						PR			
206629	206631	Epitheca princeps regina						PR		cb	sp
102035	185986	Epitheca cynosura						PR			
101797	101994	Tetragoneuria	8.5					PR			
101994	101996	Tetragoneuria cynosura						PR		sp	
101796	101797	Libellulidae				9	9	PR		sp	
101830	101831	Brachymesia gravida						PR		sp	
101797	101865	Erythemis						PR		cb	
101865	101866	Erythemis simplicicollis	7.7					PR		cb	
101797	101870	Erythrodiplax						PR		cb	
101870	101872	Erythrodiplax minuscula						PR		sp	
101797	101885	Leucorrhinia						PR			
101797	101893	Libellula	9.8	9		9	8	PR			

Appendix B: Regional Tolerance Values, Functional Feeding Groups, and Habit/Behavior Assignments for Benthic Macroinvertebrates

Parent TSN	TSN	Scientific Name	Regional Tolerance Values					Functional Feeding Group		Habit/ Behavior	
			Southeast (NC)	Upper Midwest (WI)	Midwest (OH)	Northwest (ID)	Mid-Atlantic (MACS)	primary	secondary	primary	secondary
101893	101901	Libellula auripennis						PR			
101893	101900	Libellula incesta						PR			
101893	101903	Libellula semifasciata						PR		sp	
101893	101904	Libellula vibrans						PR		sp	cb
102009	102010	Miathyria marcella						PR		sp	
101932	101933	Nannothemis bella						PR		sp	
101797	101945	Orthemis						PR		sp	
101945	101946	Orthemis ferruginea						PR		sp	
101798	101799	Pachydiplax longipennis	9.6					PR		sp	
101797	101803	Perithemis	10				4	PR		sp	
101803	101805	Perithemis seminola						PR		sp	
101803	101804	Perithemis tenera						PR		sp	cb
101808	101809	Plathemis lydia	10	8	8.2			PR			
101797	101976	Sympetrum	7.3	10			4	PR		sp	
101976	101977	Sympetrum ambiguum						PR			
101818	101820	Tramea carolina						PR			
100500	102042	Zygoptera						PR		cb	
102042	102043	Calopterygidae					5	PR		cb	
102043	102052	Calopteryx	8.3	5	3.7	6	6	PR		cb	
102052	102054	Calopteryx dimidiata						PR		cb	cn
102052	102055	Calopteryx maculata						PR			
102043	102048	Hetaerina	6.2	6	2.8			PR			
102048	102050	Hetaerina americana						PR			
102048	102049	Hetaerina titia						PR		cb	sw
102042	102077	Coenagrionidae			6.1	9	9	PR		cb	
102077	102093	Amphiagrion				5		PR		cn	cb
102077	102139	Argia			5.1	7	6	PR			
102139	102140	Argia apicalis						PR			
102139	102143	Argia fumipennis						PR			
102139	102146	Argia moesta						PR			
102139	102147	Argia sedula						PR			
102139	102148	Argia tibialis						PR		cb	
102139	102154	Argia violacea						PR		cb	
102077	102133	Chromagrion				6		PR		cb	
102077	102102	Enallagma	9	9		9	8	PR		cb	
102102	102103	Enallagma antennuatus						PR		cb	
102102	102104	Enallagma cardenium						PR		cb	
102102	102106	Enallagma daecki						PR		cb	
102102	102108	Enallagma divagans						PR		cb	
102102	102110	Enallagma dubium						PR		cb	
102102	181184	Enallagma pallidum						PR		cb	
102102	102114	Enallagma pollutum						PR		cb	
102102	102115	Enallagma signatum						PR		cb	
102102	102119	Enallagma vesperum						PR		cb	
102102	102120	Enallagma weewa						PR		cb	
102077	102078	Ischnura	9.4	9		9	9	PR		cb	
102078	206632	Ischnura hastata						PR			
102078	102082	Ischnura posita						PR		cb	

Rapid Bioassessment Protocols for Use in Streams and Wadeable Rivers: Periphyton, Benthic Macroinvertebrates, and Fish, Second Edition

B-39

Parent TSN	TSN	Scientific Name	Southeast (NC)	Upper Midwest (WI)	Midwest (OH)	Northwest (ID)	Mid-Atlantic (MACS)	Functional Feeding Group primary	secondary	Habit/ Behavior primary	secondary
102078	102084	Ischnura ramburi						PR		cb	
102077	102135	Nehalennia						PR		cb	
102135	102136	Nehalennia intergricollis						PR		cb	
102096	102099	Telebasis byersi						PR			
102077	102100	Zoniagrion				9		PR			
102058	102061	Lestes				9		PR		cb	
109215	118831	Diptera				7					
121226	121227	Blephariceridae				0		SC			
121229	121230	Agathon				0		SC		cn	
121229	121250	Bibiocephala				0		SC			
121229	121255	Blepharicera	0.2	0		0		SC		sp	bu
121229	121278	Philorus				0		SC		sp	cb
125808	127076	Ceratopogonidae			5.7	6		PR			
127277	127278	Dasyhelea						GC		sp	cn
127076	127112	Forcipomyiinae				6		PR	GC	sp	
127112	127113	Atrichopogon	6.8		4.5	6		PR	GC		
127113	127150	Atrichopogon websteri			4.4						
127112	127152	Forcipomyia				6		SC	PR	bu	
127076	127338	Ceratopogoninae				6		PR		bu	
127526	127533	Alluaudomyia						PR			
127774	127778	Bezzia				6	6	GC	PR	bu	
127526	127564	Ceratopogon					6	PR		bu	
127339	127340	Culicoides	6.5	10			10	PR	GC	bu	
127683	127720	Nilobezzia						PR			
127774	127859	Palpomyia				6		PR	GC	bu	
127859	127905	Palpomyia tibialis								bu	
127683	127729	Probezzia				6		PR		bu	
127526	127614	Serromyia				6		PR		bu	
127683	127761	Sphaeromias						PR	GC		
127526	127619	Stilobezzia						PR		sp	sw
125808	125886	Chaoboridae						PR			
125892	125904	Chaoborus						PR			
125904	125923	Chaoborus punctipennis	8.5	8				PR			
125887	125888	Eucorethra				7		PR			
125808	127917	Chironomidae				6		GC		bu	
127917	127994	Tanypodinae				7		PR		bu	
127995	127996	Clinotanypus					8	PR			
127996	127998	Clinotanypus pinguis	9.8	8	7.5						
127995	128010	Coelotanypus	6.2					PR			
128010	128012	Coelotanypus concinnus	7.7					PR			
128010	128016	Coelotanypus scapularis						PR		bu	
128010	128018	Coelotanypus tricolor						PR		bu	
	128020	Macropelopiini						PR			
127995	206646	Alotanypus									
128020	128021	Apsectrotanypus						PR		bu	
128021	128024	Apsectrotanypus johnsoni	0					PR			
128020	128026	Brundiniella				6		PR		sp	
128026	128028	Brundiniella eumorpha	3.8								

Appendix B: Regional Tolerance Values, Functional Feeding Groups,
and Habit/Behavior Assignments for Benthic Macroinvertebrates

| Parent TSN | TSN | Scientific Name | Regional Tolerance Values | | | | | Functional Feeding Group | | Habit/Behavior | |
			Southeast (NC)	Upper Midwest (WI)	Midwest (OH)	Northwest (ID)	Mid-Atlantic (MACS)	primary	secondary	primary	secondary
206647	206648	Fittkauimyia serta								sp	bu
128020	128034	Macropelopia				6		PR			
128020	128048	Psectrotanypus			8.1	10	10	PR		sp	
128048	128056	Psectrotanypus dyari	10	10	8.6						
128270	128271	Djalmabatista						PR		sp	
128271	128272	Djalmabatista pulcher						PR			
128270	128277	Procladius	9.3	9	6.5	9	9	PR	GC	sp	
128277	128285	Procladius bellus						PR			
128069	128070	Natarsia	10	8	5.9		8	PR		sp	
128070	128071	Natarsia baltimoreus			5.6						
127994	128078	Pentaneurini				6		PR			
128078	128079	Ablabesmyia			5.2		8	GC	PR		
128079	128081	Ablabesmyia annulata			4.1			OM			
128079	128083	Ablabesmyia aspera						OM			
128079	128087	Ablabesmyia cinctipes						OM			
128079	128089	Ablabesmyia hauberi						OM			
128079	128090	Ablabesmyia idei						OM			
128079	128093	Ablabesmyia janta	7.1		4.9			OM			
128079	128097	Ablabesmyia mallochi	7.6	8	5			OM			
128079	128113	Ablabesmyia peleensis	4.6					OM		sp	
128079	128121	Ablabesmyia rhamphe						OM			
128078	128130	Conchapelopia	8.7	6	4.3	6	6	PR			
		Denopelopia atria									
128161	128162	Guttipelopia guttipennis						PR			
	128237	Hayesomyia						PR		sp	
128237	128249	Hayesomyia senata			4.6						
	128131	Helopelopia			3.9		6	PR		sp	
128078	128167	Hudsonimyia						PR			
128078	128170	Krenopelopia						PR		sp	
128170	128171	Krenopelopia hudsoni						PR			
128078	128173	Labrundinia			3.8			PR			
128173	128174	Labrundinia becki						PR			
128173	128175	Labrundinia johannseni						PR			
128173	128176	Labrundinia maculata	.					PR			
128173	128177	Labrundinia neopilosella					7	PR			
128173	128178	Labrundinia pilosella	6	7	3.1			PR		sp	
128173	128182	Labrundinia virescens	4.5					PR			
128078	128183	Larsia	8.3	6	4.3	6	6	PR			
128183	128184	Larsia berneri						PR			
128183	128186	Larsia decolorata						PR			
128183	128189	Larsia indistincta						PR		sp	
	128132	Meropelopia			2.7		7				
128078	128199	Monopelopia				6		PR		sp	
128199	128200	Monopelopia boliekae						PR			
128078	128202	Nilotanypus	4	6		6		PR		sp	
128202	128203	Nilotanypus fimbriatus			2.8			PR			
128078	128207	Paramerina	2.8			6	4	PR		sp	
128207	128208	Paramerina anomala									

Rapid Bioassessment Protocols for Use in Streams and Wadeable Rivers: Periphyton, Benthic Macroinvertebrates, and Fish, Second Edition

B-41

Parent TSN	TSN	Scientific Name	Southeast (NC)	Upper Midwest (WI)	Midwest (OH)	Northwest (ID)	Mid-Atlantic (MACS)	primary	secondary	primary	secondary
			Regional Tolerance Values					Functional Feeding Group		Habit/ Behavior	
128207	128209	Paramerina fragilis			4.7						
128078	128215	Pentaneura	4.6	6		6		PR	GC		
128215	128216	Pentaneura inconspicua			4.9			PR		sp	
128215	128218	Pentaneura inculta						PR		sp	
128078	128226	Rheopelopia						PR		sp	
128226	128229	Rheopelopia paramaculipennis			2.9						
	128234	Telopelopia okoboji			4						
128078	128236	Thienemannimyia				6	6	PR		sp	
128078	128251	Trissopelopia						PR			
128078	128259	Zavrelimyia	9.3	8	4.1	8	8	PR		sp	
128259	128262	Zavrelimyia sinuosa						PR			
128323	128324	Tanypus	9.6	10	8.8		10	PR	GC		
128324	128329	Tanypus neopunctipennis			7.5			OM			
128324	128335	Tanypus carinatus						OM			
128324	128333	Tanypus punctipennis						OM		sp	
128324	128336	Tanypus stellatus						OM			
127953	127954	Boreochlus				6		GC	SC		
127917	128341	Diamesinae						GC		sp	
128342	128343	Boreoheptagyia				6		GC			
	128351	Diamesini				2		GC			
128351	128355	Diamesa	7.7	8		5		GC	SC	sp	
128351	128401	Pagastia	2.2	1		1		GC			
128351	128408	Potthastia				2		OM	GC		
128408	128409	Potthastia gaedii	2			6		GC		sp	
128408	128412	Potthastia longimana	7.4			2		GC		sp	
128351	128416	Pseudodiamesa				6		GC		sp	
128351	128426	Sympotthastia	5.7	2		2		GC	SC	sp	
128437	128440	Monodiamesa				7		GC		bu	sp
128437	128446	Odontomesa				4		GC			
128446	128447	Odontomesa fulva	5.9	4							
128437	128452	Prodiamesa				3		GC		sp	
128452	128454	Prodiamesa olivacea	7.9	3				GC		sp	
125808	128457	Orthocladiinae				5		GC		bu	
128457	128563	Corynoneura	6.2	7	3.5	7	7	GC			
128563	128565	Corynoneura celeripes			2.3			GC		sp	
128563	128567	Corynoneura lobata			3.3						
128563	128570	Corynoneura taris						GC			
128457	129182	Thienemanniella	6	6	3.7	6	6	GC			
129182	129193	Thienemanniella fusca						GC			
129182	129189	Thienemanniella similis			2.4			GC			
129182	129190	Thienemanniella xena			3.6			GC			
		Orthocladiini				6		GC			
128457	128460	Acamptocladius						GC		bu	sp
128457	128470	Antillocladius									
128457	128477	Brillia	5.2	5		5	5	SH	GC		
128477	128478	Brillia flavifrons				5		SH			
128477	128487	Brillia par								bu	cn
128477	128482	Brillia retifinis				5		SH		sp	

Appendix B: Regional Tolerance Values, Functional Feeding Groups,
and Habit/Behavior Assignments for Benthic Macroinvertebrates

Parent TSN	TSN	Scientific Name	Southeast (NC)	Upper Midwest (WI)	Midwest (OH)	Northwest (ID)	Mid-Atlantic (MACS)	Functional Feeding Group primary	secondary	Habit/Behavior primary	secondary
128457	128511	Cardiocladius	6.2	5		5		PR		cn	bu
128511	128515	Cardiocladius obscurus			2.2						
128457	128520	Chaetocladius				6		GC			
128457	128575	Cricotopus		7	4.3	7	7	SH	GC		
128575	128583	Cricotopus bicinctus	8.7		6.7		7	OM			
128575	128594	Cricotopus festivellus				7		SH			
128575	128610	Cricotopus infuscatus	9								
128575		Cricotopus Isocladius				7		SH			
128575		Cricotopus Nostococladius				7		SH			
128575	128640	Cricotopus politus						OM			
128575		Cricotopus sylvestris	10					OM			
128575	128651	Cricotopus tremulus				7	7	SH		sp	
128575	128659	Cricotopus trifascia				7		OM			
128575	128664	Cricotopus varipes	8.1								
128575	128666	Cricotopus vierriensis	4.8		4.2						
128457	128670	Diplocladius						GC		sp	
128670	128671	Diplocladius cultriger	7.7	8				GC			
128680	128681	Doncricotopus bicaudatus			4.8						
128457	128689	Eukiefferiella				8		GC	SC		
128689	128704	Eukiefferiella brehmi	3.7			8		GC			
128689	128703	Eukiefferiella brevicalcar	1.7			8		GC			
128689	128693	Eukiefferiella claripennis	5.7	8		8		GC			
128689	128695	Eukiefferiella devonica	2.6			8		GC			
128689	128705	Eukiefferiella gracei	2.7			8		GC			
128689	128706	Eukiefferiella pseudomontana				8		GC		sp	
128457	128712	Georthocladius								sp	
128457	128718	Gymnometriocnemus					7	GC		sp	bu
128457	128730	Heleniella	0			6		GC			
128457	128737	Heterotrissocladius	5.4	0		4		GC	SC	sp	
128737	128746	Heterotrissocladius subpilosus				0		GC		sp	
128457	128750	Hydrobaenus	9.6	8		8	8	SC	GC	sp	
	128771	Krenosmittia				1		GC			
128457	128776	Limnophyes			3.1	8	8	GC			
128457	128811	Lopescladius	2.2	4		6		GC		bu	sp
128457	128818	Mesosmittia								sp	
128457	128821	Metriocnemus						OM	GC		
128457	128844	Nanocladius	7.2	3	5.3	3	3	GC			
128844	128852	Nanocladius crassicornus			4.3		3	GC			
128844	128853	Nanocladius distinctus			6.1			GC			
128844	128855	Nanocladius downesi	2.6								
128844	128859	Nanocladius minimus			4.5						
128844	128860	Nanocladius rectinervis						GC		sp	bu
128844	128862	Nanocladius spiniplenus			3.5						
128457	128867	Oliveridia				6		GC			
128457	128874	Orthocladius		6	3.9			GC			
128874		Orthocladius Eudactylocladius				6		GC			
128874		Orthocladius Euorthocladius	6.3			6		GC			
128874		Orthocladius Pogonocladius				6		GC			

Rapid Bioassessment Protocols for Use in Streams and Wadeable Rivers: Periphyton, Benthic Macroinvertebrates, and Fish, Second Edition

B-43

Parent TSN	TSN	Scientific Name	Southeast (NC)	Upper Midwest (WI)	Midwest (OH)	Northwest (ID)	Mid-Atlantic (MACS)	primary	secondary	primary	secondary
			Regional Tolerance Values					**Functional Feeding Group**		**Habit/ Behavior**	
128874	128878	Orthocladius annectens						GC		sp	
128874	128882	Orthocladius carlatus			2						
128874	128885	Orthocladius clarkei	5.8								
128874	128898	Orthocladius dorenus	6.7								
128874	128913	Orthocladius lignicola						GC		sp	
128874	128920	Orthocladius nigritus	0.9								
128874	128923	Orthocladius obumbratus	8.8								
128874	128929	Orthocladius robacki	7.2								
128457	128951	Parachaetocladius	0			6	2	GC		sp	
128457	128968	Parakiefferiella	5.9		4.8	6	4	GC			
128457	128978	Parametriocnemus			2.8	5	5	GC		sp	
128978	128982	Parametriocnemus lundbecki	3.7	5				GC		sp	
128457	128989	Paraphaenocladius				5	4	GC		sp	
128457	129005	Paratrichocladius			2	6		GC		sp	bu
128457	129011	Parorthocladius				6		GC			
128457	129018	Psectrocladius	3.8	8	5.7	8	8	GC	SH		
129018	129027	Psectrocladius elatus						OM			
129018	129031	Psectrocladius limbatellus				8		GC		sp	
129018	129051	Psectrocladius sordidellus				8		GC			
128457	129052	Pseudorthocladius	0	0		0	0	GC		sp	
128457	129071	Pseudosmittia						GC		sp	
128457	129083	Psilometriocnemus						GC			
128457	129086	Rheocricotopus			4.9	6	6	GC	SH		
129086	129101	Rheocricotopus pauciseta					6	GC			
129086	129102	Rheocricotopus robacki	7.7	6	3.8						
129086	129105	Rheocricotopus tuberculatus	6.8							bu	
128457	129107	Rheosmittia						GC			
128457	129110	Smittia						GC			
128457	129152	Stilocladius						GC		sp	
128457	129156	Symbiocladius				6		PA			
	128877	Symposiocladius								sp	
128877	128915	Symposiocladius lignicola	5.4								
128457	129161	Synorthocladius	4.7	2		2		GC	SC		
129161	129162	Synorthocladius semivirens			2.5						
128457	129197	Tvetenia		5		5	5	GC			
129197	129205	Tvetenia bavarica	4			5		GC			
129197	189327	Tvetenia discoloripes	3.9			5		GC			
128457	129206	Unniella					4	GC		bu	
129206	129207	Unniella multivirga	0					GC			
128457	129208	Xylotopus	6.6	2						bu	
129208	129209	Xylotopus par					2				
128457	129213	Zalutschia					7	SH			
128457	129228	Chironominae				6		GC			
129228	129229	Chironomini				6		GC			
	206655	Apedilum									
206655	129618	Apedilum elachista								sp	bu
129231	129234	Asheum beckae						GC			
129229	129236	Axarus						GC			

Appendix B: Regional Tolerance Values, Functional Feeding Groups,
and Habit/Behavior Assignments for Benthic Macroinvertebrates

Parent TSN	TSN	Scientific Name	Southeast (NC)	Upper Midwest (WI)	Midwest (OH)	Northwest (ID)	Mid-Atlantic (MACS)	Functional Feeding Group primary	secondary	Habit/Behavior primary	secondary
129229	206657	Beardius								bu	
206657	206658	Beardius truncatus									
129229	129254	Chironomus	9.8	10	8.1	10	10	GC	SH		
129254	129280	Chironomus decorus						OM			
129254	129313	Chironomus riparius						OM		bu	
129254	129322	Chironomus stigmaterus						OM		sp	bu
129229	129350	Cladopelma	2.5	9			7	GC			
129229	129368	Cryptochironomus			4.9	8	8	PR		sp	
129368	129370	Cryptochironomus blarina	8	8							
129368	129376	Cryptochironomus fulvus	6.7	8				PR		bu	
129229	129394	Cryptotendipes	6.1	6	4.2		6	GC		bu	
129229	129421	Demicryptochironomus	2.1				8	GC			
129229	129428	Dicrotendipes	7.9		5.6	8	8	GC	FC		
129428	129436	Dicrotendipes fumidus			5.8						
129428	129441	Dicrotendipes leucoscelis						FG			
129428	129445	Dicrotendipes lobus						FG			
129428	129458	Dicrotendipes lucifer			6.3						
129428	129448	Dicrotendipes modestus	9.2	5	5.9			FG			
129428	129450	Dicrotendipes neomodestus	8.3		4.5			FG			
129428	129452	Dicrotendipes nervosus	10					FG			
129428	193743	Dicrotendipes simpsoni	10		7.4			FG			
129428	206649	Dicrotendipes thanatogratus						FG		bu	
129428	183774	Dicrotendipes tritomus						FG			
129229	129459	Einfeldia					8	GC			
129459	129460	Einfeldia austini						GC		cn	
129459	129463	Einfeldia natchitocheae						GC			
129229	129470	Endochironomus			5.6	10	10	SH	GC		
129470	129471	Endochironomus nigricans	7.5	8	5.3						
129470	129474	Endochironomus subtendens									
128457	130046	Endotribelos						GC		bu	cn
130046	130047	Endotribelos hesperium						GC			
129229	129483	Glyptotendipes	8.5	10	6.2		10	FC	GC		
129483	129484	Glyptotendipes amplus			3.2						
129483	129485	Glyptotendipes barbipes					10	FC			
129483	129493	Glyptotendipes meridionalis									
129483	129494	Glyptotendipes paripes								bu	
129483	129496	Glyptotendipes seminole									
129229	129506	Goeldichironomus					8	GC			
129506	206650	Goeldichironomus amazonicus						GC			
129506	129508	Goeldichironomus carus						GC			
129506	206651	Goeldichironomus fluctuans						GC			
129506	129512	Goeldichironomus holoprasinus	10					GC		cb	cn
129506	206652	Goeldichironomus natans						GC		bu	
129229	129516	Harnischia	7.5	8				GC	SC		
129516	129517	Harnischia curtilamellata			3.5						
129229	129522	Kiefferulus					10	GC			
129522	129523	Kiefferulus dux	10	10	5.2			GC		cn	
129525	129526	Lauterborniella agrayloides						GC			

Rapid Bioassessment Protocols for Use in Streams and Wadeable Rivers: Periphyton, Benthic Macroinvertebrates, and Fish, Second Edition

B-45

Parent TSN	TSN	Scientific Name	Regional Tolerance Values					Functional Feeding Group		Habit/ Behavior	
			Southeast (NC)	Upper Midwest (WI)	Midwest (OH)	Northwest (ID)	Mid-Atlantic (MACS)	primary	secondary	primary	secondary
129229	129535	Microtendipes	6.2	7		6		FC	GC		
129535	129540	Microtendipes caelum			2.7						
129535	129541	Microtendipes pedellus						FG			
129535	129547	Microtendipes rydalensis			2			FG			
129229	129548	Nilothauma	5.5	2	3.1		2				
129548	129551	Nilothauma bicorne						GC			
129229	129561	Pagastiella						GC		sp	
129561	206654	Pagastiella orophila						GC			
129561	129562	Pagastiella ostansa	2.6								
129229	129564	Parachironomus	9.2	10	4.1			PR	GC		
129564	129565	Parachironomus abortivus			8						
129564	129569	Parachironomus carinatus			5.3						
129564	129573	Parachironomus directus			7.9						
129564	129579	Parachironomus frequens			3.8						
129564	129595	Parachironomus hirtalatus									
129564	129581	Parachironomus monochromus	7.9								
129564	129583	Parachironomus pectinatellae			3.7						
129564	129587	Parachironomus schneideri								sp	
129564	129588	Parachironomus sublettei									
129229	129597	Paracladopelma	6.4	7				GC			
129597	129608	Paracladopelma nereis	1.8					GC		cn	
129597	129612	Paracladopelma undine	5.2					GC			
129229	129616	Paralauterborniella					8	GC		bu	
129616	129619	Paralauterborniella nigrohalterale									
129229	129623	Paratendipes	5.3	8	5.7	8	8	GC			
129623	129624	Paratendipes albimanus			4.3			GC		cn	
129623	129632	Paratendipes subaequalis						GC			
129229	129637	Phaenopsectra	6.8	7		7	7	SC	GC		
129637	129642	Phaenopsectra flavipes	8.5		5.7						
129637	129647	Phaenopsectra obediens						OM		cb	cn
129637	129652	Phaenopsectra punctipes			3.5			SC			
129229	129657	Polypedilum				6	6	SH	GC		
129657		Polypedilum Pentapedilum				6		SH			
129657	129725	Polypedilum angulum	5.6								
129657	129666	Polypedilum aviceps	4		1.9						
129657	129726	Polypedilum bergi					6	SH			
129657	129671	Polypedilum convictum	5.3		3.6						
129657	129676	Polypedilum fallax	6.7								
129657	129684	Polypedilum halterale	7.2								
129657	129686	Polypedilum illinoense	9.2		6.9						
129657	129692	Polypedilum laetum									
129657	129698	Polypedilum ontario			2.6						
129657	129708	Polypedilum scalaenum	8.7								
129657	129718	Polypedilum trigonum								bu	
129657	129719	Polypedilum tritum									
129229	129730	Robackia						GC			
129730	129731	Robackia claviger	2.4					GC		bu	
129730	129733	Robackia demeijerei	4.3			7		GC			

B-46

Appendix B: Regional Tolerance Values, Functional Feeding Groups,
and Habit/Behavior Assignments for Benthic Macroinvertebrates

Parent TSN	TSN	Scientific Name	Regional Tolerance Values					Functional Feeding Group		Habit/ Behavior	
			Southeast (NC)	Upper Midwest (WI)	Midwest (OH)	Northwest (ID)	Mid-Atlantic (MACS)	primary	secondary	primary	secondary
129229	129735	Saetheria						GC		wood	
129735	129736	Saetheria hirta						GC			
129735	129737	Saetheria tylus	8.1	4							
129229	129743	Stelechomyia					7	GC		bu	
129743	129744	Stelechomyia perpulchra	4.6					GC		bu	
129229	129746	Stenochironomus	6.4	5	3.6		5	SH	GC		
129229	129785	Stictochironomus	6.7	9	4			OM	GC	bu	
129785	129790	Stictochironomus devinctus						OM			
129229	129820	Tribelos	6.6	5			5	GC			
129820	206656	Tribelos atrum						GC			
129820	129823	Tribelos fuscicorne			5.1			GC		bu	
129820	129827	Tribelos jucundus			5.6			GC			
129229	129837	Xenochironomus						PR			
129837	129838	Xenochironomus xenolabis	7	0				PR			
129229	129842	Xestochironomus						OM			
129842	129844	Xestochironomus subletti						OM			
129872	130040	Zavreliella								bu	
130040	189328	Zavreliella marmorata									
129850	129851	Pseudochironomus	4.2	5	4.7	5		GC			
129228	129872	Tanytarsini					6	FC			
129872	129873	Cladotanytarsus	3.7	7	4.4	7	7	GC	FC	cb	sp
129872	129884	Constempellina					6	GC			
129872	129890	Micropsectra	1.4	7	3.5	7	7	GC			
129872	129932	Nimbocera					6	FC		sp	
129932	206659	Nimbocera limnetica						FG			
129872	129935	Paratanytarsus	7.7	6	4.2	6	6	GC		cn	
129935		Paratanytarsus inopterus					6	GC			
129872	129952	Rheotanytarsus	6.4	6	3.3	6	6	FC			
129952	129955	Rheotanytarsus distinctissimus						FC		cb	sp
129952	129955	Rheotanytarsus distinctissimus						FC		cb	sp
129952	129957	Rheotanytarsus exiguus						FC			
129872	129962	Stempellina	2	2		2		GC		cb	cn
129872	129969	Stempellinella	5.3	4	2.6	4	4	GC			
129872	129975	Sublettea					6	FC			
129975	129976	Sublettea coffmani	1.7		2.2						
129872	129978	Tanytarsus	6.7	6	3.5	6	6	FC	GC		
129978	130030	Tanytarsus glabrescens						FG		cb	sp
129978	129997	Tanytarsus guerlus						FG			
		Thienemanniola					6	GC			
129872	130038	Zavrelia	2.7			8		GC		sw	
125875	125877	Corethrella								sw	
125808	125930	Culicidae				8		GC		sw	
126233	126234	Aedes					8	FC			
125955	125956	Anopheles	9.1				6	FC			
126233	126455	Culex	10				8	FC			
126233	126518	Deinocerites						FC			
125931	125932	Toxorhynchites						PR			
121226	121286	Deuterophlebiidae						SC			

Rapid Bioassessment Protocols for Use in Streams and Wadeable Rivers: Periphyton, Benthic Macroinvertebrates, and Fish, Second Edition

B-47

Parent TSN	TSN	Scientific Name	Regional Tolerance Values					Functional Feeding Group		Habit/ Behavior	
			Southeast (NC)	Upper Midwest (WI)	Midwest (OH)	Northwest (ID)	Mid-Atlantic (MACS)	primary	secondary	primary	secondary
121286	121287	Deuterophlebia				0		SC		sw	cb
121287	121290	Deuterophlebia nielsoni						SC			
125808	125809	Dixidae				1	1	GC			
125809	125810	Dixa	2.8			1		GC			
125809	125854	Dixella						GC			
125809	125873	Meringodixa				2		GC		bu	
125350	125351	Psychodidae				10		GC			
125391	125392	Maruina				1		SC			
125391	125514	Pericoma			5.6	4	4	GC			
125391	125468	Psychoda	9.9		3.7	10		GC			
125468	125469	Psychoda alternata						GC		bu	
125399	125400	Telmatoscopus albipunctatus									
125762	125763	Ptychopteridae				7		GC			
125764	125765	Bittacomorpha									
125785	125786	Ptychoptera				7		GC			
125808	126640	Simuliidae				6		FC		cn	
	126658	Cnephia mutata	4		5						
126648	126674	Gymnopais						SC		cn	
126648	126687	Metacnephia				6		FC			
	126642	Parasimulium						FC			
126648	126703	Prosimulium	2.6			3		FC			
126703	126736	Prosimulium mixtum	3.3	3				FC			
126773	126774	Simulium	4.4		4.8	6	6	FC			
126774	126790	Simulium bivittatum				6		FC			
126774	126832	Simulium jenningsi					6	FC			
126774	126834	Simulium jonesi					6	FC			
126774	126841	Simulium meridionale				6		FC			
126774	126870	Simulium rivuli					6	FC			
126774	126873	Simulium slossonae						FC			
126774	126883	Simulium tuberosum					6	FC		cn	
126774	126892	Simulium venustum	7.4	5			6	FC			
126774	126903	Simulium vittatum	8.7	7		6	6	FC			
126648	126761	Stegopterna									
126648	126767	Twinnia				6		FC			
125762	125799	Tanyderidae									
	125802	Protanyderus				1				sp	bu
125799	125800	Protoplasa				5		GC			
125800	125801	Protoplasa fitchii	5								
125808	126624	Thaumaleidae						OM			
126624	126629	Thaumalea						OM			
126629	126631	Thaumalea elnora						OM			
126629	126632	Thaumalea fusca						OM			
118839	118840	Tipulidae				3		SH		bu	
118841	118905	Megistocera									
118841	119008	Prionocera				4		SH		cn	
118841	119037	Tipula	7.7	4	7.2	4	4	SH			
119037	119041	Tipula abdominalis			4						
119037		Tipula ormosia				4		OM			

Appendix B: Regional Tolerance Values, Functional Feeding Groups, and Habit/Behavior Assignments for Benthic Macroinvertebrates

Parent TSN	TSN	Scientific Name	Southeast (NC)	Upper Midwest (WI)	Midwest (OH)	Northwest (ID)	Mid-Atlantic (MACS)	Functional Feeding Group primary	secondary	Habit/Behavior primary	secondary
119655	119656	Antocha	4.6	3	2.2	3		GC			
119656	119660	Antocha monticola				3		GC			
	120488	Cryptolabis						SH	GC	bu	
121026	121027	Dicranota	0	3		3		PR			
120030	120076	Elephantomyia						SH		sp	bu
120397	120503	Erioptera				3		GC			
120397	120640	Gonomyia						GC		bu	sp
119655	119690	Helius					4	GC		bu	sp
120397	120732	Hesperoconopa				1		GC		bu	
120030	120094	Hexatoma	4.7	2	2.3	2	2	PR		bu	sp
	120095	Eriocera						PR		bu	sp
120030	120164	Limnophila				4		PR		bu	
119655	119704	Limonia	10	6		6		SH		bu	
	119706	Geranomyia					3	SH			
120397	120758	Molophilus				4		SH		bu	
120397	120830	Ormosia	6.5			3		GC		bu	
121026	121118	Pedicia				6		PR		bu	
120030	120335	Pilaria				7	7	PR			
120030	120365	Pseudolimnophila	7.3	2			2	PR			
120397	120968	Rhabdomastix				8		PR		sp	bu
120968	120977	Rhabdomastix fascigera				3		GC		bu	
120968	120995	Rhabdomastix setigera				3		GC		bu	
120030	120387	Ulomorpha									
118831	130052	Brachycera									
130928	130929	Atherix				2	2	PR			
130929	130930	Atherix lantha	2.1	2	3.1			PR			
130929	130932	Atherix variegata				2		PR			
130741	130914	Pelecorhynchidae				3		PR			
130914	130915	Glutops				3		PR			
131750	136824	Dolichopodidae	9.7	4		4		PR			
137952	137953	Dolichopus								cn	
131750	135830	Empididae	8.1	6	3.5	6		PR		sp	bu
136304	136305	Chelifera				6		GC			
135844	135849	Clinocera				6		PR			
136304	136327	Hemerodromia				6	6	PR			
136361	136377	Oreogeton				5		PA			
135844	135881	Oreothalia				6		PR			
135930	136123	Rhamphomyia				6		PR		sp	bu
135844	135920	Wiedemannia				6		PR			
130130	130150	Stratiomyidae				8		GC			
130155	130160	Allognosta				7		GC			
130408	130409	Caloparyphus				7		GC		sp	
130408	130436	Euparyphus						GC			
130685	130694	Nemotelus								sp	bu
130483	130573	Odontomyia					7	GC			
130408	130461	Oxycera								sp	bu
130483	130627	Stratiomys						FG			
130741	130934	Tabanidae				8		PR		sp	bu

Rapid Bioassessment Protocols for Use in Streams and Wadeable Rivers: Periphyton, Benthic Macroinvertebrates, and Fish, Second Edition

B-49

| Parent TSN | TSN | Scientific Name | Regional Tolerance Values | | | | | Functional Feeding Group | | Habit/ Behavior | |
			Southeast (NC)	Upper Midwest (WI)	Midwest (OH)	Northwest (ID)	Mid-Atlantic (MACS)	primary	secondary	primary	secondary
131061	131078	Chrysops	7.3	6	4.6		7	GC	PR		
131061	131062	Silvius						PR			
131318	131527	Tabanus	9.7	5		5	5	PR			
131750	148316	Canaceidae						SC		bu	
131750	146893	Ephydridae				6		GC			
131750	150025	Muscidae				6		PR			
150729	150730	Limnophora	7					PR			
138933	139013	Dohrniphora									
131750	144653	Sciomyzidae				6		PR		bu	
144770	144898	Sepedon						PR			
131750	139621	Syrphidae				10		GC			
141029	141049	Chrysogaster									
	140904	Eristalis	10		0			GC		bu	

Appendix B: Regional Tolerance Values, Functional Feeding Groups,
and Habit/Behavior Assignments for Benthic Macroinvertebrates

APPENDIX C:

TOLERANCE AND TROPHIC GUILDS OF SELECTED FISH SPECIES

APPENDIX C

Appendix C is a list of selected fishes of the United States in phylogenetic order. Included are the Taxonomic Serial Number (TSN) and the Parent Taxonomic Serial Number for each of the species listed according to the Integrated Taxonomic Information System (ITIS). The ITIS generates a national taxonomic list that is constantly updated and currently posted on the World Wide Web at <www.itis.usda.gov>. If you are viewing this document electronically, this page is linked to the ITIS web site.

Additionally, this Appendix details trophic and tolerance designations for selected fishes of the United States. To generate this list, we compiled a consensus rating for each taxon from the literature sources listed below. Exceptions are listed for each source that does not agree with the consensus of other cited literature. Exceptions are noted by first listing the designation then the literature source code in parentheses. The following is a list of the designations and literature sources used in this Appendix.

TROPHIC DESIGNATIONS

P=Piscivore

H=Herbivore

O=Omnivore

I=Insectivore (including specialized insectivores)

F=Filter feeder

G=Generalist feeder

V=Invertivore

Notes on Trophic Designations

Piscivore—although some investigators separate certain species into subcategories such as parasitic (e.g., sea lamprey) or top carnivore (e.g., walleye), we have grouped these together as piscivores for this list.

TOLERANCE DESIGNATIONS (relevant to non-specific stressors)

I = Intolerant

M = Intermediate

T = Tolerant

Notes on Tolerance Designations

Intolerant—although some investigators separate certain species into subcategories such as rare intolerant, special intolerant or common intolerant, we have grouped these together as intolerant for this list.

Literature Sources For Trophic/Tolerance Designations

(A) = Midwestern United States (Karr et al. 1986)

(B) = Ohio (Ohio EPA 1987)

(C) = Midwestern United States (Plafkin et al. 1989)

(D) = Central Corn Belt Plain (Simon 1991)

(E) = Wisconsin Warmwater (Lyons 1992)

(F) = Maryland Coastal Plain (Hall et al. 1996)

(G) = Northeastern United States (Halliwell et al. 1999)

APPENDIX D:

SURVEY APPROACH FOR COMPILATION OF HISTORICAL DATA

Rapid Bioassessment Protocols For Use in Streams and Wadeable Rivers: Periphyton, Benthic Macroinvertebrates, and Fish, Second Edition

D-1

This Page Intentionally Left Blank

QUESTIONNAIRE SURVEY FOR EXISTING BIOSURVEY DATA AND BIOASSESSMENT INFORMATION

Ecological expertise and knowledge of the aquatic ecosystems of a state can reside in agencies and academic institutions other than the water resource agency. This expertise and historical knowledge can be valuable in problem screening, identifying sensitive areas, and prioritizing watershed-based investigations. Much of this expertise is derived from biological survey data bases that are generally available for specific surface waters in a state. A systematic method to compile and summarize this information is valuable to a state water resource agency.

The questionnaire survey approach presented here is modified from the methods outlined in the original RBP IV (Plafkin et al. 1989) and is applicable to various types of biological data. The purpose of this questionnaire survey is to compile and document historical/existing knowledge of stream physical habitat characteristics and information on the periphyton, macroinvertebrate, and fish assemblages.

The template questionnaire is divided into 2 major sections: the first portion is modeled after RBP IV and serves as a screening assessment; the second portion is designed to query state program managers, technical experts, and researchers regarding existing biosurvey and/or bioassessment data. This approach can provide a low cost qualitative screening assessment (Section 1) of a large number of waterbodies in a relative short period. The questionnaire can also prevent a duplication of effort (e.g., investigating a waterbody that has already been adequately characterized) by polling the applicable experts for available existing information (Section 2).

The quality of the information obtained from this approach depends on survey design (e.g., number and location of waterbodies), the questions presented, and the knowledge and cooperation of the respondents. The potential respondent (e.g., agency chief, program manager, professor) should be contacted initially by telephone to specifically identify appropriate respondents. To ensure maximum response, the questionnaire should be sent at times other than the peak of the field season and/or the beginning or end of the fiscal year. The inclusion of a self-addressed, stamped envelope should also increase the response rate. A personalized cover letter (including official stationary, titles, and signatures) should accompany each questionnaire. As a follow-up to mailings, telephone contact may be necessary.

Historical data may be limited in coverage and varied in content on a statewide basis, but be more comprehensive in coverage and content for specific watersheds. A clearly stated purpose of the survey will greatly facilitate evaluation of data from reaches that are dissimilar in characteristics. The identification of data gaps will be critical in either case. Regardless of the purpose, minimally impaired reference reaches may be selected to serve as benchmarks for comparison. The definition of minimal impairment varies from region to region. However, it includes those waters that are generally free of point source discharges, channel modifications, and/or diversions, and have diverse habitats, complex substrates, considerable instream cover and a wide buffer of riparian vegetation. Selection of specific reaches for consideration (e.g., range and extent) in the questionnaire survey is ultimately dependent on program objectives and is at the discretion of the surveyor. The questionnaire approach and the following template form allows considerable flexibility. Results can be reported as histograms, pie graphs, or box plots.

Rapid Bioassessment Protocols For Use in Streams and Wadeable Rivers: Periphyton, Benthic Macroinvertebrates, and Fish, Second Edition

D-3

Questionnaire design and responses should address, when possible, the:

- ! extent of waterbody or watershed surveyed

- ! condition of the periphyton, macroinvertebrate and/or fish assemblage

- ! quality of available physical habitat

- ! frequency of occurrence of particular factors/causes limiting the biological condition

- ! effect of waterbody type and size on the spatial and temporal trends, if known

- ! likelihood of improvement or degradation based on known land use patterns or mitigation efforts

BIOASSESSMENT/BIOSURVEY QUESTIONNAIRE

Date of Questionnaire Survey _____

This questionnaire is part of an effort to assess the biological condition or health of the flowing waters of this state. Our principle focus is on the biotic health of the designated waterbody as indicated by its periphyton, macroinvertebrate and/or fish community. You were selected to participate in this survey because of your expertise in periphyton, macroinvertebrate, and/or fish biology and your knowledge of the waterbody identified in this questionnaire.

Please examine the entire questionnaire form. If you feel that you cannot complete the form, check here [ ] and return it. If you are unable to complete the questionnaire but are aware of someone who is familiar with the waterbody and/or related bioassessments, please identify that person's name, address, and telephone number in the space provided below:

Contact: Name_____

                 Address_____

                 Agency/Institution_____

                 Phone_____Fax_____

                 Email_____

- -

This questionnaire is divided into two major sections. Section 1 serves as a screening assessment and Section 2 is a request for existing biosurvey data and/or bioassessment results.

This form addresses the following waterbody:

_____
Waterbody

State:_____   County:_____   Lat./Long.:_____   Waterbody code:_____

Ecoregion:_____   Subecoregion:_____   Description of site/reach:_____

Drainage size:_____   Flow: <1cfs; 1-10cfs; >10cfs   _____

- -

Description of data set (i.e., years, seasons, type of data, purpose of survey)_____

_____
_____
_____
_____
_____

SECTION 1. SCREENING ASSESSMENT

Using the scale of biological conditions found in the following text box, please circle the rank that best describes your impression of the condition of the waterbody.

SCALE OF CONDITIONS

5   Species composition, age classes, and trophic structure comparable to non (or minimally) impaired waterbodies of similar size in that ecoregion or watershed.

4   Species richness somewhat reduced by loss of some intolerant species; less than optimal abundances, age distributions, and trophic structure for waterbody size and ecoregion.

3   Intolerant species absent; considerably fewer species and individuals than expected for that waterbody size and ecoregion; trophic structure skewed toward omnivory.

2   Dominated by highly tolerant species, omnivores, and habitat generalists; top carnivores rare or absent; older age classes of all but tolerant species rare; diseased fish and anomalies relatively common for that waterbody size and ecoregion.

1   Few individuals and species present; mostly tolerant species; diseased fish and anomalies abundant compared to other similar-sized waterbodies in the ecoregion.

0   No fish, depauperate macroinvertebrate and/or periphyton assemblages.

(Circle one number using the scale above.)

1.   Rank the current conditions of the reach

        5   4   3   2   1   0

If impairment noted (i.e., scale of 1-3 given), complete each subsection below by **checking off** the most appropriate limiting factor(s) and probable cause(s). Clarify if reference is to past or current conditions.

PHYSICOCHEMICAL

(a.) WATER QUALITY

Limiting Factor	Probable Cause

□ Temperature too high
□ Temperature too low
□ Turbidity
□ Salinity
□ Dissolved oxygen
□ Gas supersaturation
□ pH too acidic
□ pH too basic
□ Nutrient deficiency
□ Nutrient surplus
□ Toxic substances
□ Other (specify below)
_____
 	
□ Not limiting
	□ Primarily upstream
	□ Within reach
	  Point source discharge
	    □ Industrial
	    □ Municipal
	    □ Combined sewer
	    □ Mining
	    □ Dam release
	  Nonpoint source discharge
	    □ Individual sewage
	    □ Urban runoff
	    □ Landfill leachate
	    □ Construction
	    □ Agriculture
	    □ Feedlot
	    □ Grazing
	    □ Silviculture
	    □ Mining
	□ Natural
	□ Unknown
	□ Other (specify below)
	_____

(b.) WATER QUANTITY

Limiting Factor	Probable Cause

□ Below optimum flows
□ Above optimum flows
□ Loss of flushing flows
□ Excessive flow fluctuation
□ Other (specify below)
_____
 	
□ Not limiting
	□ Dam
	□ Diversion
	  Watershed conversion
	    □ Agriculture
	    □ Silviculture
	    □ Grazing
	    □ Urbanization
	    □ Mining
	□ Natural
	□ Unknown
	□ Other (specify below)
	_____

Rapid Bioassessment Protocols For Use in Streams and Wadeable Rivers: Periphyton, Benthic Macroinvertebrates, and Fish, Second Edition

D-7

BIOLOGICAL/HABITAT
(Check the appropriate categories)

(a.) Limiting Factor	HABI	PERI	MACR	FISH
Insufficient instream structure				
Insufficient cover				
Insufficient sinuosity				
Loss of riparian vegetation				
Bank failure				
Excessive siltation				
Insufficient organic detritus				
Insufficient woody debris for organic detritus				
Frequent scouring flows				
Insufficient hard surfaces				
Embeddedness				
Insufficient light penetration				
Toxicity				
High water temperature				
Altered flow				
Overharvest				
Underharvest				
Fish stocking				
Non-native species				
Migration barrier				
Other (specify) _____				
Not limiting				

Key:

HABI - Habitat   PERI - Periphyton
MACR - Macroinvertebrates   FISH - Fish

(b.) Probable Cause	HABI	PERI	MACR	FISH
Agriculture				
Silviculture				
Mining				
Grazing				
Dam				
Diversion				
Channelization				
Urban encroachment				
Snagging				
Other channel modifications				
Urbanization/impervious surfaces				
Land use changes				
Bank failure				
Point source discharges				
Riparian disturbances				
Clear cutting				
Mining runoff				
Stormwater				
Fishermen				
Aquarists				
Agency				
Natural				
Unknown				
Other (specify) _____				

Key:

Rapid Bioassessment Protocols For Use in Streams and Wadeable Rivers: Periphyton, Benthic Macroinvertebrates, and Fish, Second Edition

D-9

HABI - Habitat   PERI - Periphyton
MACR - Macroinvertebrates   FISH - Fish

SUMMARY: ASPECT OF PHYSICOCHEMICAL OR BIOLOGICAL CONDITION AFFECTED

☐ Water quality
☐ Water quantity
☐ Habitat structure
☐ Periphyton assemblage
☐ Macroinvertebrate assemblage
☐ Fish assemblage
☐ Other (specify) _____

Rapid Bioassessment Protocols For Use in Streams and Wadeable Rivers: Periphyton, Benthic Macroinvertebrates, and Fish, Second Edition

D-11

SECTION 2. AVAILABILITY OF DATA

Please complete this section with applicable response(s) and fill in the blanks with appropriate information based on your knowledge of available biosurvey and bioassessment information.

Reach characterized by:

- ☐ Stream habitat surveys
- ☐ Periphyton surveys   assemblage ☐   key species ☐
- ☐ Macroinvertebrate surveys   assemblage ☐   key species ☐
- ☐ Fish surveys   assemblage ☐   key species ☐

Sampling gear(s) or methods   Sampling frequency (spatial and temporal)

_____   _____
_____   _____
_____   _____
_____   _____

Data analysis/interpretation based on: Electronic file available:

Tabulated data   ☐   Format _____
Graphical data   ☐   _____
Multivariate analyses.   ☐   _____
Multimetric approach.   ☐

Statistical routines include: Metrics include:

_____   _____
_____   _____
_____   _____
_____   _____

www.ingramcontent.com/pod-product-compliance
Lightning Source LLC
Chambersburg PA
CBHW080634180526
45168CB00008B/3164